Y0-BSM-209

Air Pollution Impacts on Crops and Forests

A Global Assessment

AIR POLLUTION REVIEWS

Series Editor: Robert L. Maynard
(Department of Health, Skipton House, London, U.K.)

Vol. 1: The Urban Atmosphere and Its Effects
Peter Brimblecombe & Robert L. Maynard

Vol. 2: The Effects of Air Pollution on the Built Environment
(Peter Brimblecombe)

Forthcoming:

Vol. 3: Air Pollution and Health
(Jon Ayres, Robert L. Maynard & Roy Richards)

Vol. 5: Indoor Air Pollution
(Robert L. Maynard, Mike Ashmore & Peter Brimblecombe)

Air Pollution Reviews – Vol. 4

Air Pollution Impacts on Crops and Forests

A Global Assessment

Editors

Lisa Emberson
Stockholm Environment Institute-York, University of York, UK

Mike Ashmore
University of Bradford, UK

Frank Murray
Murdoch University, Australia

ICP

Imperial College Press

Published by

Imperial College Press
57 Shelton Street
Covent Garden
London WC2H 9HE

Distributed by

World Scientific Publishing Co. Pte. Ltd.

5 Toh Tuck Link, Singapore 596224

USA office: Suite 202, 1060 Main Street, River Edge, NJ 07661

UK office: 57 Shelton Street, Covent Garden, London WC2H 9HE

Library of Congress Cataloging-in-Publication Data
Air pollution impacts on crops and forests / [edited by] Mike Ashmore, Lisa Emberson, Frank Murray.
p. cm. -- (Air pollution reviews ; vol. 4)
Includes bibliographical references.
ISBN 1-86094-292-X
1. Crops--Effect of air pollution on. 2. Trees--Effect of air pollution on. 3. Crops--Wounds and injuries. 4. Trees--Wounds and injuries. 5. Crop losses. 6. Trees--Losses. 7. Air quality management. I. Ashmore, M. R. (Mike R.) II. Emberson, Lisa. III. Murray, Frank. IV. Series.

SB745 .A388 2003
632'.1--dc21 2002038711

British Library Cataloguing-in-Publication Data
A catalogue record for this book is available from the British Library.

Copyright © 2003 by Imperial College Press

All rights reserved. This book, or parts thereof, may not be reproduced in any form or by any means, electronic or mechanical, including photocopying, recording or any information storage and retrieval system now known or to be invented, without written permission from the Publisher.

For photocopying of material in this volume, please pay a copying fee through the Copyright Clearance Center, Inc., 222 Rosewood Drive, Danvers, MA 01923, USA. In this case permission to photocopy is not required from the publisher.

Printed in Singapore by World Scientific Printers (S) Pte Ltd

CONTENTS

Air Pollution Impacts on Vegetation in Developing Countries

CONTRIBUTORS

Nassar Abdel-Latif
Air Pollution Research Department
National Research Centre
El-Bohos St., Dokki 12311
Cairo, Egypt

Madhoolika Agrawal
Department of Botany
Banaras Hindu University
Varanasi 221005, India

Mike Ashmore
Department of Environmental Science
University of Bradford
West Yorkshire, BD7 1DP, UK

Maria Lourdes de Bauer
Instituto de Recursos Naturales, CP
Colegio de Postgraduados
Comision de Estudios Ambientales
56230 Montecillo, Mexico

Marisa Domingos
Institute of Botany
Seção de Ecologia
Instituto de Botânica
Caixa Postal 4005
01061 970 São Paulo, SP, Brazil

Lisa Emberson
Stockholm Environment Institute at York (SEI-Y)
Biology Department
Box 373
University of York
York YO10 5YW, UK

Takeshi Izuta
Department of Environmental and Natural Resource Science
Faculty of Agriculture
Tokyo University of Agriculture and Technology
Fuchu, Tokyo 183-8509, Japan

Johan Kuylenstierna
Stockholm Environment Institute at York (SEI-Y)
Biology Department
Box 373
University of York
York YO10 5YW, UK

Chin-pin Liu
Division of Watershed Management
Taiwan Forestry Research Institute
Taipei, Taiwan

Fiona Marshall
Department of Environmental Science and Technology
Faculty of Life Sciences
Imperial College London
Silwood Park, Ascot, Berkshire, UK

Frank Murray
School of Environmental Science
Division of Science and Engineering
Murdoch University
Murdoch, WA 6150, Australia

Kevin Percy
Natural Resources Canada
Canadian Forest Service — Atlantic Forestry Centre
PO Box 4000, Fredericton, NB
Canada E3B 5P7

Bor-Hung Sheu
Department of Forestry
National Chung Hsing University
Taichung, Taiwan

Mieke van Tienhoven
CSIR — Division of Water, Environment and Forestry Technology
PO Box 17001
Congella, Durban 4013,
South Africa

Abdul Wahid
Environmental Pollution Research Laboratory
Government College University (GCU)
Lahore, Pakistan

Youbin Zheng
Controlled Environment Systems Research Facility (CESRF)
Environmental Biology Department
University of Guelph
Bovey Building, RM1212
Guelph, Ontario N1G 2W1
Canada

PREFACE

Air pollution is a problem which now affects every part of our planet. While the effects of air pollution on human health are the most important concern, the impacts on crop production, forest vitality and biodiversity may also have considerable implications for human welfare. The most serious and most visible air pollution problems across the planet are found in and around major cities and industrial areas. However, secondary pollutants such as ozone tend to reach their highest concentrations at some distance from the urban and industrial centres that are the source of their precursor pollutants. Hence, secondary pollutants may cause significant environmental impacts over large expanses of rural regions.

The major local air pollution problems arising from domestic coal combustion, industry and power generation which historically affected Europe and North America are now largely a thing of the past. However, the pollution problems associated with the dramatic increase in vehicle numbers and emissions over the last 50 years still remain a significant issue which require further action. Over the same period, emissions from industry, energy production and vehicles have increased in many regions of Asia, Africa and Latin America. Furthermore, the climate in these regions is often more favourable than that in Europe and North America for the formation of ozone. Field observations and experiments that are described in this book demonstrate visible leaf injury, declines in forest health, and loss of crop yield as a result of air pollution problems, in countries such as China, Mexico, Egypt and India.

The effects on crop production and forest vitality due to this shift in the global patterns of air pollutant emissions and concentrations remain poorly understood. In many countries, the amount of research is extremely limited and awareness of the impacts caused by air pollution is very low. Air pollutants are known to cause significant effects on crop and forest yields in the absence of any recognisable injury symptoms. Hence, it is quite probable that significant yield losses are occurring in many regions of the world without the problem having been recognised. Such impacts may have very serious consequences, both for the national economy and for the livelihoods of individual farmers, in many countries where industrial and urban development are associated with a rapidly increasing population.

However, policies to reduce emissions are often technically demanding and costly, and need to be balanced against a range of other economic and social priorities. This means that information is urgently needed in order to fully assess the possible benefits of reduced pollutant emissions in terms of impacts on crops, forests and the wider natural environment, as well as human health. We hope that this book will contribute both to an increased awareness of the problem and to the development of tools to better assess the impacts of air pollution on crops and forests in all parts of the world.

Our aim in this book is to provide the first truly global assessment of the scale of the direct impacts of air pollution on crops and forests. We felt that such an assessment had to be based on the perspectives on the ground from individual countries. Therefore, the core of this book comprises assessments of the problem by experts from 12 different countries on every continent. These contributions describe the evidence of air pollution effects on crop yields and forest vitality in the context of current air pollution emissions and concentrations, and in the context of policies for the management of air quality and natural resources. We are most grateful for the contributions of all these experts in providing a range of different perspectives and views on air pollution problems.

We also wanted to place these studies in a broader global framework. Therefore, the book provides an overview of how emission patterns have changed over the last 30 years, and attempts to present a global assessment of the scale of both current and future impacts of ozone and sulphur dioxide. This assessment clearly indicates the growing significance of ozone as a global constraint on crop production over the next two decades based on current projections and policies. Finally, we aim to place this issue in a wider socio-economic context. This is achieved by evaluating methods to translate from biological to socio-economic impacts by quantifying the benefits of air pollution abatement, including non-marketable environmental goods and services.

Lisa Emberson
Mike Ashmore
Frank Murray

Acknowledgements

The work contained in this book was coordinated by the Stockholm Environment Institute (SEI). The work has been developed as part of an activity investigating the impacts of air pollution on crops and forests which forms part of the Programme on Regional Air Pollution In Developing Countries (RAPIDC), funded by the Swedish International Development Cooperation Agency (Sida). We are also indebted to Steve Cinderby, Erik Willis and Lisetta Tripodi for their help with map production graphics and proofreading.

Air Pollution Impacts on Crops and Forests

CHAPTER 1

AIR POLLUTION IMPACTS ON CROPS AND FORESTS: AN INTRODUCTION

L. Emberson

1. Introduction

In Europe and North America, air pollution has been recognised as a cause of injury to vegetation over the past few centuries. Early air pollution impacts characteristically caused severe, but localised, effects close to emission sources and in urban and industrialised areas. One of the best known case studies of such localised pollutant effects occurred at a smelter complex in Sudbury, Canada during the 1900s. Entire forest communities were lost up to a distance of 15 km downwind of the complex as a result of sulphur dioxide (SO_2) and metal emissions, with other ecological effects extending to a greater distance. The effects are still in evidence today even after the introduction of significant emission reductions during the 1980s (Winterhalder, 1996).

More recently air pollution problems have become regional as well as local in extent. Regional problems associated with photochemical air pollution became apparent with observations of severe forest decline in parts of North America during the early 1960s. Chlorosis (yellowing) and premature senescence of needles of ponderosa pine trees in the San Bernardino National Forest were found to be principally a result of elevated tropospheric ozone (O_3) levels in the region (Miller and McBride, 1999). Similar effects of long-term O_3 exposure on conifer forest ecosystems were experienced throughout this southern Californian mountain range region and were related to the transport of polluted air masses for considerable distances from urban source areas (Hoffer *et al.*, 1982). Similar instances of regional forest decline which have been recorded in other regions of both North America and Europe have been attributed, at least in part, to regionally elevated O_3 concentrations (Chappelka and Samuelson, 1998; Skärby *et al.*, 1998). As well as effects on forest trees, detrimental effects of photochemical oxidants were observed on agricultural crops in the U.S. during the late 1940s and 1950s (Middleton *et al.*, 1950). Ambient concentrations of O_3 are now thought to be capable of decreasing the productivity of a wide range of crops in Europe (Fuhrer *et al.*, 1997) and North America (Tingey *et al.*, 1993).

The increased understanding of the damage that air pollutants such as O_3, SO_2 and nitrogen oxides (NO_x) were causing by acting directly on vegetation resulted in an awareness of the need for air quality management strategies to reduce ambient pollution levels. These strategies were based on the identification of air quality guidelines for the pollutant receptors of interest (i.e. human health, vegetation and materials) which could then be applied to assess the need for, and benefits of, pollution control measures. For example, in order to establish causal relationships between air pollution and regional scale vegetation damage, numerous field surveys and experimental studies were performed in both North America and Europe, as discussed in Chaps. 2 to 5. These studies enabled the development of dose-response relationships for different vegetation types that were used to define acceptable levels of air quality and provide targets for emission reduction strategies. Dose-response relationships were also used to indicate the current scale of the problem; for example, estimates of economic crop losses due to O_3 in the U.S. were calculated as being in excess of US\$3 $\times$ 10^9 per year (Adams *et al.*, 1988). The application of dose-response relationships in economic assessments of crop loss is discussed further in Chap. 15.

The use of air quality guidelines to inform emission reduction strategies has gone some way to improve air quality in both Europe and North America. However, much of the reduction in levels of some atmospheric pollutants in these regions has also been due to other factors such as economic restructuring and changes in energy policy. In contrast to the situation in Europe and North America, air pollutant emissions have been increasing over the last two decades in many developing countries of Latin America, Africa and Asia (UNEP, 1997). This has been primarily due to rapid economic growth in many of these countries resulting in increases in urbanisation, transportation, industrialisation, and energy generation (World Resources Institute, 1996). The contribution of different sector emissions to air pollution problems varies considerably across regions resulting in variable pollutant climates in different countries. However, it is possible to identify the major pollutants produced by these human activities as SO_2, O_3, suspended particulate matter (SPM) and NO_x (World Resources Institute, 1994). It is the impact of these pollutants in developing countries which is the prime focus of this volume.

Air pollution impacts can be divided into two classes — direct and indirect injury. In this book we deal primarily with the former, which constitutes injury resulting directly from exposure to the pollutant (e.g. gaseous uptake of the pollutant by vegetation resulting in internal cellular damage or changes to biochemical or physiological processes). SO_2 and NO_x can also contribute to acidification of sensitive soils, which may be accompanied by a depletion of base cations, affecting the local vegetation over relatively long timescales (e.g. Grennfelt *et al.*, 1994). NO_x and ammonia emissions can also cause long-term eutrophication of nutrient-poor terrestrial ecosystems, although the additional nitrogen deposition may also lead to short-term stimulation of growth. Given the scarcity of information describing responses of vegetation in developing regions, and the relatively short period of elevated emissions in many regions, it is crucial to ascertain the immediate effects of air pollutants before considering these longer-term impacts.

Furthermore, a recent global assessment by Kuylenstierna *et al.* (2001) showed that modelled deposition of acidity in 1990 only exceeded critical loads for soil acidification in small areas of China and Southeast Asia outside Europe and North America, although this situation was predicted to change significantly by 2050. There is also little evidence that productive crop systems are sensitive to acidification, while deposition of atmospheric nitrogen deposition is usually significantly smaller than the inputs from organic and/or inorganic fertilisers in agricultural systems. Since it is the impact of air pollutants on crop production which is likely to be of greatest concern in most developing countries, the focus on direct impacts of atmospheric pollutants in this volume reflects the likely economic significance of the issue.

2. Pollutant Emissions

The pollutants considered in this book are emitted from a number of different sources that can be grouped according to different sectors (e.g. industry, power generation, transport, biomass burning, etc.). Knowledge of the sources of different pollutants is crucial to an understanding of the variable nature of the changes in pollutant concentrations, and the resulting pollutant mixtures, found in different countries. The following section briefly describes the sectors primarily responsible for the emissions of each of the different pollutants in turn. It then considers data on emission trends from different sources in various regions of the world.

2.1. *Pollutant Sources*

Sulphur dioxide (SO_2)

The main source of anthropogenic SO_2 is from the combustion of fossil fuels containing sulphur. These are predominantly coal (especially poor quality, high sulphur content brown coal or lignite) and fuel oil, since natural gas, petrol and diesel fuels have a relatively low sulphur content. In general, combustion of coal in power stations is the most important source of SO_2 emissions. Other sectors that contribute significantly to SO_2 emissions include other industrial processes, as well as domestic and commercial heating, although sector contributions will vary by country and region. SO_2 is a primary pollutant (i.e. one that is released directly into the atmosphere from pollutant sources) and therefore concentrations tend to be directly related to the extent of local emissions and the height of emissions; as such, high levels of SO_2 are typically associated with urban or industrial areas. Use of tall stacks for major industrial processes and power stations can effectively reduce ground-level concentrations, although they are less beneficial in terms of long-range transport. Levels of SO_2 in the atmosphere are highest under meteorological conditions that lead to poor dispersal of low-level emissions, for example under temperature inversions with low wind speeds.

Nitrogen oxides (NO_x)

Both nitric oxide (NO) and nitrogen dioxide (NO_2) are known to have impacts on vegetation. NO_2 is predominantly a secondary pollutant formed mainly by reaction between emissions of the primary pollutant nitric oxide (NO) and O_3. The rapid conversion of NO to NO_2 results in the atmospheric burden of NO_x (as the sum of the two compounds is known) being predominantly NO_2 at locations away from sources. As for SO_2, these pollutants are produced during combustion processes but whilst SO_2 emissions increase with the sulphur content of the fuel, NO_x emissions depend upon other factors. The major source of NO_x is from high temperature (>1400°C) combination of atmospheric nitrogen and oxygen during combustion. There is also a smaller contribution from the combustion of nitrogen contained in the fuel. Combustion of fossil fuels from stationary sources (heating and power generation) and from motor vehicles are the main sources of NO_x.

Ozone (O_3)

Unlike other gaseous pollutants, O_3 occurs naturally at relatively high background concentrations. Background surface O_3 concentrations have been steadily rising from between 10 to 20 ppb at the beginning of the 20th century to values in the range 20 to 40 ppb in recent years (Volz and Kley, 1998). These concentrations are the result of two different processes; O_3 transfer from the stratosphere into the troposphere and photochemical reactions. The latter involve O_3 formation from the recombination of atomic and molecular oxygen via the phytolosis of NO_2, and O_3 destruction by reaction with NO to reform NO_2. Net production of O_3 leading to elevated O_3 concentrations occurs in atmospheres polluted with hydrocarbons and NO_x. In such situations, hydrocarbon degradation produces free radical species that react with NO producing NO_2, thereby facilitating net production of O_3. O_3 concentrations tend to be higher in suburban and rural locations as reaction with NO leads to local-scale depletion of O_3. O_3 can cause damage to ecosystems located considerable distances from the source of the primary pollutants. The main sources of NO_x have been described above; emission sources of volatile hydrocarbons include natural sources such as release from forest trees. In urban areas, road transport tends to be the major contributor, although use of solvents, for example in paints and adhesives, can be a significant source. Emissions from road transport include both the evaporation of fuels and the emission of unburned and partially combusted hydrocarbons and their oxidation products from vehicle exhausts.

Suspended particulate matter (SPM)

The term suspended particulate matter (SPM) includes finely divided solids or liquids that range in size from 0.1 to approximately 25 μm in diameter. However, the most common measure of particles used to quantify pollutant concentrations in developed countries is now PM_{10}, the abbreviation for particulate matter having an aerodynamic diameter less than 10 μm. This is the size range that includes the majority of the total suspended particles in the atmosphere by mass (Beckett *et al.*, 1998). SPMs can be categorised into two groups. Primary particles are emitted directly from source (e.g. from vehicle exhausts), while secondary particles are formed by interactions with other compounds, e.g. nitrate formation from the photo-oxidation of NO_x. In general, most coarse particles (i.e. those between 2.5 to 10 μm) are made up of both natural and organic particles whilst the fine fraction (i.e. < 2.5 μm) tend to mostly be of anthropogenic origin, including nitrate and sulphate aerosols. SPM tends to be an urban pollutant; however this does depend on the size of the particles under consideration. This is the reason for higher rates of deposition closer to particle emission sources and the classification of SPM as a local pollutant. Emission sources of SPM in developing regions may well differ from those in developed regions; for example forest fires, local industry and less developed infrastructure may contribute significantly to atmospheric SPM load.

Fluorides

Fluoride (F) compounds are recognised as common gaseous or particulate pollutants produced from many industrial processes. The most important F-emitting industrial sources are aluminium smelters and fertiliser phosphate factories. As such, fluoride [as hydrogen fluoride (HF)] did not become a major air pollutant problem in North America and Europe until the expansion during the 1940s of the aluminium smelting and phosphate fertiliser industries (Wellburn, 1994). Coal combustion, steel works, brickyards, and glass works may also significantly contribute to F emissions.

2.2. *Regional Variations and Trends*

A major driving force behind the increases in air pollution in many developing regions is the recent rapid increases in urban populations in these regions, leading to increased energy demand, industrial activity and road traffic. The different rates of urbanisation across the world are shown in Fig. 1. Since 1950, the number of people living in urban areas has risen dramatically from 750 million to more than 2500 million people (UNEP, 2000). The Asian and Pacific region has experienced the greatest absolute increase, with the urban population almost doubling between 1975 and 1995. Rapid increases in urbanisation are also occurring in Latin America and Africa, albeit rather more slowly compared to Asia. However, it is worth noting that the percentage of the population in urban areas is still highest in North America and Europe.

Similarly, Fig. 2 shows that the number of motor vehicles are also starting to increase in developing regions; the Asian and Pacific region has experienced the greatest increase amongst developing country regions with an increase of 40% between 1980 and 1995. This represents a greater rate of increase than in North America, though not as large

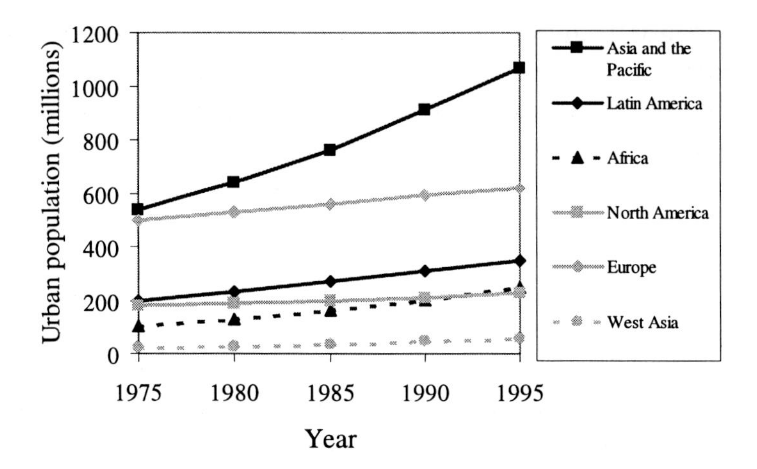

Fig. 1. Increases in regional urban populations from 1975 to 1995 (UNEP, 2000).

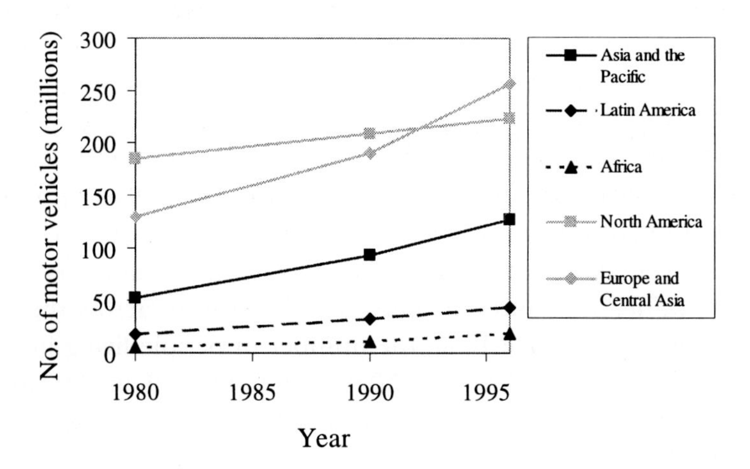

Fig. 2. Increases in regional numbers of motor vehicles from 1980 to 1995 (UNEP, 2000).

as increases experienced in Europe. If current rates of expansion continue there will be more than 1000 million vehicles on the road worldwide by 2025 (UNEP, 2000). In many urban areas, motor vehicles are the most important source of air pollutants; for example in Delhi and Beijing, emissions from motor vehicles represent 57% and 75% of total urban pollutant emissions, respectively (World Resources Institute, 1994).

Air pollution from industry has also become a feature of a number of developing countries. Although the sizes of many industrial plants have tended to be relatively small by western standards, the cumulative effect of many small industrial sources of pollution may be considerable. In addition, the displacement of polluting industries to countries where less emphasis is placed on emission control could cause significant regional problems in the future (Abdul Rahim, 2000).

The growth in energy demand in recent decades has been especially rapid in developing countries and regions. For example, growth in energy demand for the Asian and Pacific region was 3.6% year^{-1} between 1990 and 1992 compared with an average of 0.1% growth for the whole world (ADB, 1994). This has resulted in an increase in the consumption of coal; for example, the Asia Pacific region accounted for 41% of world coal consumption in 1993 (EIA, 1995). Large power plants in many developing countries often still use fossil fuels with a high sulphur content. As a result, these power plants comprise the greatest contributors to SO_2 emissions in these regions. Figure 3 shows that at the same time as SO_2 emissions have been decreasing in Europe and North America, the increased burning of fossil fuels in Asia has resulted in a dramatic increase in SO_2 emissions; these trends are projected to continue over the next decade.

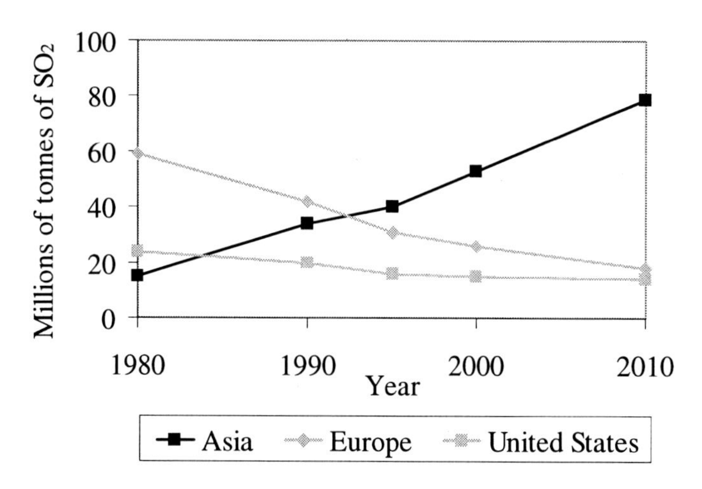

Fig. 3. Regional SO_2 emissions from fossil fuel burning (UNEP, 2000).

In addition to air pollution emissions from road traffic, energy demand, industry, and power generation, deterioration in air quality in developing regions also results from a number of other activities including forest conversion and agricultural activities, refuse disposal and household heating and cooking. The widespread use of fire to clear previously logged forests has received recent international attention as a source of air pollution, and specifically of increasing SPM concentrations. Forest fires have become a major cause of air quality problems in countries such as Brazil, Indonesia and Malaysia (Glover and Jessup, 1999) and result in both urban and regional scale deterioration in air quality.

The broad global trends discussed above demonstrate a shift in air pollution problems from developed to developing countries. However, it is also apparent (e.g. from the comparatively large increases in numbers of motor vehicles, energy demand and industrial growth in the Asian and Pacific regions, compared to African and Latin American regions) that pollutant emissions and resulting pollution loads vary within the developing world. Some of the key regional differences are discussed below.

The **Asia and Pacific** region has seen energy demand rise faster than in any other part of the world; per capita commercial energy use more than doubled in most parts of the region between 1975 and 1995 (UNEP, 2000). Fossil fuels account for about 80% of energy generation; since both China and India rely heavily on coal, this has become the most utilised fossil fuel within the region (EIA, 1995) and has resulted in rapidly increasing emissions of SO_2. NO_x emissions from fossil fuel combustion have been shown to increase in line with SO_2 emissions (Hameed and Dignon, 1992). Transportation contributes the largest share of air pollutants to the urban environment; urban levels of smoke and dust are generally twice the world average and more than five times as high as in industrial countries and Latin America (ADB, 1997). The concentrations of air pollutants in this region are expected to increase in the future. For example, a study performed to characterise regional emissions in China estimated that SO_2 emissions are projected to increase from 25.2 million metric tonnes (Mt) in 1995 to 30.6 Mt in 2020; these projections allow for the fact that emission controls will be implemented on major power plants.

Emissions of NO_x are also projected to increase in China from 12.0 Mt in 1995 to between 26.6 and 29.7 Mt by 2020 (Streets and Waldhoff, 2000). These emissions will be concentrated in the populated and industrialised areas of China, which tend to coincide with areas where agricultural productivity is relatively high.

Atmospheric pollution has only emerged as a problem in **African** countries over the past few decades. This is in part related to the dramatic increase in commercial energy consumption, which has risen by 145% since 1973 (McCormick, 1997). Of anthropogenic SO_2 emissions estimated for 1990 at 6.9 Mt for the whole of Africa, 3.5 Mt resulted from fossil fuel burning. In contrast, industrial processes such as smelting contributed 1.6 Mt SO_2, and land-use changes and waste treatment 1.7 Mt SO_2 (Olivier *et al.*, 1996). The highest concentrations of heavy industries are found in Zambia, South Africa and Nigeria, with industry occurring on much smaller scales in other African countries (van Tienhoven, 2000). The most important sources of industrial emissions include thermal power stations, copper smelters, ferro-alloy works, steel works, foundries, fertiliser plants and pulp and paper mills (UNEP, 2000). In southern Africa, air pollution originates largely from thermal power stations; estimates for 1994 indicated that about 89% of the electricity generation in this region is from coal. Approximately 97% of this coal generated electricity is produced in South Africa where the coal sulphur content is about 1% (Siversten *et al.*, 1995). This dependence on coal-based thermal power is likely to persist into the future indicating that SO_2 pollution will remain a problem for many years to come. In addition, the world's richest mineral field runs through most of the southern African countries. Smelters processing ores from these mineral deposits represent one of the major sources of air pollution in southern Africa and future exploitation of these deposits make this industry an emission sector of growing concern.

In the countries of **Latin America and the Caribbean**, nearly three-quarters of the population are urbanised, many in megacities, with the air quality in most major cities rising to levels that pose a threat to human health. Reliable emission inventories for these regions are not readily available. Trends emerging from completed (Uruguay and Argentina) and preliminary (Costa Rica, Mexico and Venezuela) inventories suggest that more than 50% of emissions come from industrial production and energy generation (UNEP, 2000). In Central America, more than 50% of the energy produced is generated by hydropower. However, as exemplified by Brazil, there has recently been a re-direction of energy policy from hydroelectricity toward fossil-fuelled electricity generation, changes that are being supported by international and bilateral funding agencies (Rosa and Schaeffer, 1995). As such, energy-related pollutant emissions from the Latin America region might be expected to increase significantly in the future.

3. Pollutant Impacts

The air pollutants currently considered to be most important in causing direct damage to vegetation are SO_2, NO_x, O_3, F and SPM. Direct effects of air pollution can be further classified into visible and invisible injury. Visible injury normally takes the form of discolourations of the leaf surface caused by internal cellular damage. Such injury can reduce the market value of agricultural crops for which visual appearance is important (e.g tobacco and spinach). It can also lead to yield reductions, while the damaged parts of the leaf surface can provide points of entry for plant pathogens. Invisible injury results from pollutant impacts on plant physiological or biochemical processes and can lead to significant loss of growth or yield and changes in nutritional quality (e.g. protein content) (Ashmore and Marshall, 1999). Visible injury tends to be associated with short-term exposures to high pollutant levels whilst invisible injury is generally a consequence of longer-term exposures to moderately elevated pollution concentrations. While visible injury can be identified in the field, loss of yield can only be identified with suitable control plants, and so can go undetected especially if there is little awareness of air pollution issues.

A brief description of the process by which each of the five different pollutants cause injury to vegetation is given below; this information is summarised in Table 1. A small number of photographs from different countries, illustrating characteristic symptoms of visible injury caused by SO_2, O_3 and F, are distributed through the text. These are described in more detail in the relevant chapter. Appendix 1 also

Table 1. Summary of the major sources, impacts and scale of effects of the major pollutants considered in this book.

Pollutant	Major sources	Major impacts	Major scale of effects
Sulphur dioxide (SO_2)	Power generation; industry; commercial and domestic heating	Visible foliar injury; altered plant growth; elimination of lichens and bryophytes; forest decline	Local
Nitrogen oxides (NO_x)	Power generation; transport	Altered plant growth; enhanced sensitivity to secondary stresses; eutrophication	Local
Ozone (O_3)	Secondary pollutant formed from NO_x and hydrocarbons	Visible foliar injury; reduced growth; forest decline	Regional
Suspended particulate matter (SPM)	Transport; power generation; industry; domestic heating	Altered plant growth; enhanced sensitivity to secondary stresses	Local
Fluorides	Manufacturing and smelting industries	Reduced plant growth; fluorosis in grazing animals	Local

provides a brief reference table for inter-conversion of measurement units for the different pollutants.

Sulphur Dioxide (SO_2)

SO_2 enters leaves through the stomata, but is also deposited at significant rates to wet surfaces, where it may dissociate to form sulphite or bisulphite and react with cuticular waxes. This can affect the cuticle to such an extent that a certain amount of SO_2 can enter via the damaged cuticle (Wellburn, 1994). Critical to the impact of the internal SO_2 dose are the buffering capacities of the cellular fluids. Higher plants have some capacity to control intracellular gradients of acidity and alkalinity, but in lichens, such control mechanisms are relatively primitive. This is considered to be the reason why lichens are most severely affected by SO_2, resulting in "lichen deserts" in regions where SO_2 emissions are especially high (e.g. Nash and Worth, 1988).

SO_2 causes visible injury characterised by chlorosis of leaf tissue (whitened areas of dying tissue where the pigments have been broken down). Even when no visible injury is apparent, SO_2 can cause a reduction in growth and yield. However, in sulphur (S) deficient areas, low levels of SO_2 may actually be beneficial. Changes in fertiliser practice coupled with SO_2 emission reductions over the last three decades in Western Europe and North America have resulted in the occurrence of S deficiency in some agricultural species growing under specific soil types and climatic conditions (Blake-Kalff *et al.*, 2000). SO_2 can also indirectly affect crop yields through effects on the prevalence of plant pathogens and insect pests (Bell *et al.*, 1993), as can NO_x.

Nitrogen oxides (NO_x)

The predominant pathway of NO_x entry into plant leaves is through the stomata, although cuticular resistances to NO_2 entry are lower than for both SO_2 and O_3. The biochemical effects of NO and NO_2 are quite different and there is some uncertainty over which oxide is more toxic (CLAG, 1996). NO_x can reduce plant growth at high concentrations, although growth stimulations can be caused by low NO_x concentrations, generally under situations of low soil nitrogen (CLAG, 1996). However, even if growth is stimulated, exposure to NO_x can have adverse effects such as heightened sensitivity to drought, pests, and in some cases, to frost (CLAG, 1996). Rare instances of visible injury caused by exposure to very high concentrations of NO_2 are characterised by chlorotic areas on leaves associated with necrotic patches. Prolonged exposure to NO_x has been shown to suppress plant growth via inhibition of photosynthesis. The combination of NO_x with other pollutants has been found to cause synergistic effects on plants (i.e. affect vegetation to a greater extent in combination than individually). This is particularly true of NO_2 and SO_2 (Ashenden and

Mansfield, 1978) but synergistic effects have also been observed between NO_x and O_3 (CLAG, 1996).

Ozone (O_3)

Unlike SO_2 and NO_x, consideration of the toxicity of O_3 is not complicated by its role as a source of an essential nutrient. O_3 transfer via the leaf cuticle is negligible and O_3 uptake is almost entirely through the stomata. On entry to the sub-stomatal cavity, O_3 reacts with constituents of the aqueous matrix associated with the cell wall to form other derivatives which result in the oxidation of the sensitive components of the plasmalemma, and subsequently the cytosol. The inability to repair or compensate for altered membrane permeability can manifest itself as symptoms of visible injury, which are generally associated with short-term exposures to high O_3 concentrations. Symptoms of acute injury on broad-leafed plants include chlorosis, bleaching, bronzing, flecking, stippling and uni- and bifacial necrosis. On conifers, tip necrosis, mottling and banding are all common symptoms (Kley *et al.*, 1999). Chronic exposures may or may not result in visible foliar symptoms (usually characterised by chlorosis, premature senescence and leaf abscission). However, reductions in growth from chronic exposures are well documented and can result in crop yield losses, reductions in annual biomass increments for forest trees and shifts in species composition of semi-natural vegetation (Fuhrer *et al.*, 1997).

Suspended particulate matter (SPM)

SPM can produce a wide variety of effects on the physiology of vegetation that in many cases depend on the chemical composition of the particles. Heavy metals and other toxic particles have been shown to cause damage and death of some species as a result of both the phytotoxicity and the abrasive action during turbulent deposition. Heavy loads of particles can also result in reduced light transmission to the chloroplasts and the occlusion of stomata, decreasing the efficiency of gaseous exchange (and hence water loss). They may also disrupt other physiological processes such as bud-break, pollination and light absorption/reflectance. Indirect effects of particulate deposition such as predisposition of plants to infection by pathogens and the long-term alteration of genetic structure have also been reported. In contrast, a few instances of particle deposition producing positive growth responses have also been observed and related to the capture and utilisation of nutrient particles from the atmosphere (Beckett *et al.*, 1998).

Fluorides

Both gaseous and particulate fluorides are deposited on plant surfaces and some penetrate directly into the leaf if the cuticle is old or weathered. Fluoride dissolved in water on the leaf surface can be absorbed by diffusion through the cuticle (Brewer *et al.*, 1960). Gaseous fluoride is absorbed via the stomata and transported by the transpirational flow in the apoplast, and can accumulate at toxic levels in the tips and margins of the leaves (Jacobsen *et al.*, 1996). Plant species show a wide range of susceptibilities to fluoride; for example, studies have shown that young conifers and vines are especially sensitive, whilst tea and cotton are resistant to fluoride pollution (Wellburn, 1994). Leaf injury in the form of chlorosis and necrosis of leaf tips and margins has been described for a number of species in relation to emissions from aluminium smelters (Treshow and Anderson, 1989). Studies have also shown fluorides to cause reductions in photosynthesis, respiration and metabolism of amino acids and proteins. Accumulation of fluoride in plants has been associated with fluorosis in grazing animals. Animals grazing on pasture close to brickworks, smelters, and phosphate fertiliser factories, or fed forage gathered from such areas, may develop fluorosis, a condition characterised by damage to the musculoskeletal system including softening of teeth, difficulty in mastication, lameness and painful gait (e.g. Patra *et al.*, 2000).

4. Significance of Agriculture and Forestry

This book is primarily concerned with assessing the impacts of air pollution on crops and forests in developing regions. Therefore, it is worth considering briefly how agriculture and forestry in these regions has changed in recent years and the impact this may have, both now and in the future, on sustainable food production, and on forest provision for the needs of the human population. It is especially important to consider air pollution impacts in the context of other environmental constraints on agriculture and forest ecosystems, since the added stress of air pollution may well compound and exacerbate environmental problems that are perhaps currently more widely acknowledged.

Developing countries currently comprise about 52% of the world arable land and 75% of the world population. In the future, the area covered by arable land is expected to stay the same or decline, in contrast the population of developing countries is expected to increase to 90% of the world total by 2050. Furthermore, at least three billion (about 30% of the population) will live in arid and semi-arid regions with severe water shortage problems and desertification (Lal, 2000). Figure 4 clearly shows that the area of arable land available per head of population is decreasing in all regions (UNEP, 2000) and that the global availability of cropland has now fallen by 25% over the past two decades. This is in part due to the rapidly increasing global population but is also exacerbated by other factors that affect the availability and productivity of agricultural land. These include conversion of agricultural land to other uses (i.e. urbanisation and infrastructure development), human-induced degradation of soils and the shortage of renewable fresh water.

Soil degradation occurs for a number of reasons, some of which are briefly described here. The cultivation of marginal lands frequently results in nutrient depletion as a result of soil inputs being lower than the products harvested. Pollution and contamination of soils by disposal of urban and industrial wastes also results in soil degradation. Degradation of irrigated land by salinisation is a serious problem, with areas of land affected by soil salinity estimated at 14% of the total irrigated land area in India, 13.5% in China, 24.2% in Pakistan, 26.2% in Mexico, and 27.3% in Egypt (Lal, 2000). Farmers have traditionally satisfied increasing demand by ploughing new land (extensification) but opportunities for expansion are now limited primarily to marginal areas. Raising productivity (intensification) has therefore been central to increased grain production. Fertiliser use continues to rise in many

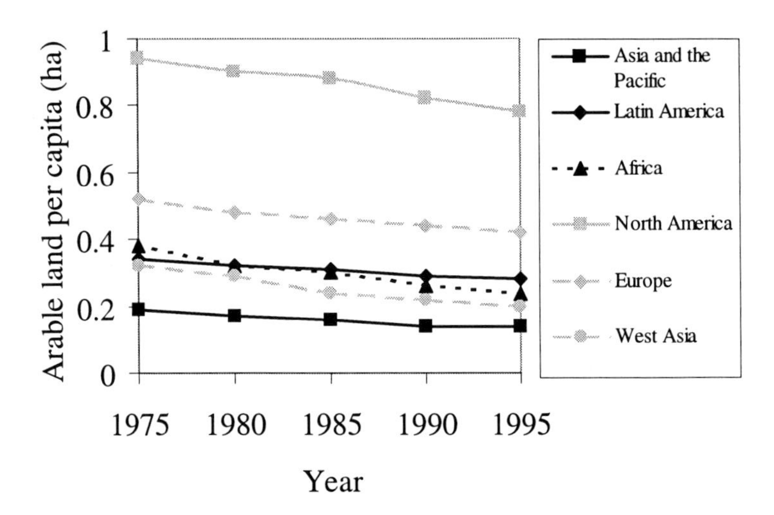

Fig. 4. Regional decrease in arable land per capita (UNEP, 2000).

developing countries, although there is concern about diminishing returns from increased applications and the threat of nitrate pollution of freshwater supplies and eutrophication. Irrigation has also been an important factor in providing increased grain yields with irrigated areas expanding at 2.3% yearly from 1950 to 1995 (FAO, 1997). However, such expansion is not sustainable over the long term and further increases in yield are likely to be limited by the availability of freshwater, with populous and arid areas most likely to be hardest hit in the future.

Breeders have significantly boosted the yield potential of cereals over recent decades, and the currently controversial use of genetically modified strains may do so further. However, crop-breeding programmes may also tend to select for higher values of stomatal conductance since this will increase rates of CO_2 assimilation. This may inadvertently be increasing the potential sensitivity of plants to air pollutants since the main pathway of pollutant uptake is through the stomata. This was suggested to be the reason for the greater sensitivity to O_3 of more recent wheat cultivars observed in Greece (Velissariou *et al.*, 1992).

In **Asia and the Pacific**, the extent of land conversion for agricultural activities varies within the region. For example, China now has

0.12 ha capita^{-1} of arable land, which is less than half the world average. In addition, 60% of the Chinese arable land has some degree of quality constraints in terms of soil type, fertility, slope or climate. The most marked reduction in cultivated area is seen in the eastern coastal provinces where agricultural land has been lost due to the development of new industries, roads and housing, as a consequence of the high economic growth rates. This decline in total arable area has so far been negated with an increase in fertiliser use; for example, during the period 1980 to 1995 the total use of fertilisers increased by a factor of three (Aunan *et al.*, 2000). This increased use of fertiliser is likely to contribute significantly to increased emissions of NO_x, and it has been estimated that the N-fertiliser induced soil emission of NO_x may represent as much as 14% of present NO_x emissions in Eastern Asia (Chameides *et al.*, 1999).

In **Africa**, agriculture contributes about 40% of regional GDP and employs more than 60% of the labour force (World Bank, 1988). Land degradation caused by soil erosion, declining fertility, salinisation, soil compaction, agrochemical pollution and desertification poses serious problems throughout Africa. One estimate suggests that African crop yields could be halved within 40 years if the degradation of cultivated lands were to continue at present rates (UNEP, 2000). As a result of declining food security, the number of undernourished people in Africa nearly doubled from 100 million in the late 1960s to nearly 200 million in 1995 (UNEP, 2000). However, the agricultural potential of the country has not yet been fully realised; of an estimated 632 million hectares of arable land in Africa, only 179 million hectares are currently cultivated (FAO, 1997). Unfortunately, much of the land not currently utilised is often far from centres of population where infrastructure is poor. In addition, nearly 40% of this land is found in only three countries (the Democratic Republic of the Congo, Nigeria and the Sudan) so the potential for agricultural extensification and associated benefits are likely to be unevenly distributed across the region.

Latin America has the world's largest reserves of cultivatable land with the agricultural potential of the region estimated at 576 million hectares. However, almost 250 million hectares of land are affected by land degradation with soil erosion constituting the major threat (UNEP,

2000). In addition, the environmental costs of improved farm technologies have been high. For example, during the 1980s Central America increased production by 32% and its cultivated area by 13%, but doubled its consumption of pesticides (FAO, 1997). The depletion and destruction of forest resources to clear land for agricultural activities, especially in the Amazon region, is also of regional and global concern.

In terms of forestry, much of the remaining intact natural forested areas are to be found in the Amazon basin, Canada, central Africa, southeast Asia and the Russian Federation. These areas are of particular value since they contribute to local and national economic growth, provide recreational areas and are rich in biodiversity. For example, throughout Africa there has been an increasing demand for wood products, with consumption almost doubling during 1970 to 1994. At least 90% of Africans depend on firewood and other biomass for energy needs (FAO, 1997). Domestic wood shortage is occurring in some areas of Asia and the Pacific, such as the Philippines, Thailand and in South Asia. In many developing countries, forested areas are cleared as populations expand and pressures to exploit natural resources increase, with the cheapest means of clearing forested areas being by fire.

5. Issues and Lack of Current Knowledge

Most scientific investigations into the impacts of air pollution on vegetation have been conducted during the last 40 years in North America and western Europe. These studies have resulted in the definition of dose-response relationships or threshold concentrations for adverse effects for a number of different air pollutants and for a number of different species and cultivars. These dose-response relationships have been used in a variety of ways (discussed in detail in Chaps. 2 to 5 and 14) to assess the risk to vegetation of ambient pollution levels, with the overall aim of informing appropriate emission abatement policy. The information presented in Chaps. 6 to 13 clearly shows that current ambient levels of pollution in several developing countries are causing visible injury and losses in productivity of many species of agricultural crops and forest trees. However, it is very difficult to assess the full magnitude of damage and productivity loss due to air pollution in developing regions.

This introduction has described the recent increases in pollutant emissions in certain developing regions and shown that such trends are likely to continue into the future. In addition, evidence of both environmental and socio-economic constraints on the expansion of agriculture, and statistics describing decreases in arable land unit per capita, give serious cause for concern as to the ability of agriculture to satisfy future human needs. Chameides *et al.* (1994) emphasise the proximity of agriculture to fossil-fuel burning centres by identifying continental scale metro-agro complexes (CSMAPs) namely in North America, Europe, Eastern China and Japan. The proximity of agriculture to urban-industrial emission sources also occurs on a smaller scale in other developing country regions as described, for example, around the Cairo region in Egypt (Abdel-Latif, this volume) the Punjab region of Pakistan (Wahid, this volume) and peri-urban areas in India (Agrawal, this volume). These examples emphasise the potential threat to agricultural production caused by both present and future air pollution concentrations.

In addition, the clearance of forested areas has increased the importance of the remaining intact forest areas for wood products, economic growth, recreation and biodiversity. Carbon sequestration in forests is also a major component of the global carbon cycle. Any additional stresses caused by air pollution impacts could have serious implications for a wide range of forest-related socio-economic factors, especially in those forested areas located close to pollution sources. From the information collated in Chaps. 6 to 13, it is clear that most air pollution studies have focussed on investigating effects on agricultural crops rather than forest trees, even though severe local impacts have been identified for forests in some areas.

It is therefore crucial that stakeholders (e.g. in agriculture, forestry, industry and government) are made aware of the potential impacts of air pollution in different regions. This can only be achieved by identification of the magnitude and spatial extent of impacts resulting from current ambient air pollution levels, and assessment of how such impacts are likely to change in the future. This requires a number of different steps. Firstly, the impacts of air pollutants on different species and cultivars need to be clearly understood and reliable dose-response relationships defined for local varieties of important species. Secondly, monitoring networks must be established to record pollutant concentrations in rural agricultural/forested areas as well as urban/industrial locations. Thirdly, reliable air pollution transport models need to be developed to estimate current and future regional pollution levels. Finally, suitable methodologies must be developed to relate pollution to impacts, for use in risk assessment by policy-makers. For these policy assessments, it is of course important to consider impacts on crops and forests alongside those on human health and on the built environment.

A number of studies have already attempted to apply European and North American dose-response relationships to estimate crop yield reductions in developing countries (e.g. Chameides *et al.*, 1999; Chameides *et al.*, 1994; Aunan *et al.*, 2000). The fundamental uncertainty associated with these assessments lies in the use of dose-response relationships established for North American or European species and cultivars, and the assumption that local species or cultivars will respond in a similar way to pollutant doses. However, there may be significant differences in the responses of local varieties from developing country regions. Further research to establish local species and cultivar responses to air pollution should be a high priority for the future. Additional uncertainties that may alter plant response to air pollution have been identified as those associated with climate, agricultural management practices (e.g. the use of fertilisers, pesticides, herbicides and irrigation); crop phenology; and pollutant exposure patterns (Massman *et al.*, 2000). These and other complicating factors related to the potential transferability of dose-response relationships and impact assessment methodologies are discussed in detail in Chap. 14.

6. Aims and Objectives

The objective of this book is to assess existing evidence describing the impacts of air pollutants on vegetation in developing countries and regions. However, in order to put this evidence into context, Chaps. 2 to 5 first provides a summary of the effects of air pollution on vegetation in industrialised countries/regions (i.e. North America, Europe,

Australia and Japan), where information is more extensive. This summary involves, for each region, a description of emission trends, vegetation composition and key observational and experimental evidence of air pollutant impacts on vegetation. Current dose-response and risk assessment methods are also discussed in relation to the scale of the air pollution problem. Chapters 6 to 13 is the central section of this book, describing evidence from developing regions of air pollutant impacts on vegetation, using country-specific case studies. Regionally selected experts (from China, Taiwan, India, Pakistan, Egypt, South Africa, Brazil and Mexico) collected evidence describing regional emissions, vegetation distributions and local pollutant concentrations. This information was related to observations and experimental evidence of injury to crops and forests in the field, in an attempt to define levels of ambient pollutant concentrations that cause vegetation damage. Key observations of damage, including instances of visible injury in the field, transect studies along pollution gradients and controlled experimental investigations of impacts on selected crop and forest species, were described for five pollutants. These pollutants, chosen for their prevalence and known phytotoxicity, were SO_2, O_3, NO_x, SPM and F.

The potential of using this information to define dose-response relationships is considered in Chap. 14. Information collected for SO_2 and O_3 has been collated and is compared with North American, European, Australian and Japanese exposure-response relationships in order to investigate the variability in response of different vegetation types to pollutants by region. Regional and global models describing pollutant concentrations have been used to give some indication of the spatial extent of air pollution problems. Results from a global O_3 and regional SO_2 model are presented to compare predicted air pollutant concentrations with the location of field observations of damage. This information is used to indicate the probable extent of vegetation damage both for the present day and in the future. Finally, Chap. 15 discusses how assessments of air pollution damage to crops and forests have been used to guide pollution abatement policies in industrialised countries, and considers how applicable these techniques are to the developing world. Chapter 15 also includes a case study describing research in peri-urban areas of India in which an innovative interdisciplinary approach has been used to assess the social and economic policy implications of air pollution damage.

References

Abdul Rahim N. (2000) General problems associated with air pollution in developing countries. In *Air Pollution and the Forests of Developing and Rapidly Industrializing Countries,* eds. Innes J.L. and Haron A.H. IUFRO Research Series 4. DCABI Publishing, Oxfordshire, U.K.

Adams R.M., Glyer J.D. and McCarl B.A. (1988) The NCLAN economic assessment: approaches and findings and implication. In *Assessment of Crop Losses from Air Pollutants,* eds. W.W. Heck, O.C. Taylor and D.T. Tingey. Elsevier, London, pp. 473–504.

ADB (1994) *The Environment Programme: Past, Present and Future.* Asian Development Bank, Manila.

ADB (1997) *Emerging Asia. Changes and Challenges.* Asian Development Bank, Manila.

Ashenden T.W. and Mansfield T.A. (1978) Extreme pollution sensitivity to grasses when SO_2 and NO_2 are present in the atmosphere together. *Nature* **273**, 142–143.

Ashmore M.R. and Marshall F.M. (1999) Ozone impacts on agriculture: an issue of global concern. *Adv. Bot. Res.* **29**, 32–52.

Aunan K., Bernsten T.K. and Seip H.M. (2000) Surface ozone in China and its possible impact on agricultural crop yields. *Ambio* **29**, 294–301.

Beckett K.P., Freer-Smith P.H. and Taylor G. (1998) Urban woodlands: their role in reducing the effects of particulate pollution. *Env. Pollut.* **99**, 347–360.

Bell J.N.B., McNeill S., Houlden S., Brown V.C., Mansfield P.J. (1993) Atmospheric change: effect on plant pests and diseases. *Parasitology* **106**, S11–S24.

Blake-Kalff M.M.A., Hawkesford M.J., Zhao F.J. and McGrath S.P. (2000) Diagnosing sulfur deficiency in field-grown oilseed rape (*Brassica napus* L.) and wheat (*Triticum aestivum* L.). *Plant and Soil* **225**, 95–107.

Brewer R.F., Sutherland F.H. and Guillemet F.B. (1960) Sorption of fluorine by citrus foliage from equivalent solutions of HF, NaF, NHF_4 and H_2SiF_6. *Am. Soc. Horti. Sci.* **76**, 215–219.

Chameides W.L., Kasibhatla P.S., Yienger J. and Levy H. (1994) Growth of continental-scale metro-agro-plexes, regional ozone pollution, and world food production. *Science* **264**, 74–77.

Chameides W.L., Xingsheng L., Xiaoyan T., Xiuji Z., Chao L., Kiang C.S., St. John J., Saylor R.D., Liu S.C., Lam K.S., Wang T. and Giorgi F. (1999)

Is ozone pollution affecting crop yields in China? *Geophys. Res. Lett.* **26**, 867–870.

Chappelka A.H. and Samuelson L.J. (1998) Ambient ozone effects on forest trees of the eastern United States: a review. *New Phytol.* **139**, 91–108.

CLAG (1996) *Critical Levels of Air Pollutants for the United Kingdom.* Critical Loads Advisory Group, Institute of Terrestrial Ecology, Penicuik.

EIA (1995) *International Energy Annual: 1993.* Energy Information Agency, US Department of Energy, Washington DC, United States.

FAO (1997) *State of the World's Forests 1997.* FAO, Rome.

Fuhrer J., Skärby L. and Ashmore M.R. (1997) Critical levels for ozone effects on vegetation in Europe. *Env. Pollut.* **97**, 91–106.

Glover D. and Jessup T. (1999) *Indonesia's Fires and Haze. The Cost of Catastrophe.* ISEAS Publications, Singapore.

Grennfelt P., Hov Ø. and Derwent D. (1994) Second generation abatement strategies for NO_x, NH_3, SO_2 and VOCs. *Ambio* **23**, 425–433.

Hameed S. and Dignon J. (1992) Global emissions of nitrogen and sulfur oxides in fossil fuel combustion 1970–86. *J. Air Waste Manag. Assoc.* **42**, 159–163.

Hoffer T.E., Farber R.J. and Ellis E.C. (1982) Background continental ozone levels in the rural US southwestern desert. *Sci. Total Env.* **23**, 17–30.

Jacobsen J.S., Weinstein L.H, McCune D.C. and Hitchcock A.E. (1966) The accumulation of fluorine by plants. *J. Air Pollut. Control Assoc.* **16**, 412–417.

Kley D., Kleinmann M., Sanderman H. and Krupa S. (1999) Photochemical oxidants; state of the science. *Env. Pollut.* **100**, 19–42.

Kuylenstierna J.C.L., Rodhe H., Cinderby S. and Hicks K. (2001) Acidification in developing countries: ecosystem sensitivity and the critical load approach on a global scale. *Ambio* **30**, 20–28.

Lal R. (2000) Soil management in the developing countries. *Soil Sci.* **165**(1), 57–72.

Massman W.J., Musselman R.C. and Lefohn A.S. (2000) A conceptual ozone dose-response model to develop a standard to protect vegetation. *Atmos. Env.* **34**, 745–759.

McCormick J. (1997) *Acid Earth — The Politics of Acid Pollution,* 3rd Ed. Earthscan Publications, London.

Middleton J.T., Kendrick J.B. and Schwalm H.W. (1950) Injury to herbaceous plants by smog or air pollution. *Plant Dis. Rep.* **34**, 245–252.

Miller P.R. and McBride J., eds. (1999) *Oxidant Air Pollution Impacts in the Montane Forests of Southern California: The San Bernadino Mountain Case Study.* Springer-Verlag, New York.

Nash T.H. and Worth V. (1988) *Lichens, Bryophytes and Air Quality.* Cramer, Berlin.

Olivier J.G.J., Bouwman A.F., van der Maas C.W.M., Berdowski J.J.M., Veldt C., Bloos J.P.J., Visschedijk A.J.H., Zandveld P.Y.J. and Haverlag J.L. (1996)

Description of EDGAR Version 2.0: A set of global emission inventories of greenhouse gases and ozone-depleting substances for all anthropogenic and most natural sources on a per country basis and on 1° × 1° grid. National Institute of Public Health and the Environment (RIVM) report no. 771060 002/TNO-MEP report no. R96/119. http://www.rivm.nl/env/int/coredata/edgar/

Patra R.C., Dwivedi S.K. and Bhardwaj B. (2000) Industrial fluorosis in cattle and buffalo around Udaipur, India. *Sci. Total Env.* **253**, 145–150.

Rosa L.P., and Schaeffer R. (1995) Global warming potentials: the case of emissions from dams. *Energy Policy* **23**, 149–158.

Siversten B., Matale C. and Pereira L.M. (1995) *Sulphur Emissions and Transfrontier Pollution in Southern Africa.* SADC, Maseru, Lesotho.

Skärby L., Ro-Poulsen H., Wellburn A.R. and Sheppard L.J. (1998) Impacts of ozone on forests: a European perspective. *New Phytol.* **139**, 109–122.

Streets D.G. and Waldhoff S.T. (2000) Present and future emissions of air pollutants in China: SO_2, NO_x, and CO. *Atmos. Env.* **34**, 363–374.

Tingey D.T., Olsyk D.M., Herstrom A.A. and Lee E.H. (1993) *Effects of Ozone on Crops.* In *Tropospheric Ozone: Human Health and Agricultural Impacts,* ed. McKee D.J. Lewis Publishers, Boca Raton, pp. 175–206.

Treshow M. and Anderson F.K. (1989) *Plant Stress from Air Pollution.* John Wiley, Chichester.

UNEP (1997) *Global Environment Outlook — 1.* United Nations Environmental Programme, Global State of the Environment Report 1997. Oxford University Press, New York.

UNEP (2000) *Global Environment Outlook — 2.* United Nations Environmental Programme, Global State of the Environment Report 2000. Oxford University Press, New York.

van Tienhoven A.M. (2000) Forestry problems in Africa. In *Air Pollution and the Forests of Developing and Rapidly Industrializing Countries,* eds. Innes J.L. and Haron A.H. IUFRO Research Series 4. DCABI Publishing, Oxfordshire, U.K.

Velissariou D., Barnes J.D. and Davison A.W. (1992) Has inadvertent selection by plant breeders affected the O_3 sensitivity of modern Greek cultivars of spring wheat? *Agriculture Ecosystems Env.* **3**, 79–89.

Volz A. and Kley D. (1998) Evaluation of the Montsouris series of ozone measurements made in the nineteenth century. *Nature* **332**, 240–242.

Wellburn A. (1994) *Air Pollution and Climate Change,* 2nd Ed. Longman Scientific and Technical. New York.

Winterhalder K. (1996) Environmental degradation and rehabilitation of the landscape around Sudbury, a major mining and smelting area. *Env. Rev.* **4**, 185–224.

World Bank (1998) *The World Bank and Climate Change*: Africa. http://www.worldbank.org/html/extdr/climchng/afrclim.htm
World Resources Institute (1994) *World Resources 1994–1995*. A Joint Report by WRI, UNEP and UNDP. Oxford University Press, New York.
World Resources Institute (1996) *World Resources 1996–1997*. A Joint Report by WRI, UNEP and UNDP. Oxford University Press, New York.

Air Pollution Impacts on Vegetation in Industrialised Countries

The effects of air pollution on crops and forests have been studied intensively over much of the last century in Europe and North America, and other industrialised regions. However, it is only in the last 20 years that methods have begun to be developed to assess the geographical distribution of the risk of significant impacts. This has become particularly important as acute local damage has become less significant compared with the wide-scale impacts of chronic exposures to regional air pollutants, such as ozone. Such continental-scale evaluations offer many challenges, because of the diversity of vegetation types, climatic conditions, soils and pollution exposure patterns. Indeed, despite the extent of research into the issue, many problems remain to be resolved in linking local field observations and experimental studies to wider-scale assessments. The sections within this chapter consider pollutant emissions, pollutant concentrations in relations to standards, evidence of significant impacts on crops and forests, and risk evaluation for four industrialised countries or regions. The individual authors commissioned to collate this information have chosen to focus on specific aspects of air pollution within their country or region that are broadly complementary.

The first section of this chapter, on North America, provides an overview of the evidence for adverse effects on vegetation across the continent. This section emphasises the importance of major collaborative research programmes, both nationally and regionally, which have provided significant advances in both the understanding of fundamental mechanisms of pollution impact and in developing risk assessment tools. These programmes initially focused on arable crops, but since the mid-1980s the more difficult problem of evaluating long-term impacts on major forest communities has been addressed. The results of these programmes, integrating field observations and experimental studies, have emphasised the importance of long-term impacts of air pollution in reducing carbon reserves, ▶

increasing the impacts of water stress and reducing nutrient availability.

The second section, on Europe, places particular emphasis on the development and application of air quality criteria, in the form of critical loads and levels. This is because, uniquely, these values have played an important role in international negotiations to reduce the impacts of air pollutants across Europe on the natural environment. This approach has led to a number of international agreements to reduce pollutant emissions which are based on assessment of effects, the most recent being the Gothenburg Protocol, which includes acidification, eutrophication and ground-level ozone. Both the North American and European contributions emphasise the importance of developing more mechanistic process-based models as the key step in providing risk evaluations that are more closely related to actual effects in the field.

The final two sections consider, more briefly, the extent of air pollution impacts in two contrasting parts of the world: Japan, a country with a very high population density and a subtropical climate, and Australia, a country with a low population density and a largely arid climate. While regional-scale pollutants, such as ozone and acid rain, are, as in Europe and North America, the major concerns in Japan, the focus in Australia is more on the local impacts of SO_2 and fluoride in industrial zones. Both contributions emphasise the development of dose-response functions for a range of species as key tools in assessing the impacts of ambient pollutant concentrations.

CHAPTER 2

AIR POLLUTION IMPACTS IN NORTH AMERICA

K. Percy

1. Introduction

There is a long record of air pollution damage to forests and crops in North America. Regional levels of air pollution have been positively linked to decline in at least four diverse and widely separated forest types (McLaughlin and Percy, 1999). Pollutants of concern to forests at the regional scale include wet- (not discussed in this paper) and dry-deposited acids and ozone (O_3). In the past, sulphur dioxide (SO_2) emissions from industrial processes such as sulphide-ore smelting and fossil fuel combustion (petroleum/coal) have been sufficient to induce acute injury to foliage on a large number of tree/crop species across North America (cf. Legge *et al.*, 1998). Plate 1 show examples of SO_2 injury to two tree species. The best known case history (SO_2, particulate, Cu, Ni) in North America occurred around the smelter complex in Sudbury, Ontario, Canada. By 1979, the areal extent of damage due to long-term exposure (early 1900s onwards) of the surrounding boreal forest ecosystem was manifested in no forest communities within 3 km SSE of the smelters, only pockets of remnant forest up to 8 km, and bare hilltops as far as 15 km (Freedman and Hutchinson, 1980a). Following installation of a 380 m smokestack emitting over three million tons of SO_2 in 1973, only 3% of sulphur

Plate 1. Field injury to Sumac (*Rhus* species) in the U.S., caused by exposure to SO_2 (source: L. Weinstein).

was calculated to be deposited within a 60 km radius and high levels of Ni and Cu were found in soils up to 70 km away (Freedman and Hutchinson, 1980b). With significant reduction in SO_2 and metal emissions from the mid-1980s onwards, rehabilitation of 10,000 ha of barren land and 36,000 ha of stunted, open birch-maple woodland is underway (Winterhalder, 1996).

With successful implementation of SO_2 emission control strategies in North America, vegetation injury is now restricted to localised geographic areas (Legge *et al.*, 1998) around smaller point sources. Legge *et al.* (1996) summarised boreal forest response to sour gas emissions in a 25-year study and demonstrated a relationship between ambient SO_2 exposure and boreal tree growth. The pollutant of most concern to crops remains O_3 with large economic impacts occurring over wide geographic areas in both the United States (EPA, 1996) and southern Ontario, Canada.

1.1. *Recent Emission and Concentration Trends*

Emissions of SO_2 in eastern North America have declined from over 20 million tons in 1980 to under 17 million tons in 1993 (EC, 1997a). U.S. air concentrations of SO_2 decreased 37% between 1985 and 1995 and particulate SO_4 concentration reductions have been widespread, except at several mid- to high-elevation, forested areas (NAPAP, 1998). U.S. emissions of anthropogenic NO_x decreased 6.5% from 23.3 million tons in 1980 to 21.8 million tons in 1995 (NAPAP, 1998). Trends in nitrogen species (HNO_3, NO_3) were generally more variable than for sulphur. Inputs of dry deposition may add 8–37% more S, and 15–65% more N, to that received via wet deposition (rain, cloud, fog), the dominant form of S and N deposition, depending on the region (EC, 1997a).

Composite national daily maximum 1h O_3 concentrations in the U.S. decreased 15% between 1987 and 1996. The highest national 1h maximum was in 1988. Ozone levels have declined 10% since 1987 at 194 rural monitoring sites (EPA, 1998). In Canada, time series analysis (Dann and Summers, 1997) identified a significant declining trend in daily maximum O_3 concentrations ranging from −0.05% to −0.08% yr^{-1}. Average days per year (1986–1993) exceeding the pre-2000 one-hour National Ambient Air Quality Objective (NAAQO) ranged from 18 in southern Ontario to 3 in the Southern Atlantic Region and 2 in the Lower Fraser Valley, British Columbia.

In the U.S., particulate matter less than 10 μm diameter (PM_{10}) trends were published for the first time in 1997. Composite average (845 sites) PM_{10} concentrations decreased 26% between 1988 and 1997 (EPA, 1999). In Canada, total suspended particulate decreased 54% during 1974–1992 (EC, 1999). Both countries have recently revised standards for particulate matter less than 2.5 μm in diameter ($PM_{2.5}$), of particular concern to human health (Tables 1 and 2).

1.2. *Air Quality Criteria*

Canada and the U.S. differ in approach to air pollution regulation. In Canada, the federal government sets National Ambient Air Quality Objectives (NAAQO) on the basis of recommendations from the Ca-

Table 1. Canadian Ambient Air Quality Objectives and Canada-Wide Standards.

Canada-Wide Standards (CWS)			
Ozone*	65 ppb		
$PM_{2.5}$†	30 μg m^{-3}		
National Ambient Air Quality Objectives (NAAQO)‡			
	Desirable	Acceptable	Tolerable
Sulphur dioxide (ppb)			
1 Hour	172	334	
24 Hours	57	115	306
Annual	11	23	
Nitrogen dioxide (ppb)			
1 Hour		231	532
24 Hours		106	160
Annual	32	53	

Sources: Environment Canada (1999): http://www.ec.gc.ca/pdb/uaqt/obj_e.html
Environment Canada (2001): http://www.ec.gc.ca/air/gov-efforts_e.shtml

*A Canada-Wide Standard (CWS) of 65 ppb, 8-hour averaging time by 2010. Calculated as the 4th highest measurement annually, averaged over three consecutive years.

†A CWS of 30 μg/m^{-3}, 24-hour averaging time by year 2010. Achievement to be based on the 98th percentile ambient measurement annually, averaged over three consecutive years.

‡The maximum desirable level defines the long-term goal for air quality and provides the basis for an anti-degradation policy in unpolluted areas of the country. The maximum acceptable level is intended to provide adequate protection against adverse effects on humans, animals, vegetation, soil, water, materials and visibility. The maximum tolerable level is determined by time-based concentrations of air contaminants.

nadian Environmental Protection Act/Federal-Provincial Advisory Working Group on Air Quality Objectives and Guidelines. The NAAQO are listed as maximum desirable, acceptable and tolerable concentrations (Table 1). Provincial governments had the option of adopting these as objectives or enforceable standards. NAAQO are designed to provide protection to human health, vegetation, animals and materials. Recently, Canada-Wide Standards (CWS) for particulate matter less than 2.5 μm diameter ($PM_{2.5}$) and O_3 have been developed to protect human health under auspices of the Canadian Council of

Table 2. United States National Ambient Air Quality Standards (NAAQS).

Pollutant	Standard value	Standard type
Ozone*+		
1-Hour average	120 ppb	Primary and Secondary
8-Hour average	80 ppb	Primary and Secondary
$PM_{2.5}$+		
Annual arithmetic mean†	15 μg m^{-3}	Primary and Secondary
24-Hour concentration‡	65 μg m^{-3}	Primary and Secondary
PM_{10}		
Annual arithmetic mean§	50 μg m^{-3}	Primary and Secondary
24-Hour concentration¶	150 μg m^{-3}	Primary and Secondary
Sulphur dioxide (SO_2)		
Annual arithmetic mean	80 μg m^{-3}	Primary
24-Hour average	365 μg m^{-3}	Primary
3-Hour average	1300 μg m^{-3}	Secondary
Nitrogen dioxide		
Annual arithmetic mean	100 μg m^{-3}	Primary and Secondary

Source: EPA (2000) http://www.epa.gov/airs/criteria.html

+The ozone 8-hour standard and the $PM_{2.5}$ standard are pending U.S. Supreme Court reconsideration of a 1999 Federal Court ruling blocking implementation of the 1997 standards proposed by the U.S. EPA.

PM_{10} = particles with diameters ≤ 10 μm; $PM_{2.5}$ = particles with diameters ≤ 2.5 μm.

*As the three-year average of the 4th-highest daily maximum 8-hour average of continuous ambient air monitoring data over each year.

†As the three-year average of the annual arithmetic mean of the 24-hour concentrations from single or multiple population oriented monitors.

‡As the 98th percentile of the distribution of the 24-hour concentrations for a period of one year, averaged over three years, at each monitor within an area.

§As the arithmetic average of the 24-hour samples for a period of one year, averaged over three consecutive years.

¶As the 99th percentile of the distribution of the 24-hour concentrations for a period of one year, averaged over three years, at each monitor within an area.

Ministers of Environment (Table 1). The CWS of 65 ppb O_3, with an 8-hour averaging time by 2010 (Table 1), was adopted in 2000. Achievement is based on the 4th highest measurement annually, averaged over three consecutive years. The revised U.S. standard for ozone (Table 2) differs from the CWS only in attainment being based on an 8-hour 80 ppb level (EPA, 2001).

In the U.S., the United States Environmental Protection Agency (EPA) is responsible for setting National Ambient Air Quality Standards (NAAQS) for criteria pollutants (Table 2). Unlike the Canadian NAAQO, the NAAQS are legally enforced. Primary standards are designed to protect human health; secondary standards are designed to protect welfare including vegetation. NAAQS are set following a process during which a "criteria document" is developed comprising a compilation and scientific assessment of all health and welfare information available for the pollutant. EPA also develops a "staff paper" compiled by technical staff, to translate science into terms that can be used in making policy decisions. Both "papers" are based upon extensive consultation and scientific review, and are examined by the Congressionally-appointed Clean Air Scientific Advisory Committee prior to recommendations being made to the EPA Administrator for consideration in proposing revisions to the standards.

Both countries continue to make progress under the Ozone Annex to the 1991 Canada-United States Air Quality Agreement negotiated to reduce the transboundary movement of smog-causing pollutants. However, when the CWS is calculated for the 1996–1998 period using Canadian and U.S. data for eastern North America (Dann, 2001), only small portions in the extreme northeast are seen to meet the CWS. Highest 4th highest daily maximum 8-hour O_3 occurred in mid-coastal and mid-western areas, as well as the Lakes Michigan, Erie and Ontario basins with concentrations typically decreasing in a northeasterly direction (Plate 2).

1.3. *Vegetation*

North American forests are immense in range and ecological diversity. The 326 M ha of forested land in the U.S. (Powell *et al.*, 1992) and 417 M ha in Canada (NRCAN, 1998) represent a highly valuable

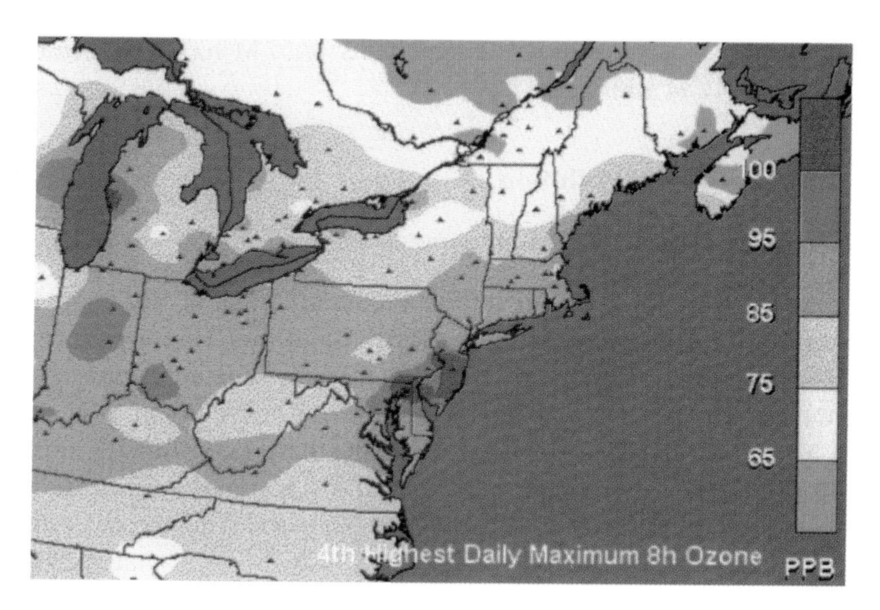

Plate 2. Northeast North American regional levels of 4th highest daily maximum O_3 calculated for the period 1996–1998 using the Canada-Wide Standard of 65 ppb, 8-hour averaging time. Triangles represent locations of Canadian and U.S. ground-level O_3 monitoring stations contributing data. Reproduced from Dann (2001).

economic resource for which maintenance of long-term productivity is a very high priority. Both actual and perceived potential responses of North American forests and crops to atmospheric pollution during recent decades have figured strongly in policy decisions on air quality regulation. During 1998 in Canada, 10.1 M ha of grains and oilseeds, along with 0.87 M ha speciality crops, were harvested (AAFC, 1998). Crops harvested and classified O_3- (Krupa *et al.*, 1998) and SO_2- (Legge *et al.*, 1998) sensitive included wheat, soybean, dry peas, dry beans and potato. SO_2-sensitive crops harvested included barley and oats. In the U.S. during 1997, 345 M ha were farmed (NASS, 1997). Crops harvested and classified as O_3- and SO_2-sensitive included wheat (23.8 M ha), soybean (26.8 M ha), cotton (5.3 M ha), potatoes (0.53 M ha) and tomato (0.15 M ha). O_3-sensitive crops harvested included edible dry beans (0.69 M ha), tobacco (0.32 M ha), rye (0.12 M ha) and red clover (0.02 M ha). SO_2-sensitive crops harvested included barley

(2.18 M ha), oats (1.09 M ha), sugar beets (0.57 M ha), and lettuce/romaine (0.12 M ha).

2. Impacts of Air Pollution

2.1. *Field Evidence of Effects on Forests*

Regional air pollution is an important component of the changing atmospheric environment in which North American forests are growing. The areal extent within which air pollutants are affecting forests at the regional level is shown in Fig. 1. High levels of mortality in northeastern hardwood forests since the early 1980s have been directly linked to air pollution. While sugar maple condition is generally improved in the U.S. (Stoyenhoff *et al.*, 1998) and Canada (Hall *et al.*, 1998), regional analysis of soil buffering in Ontario has shown that areas where critical loads of S/N were exceeded had higher levels of branch

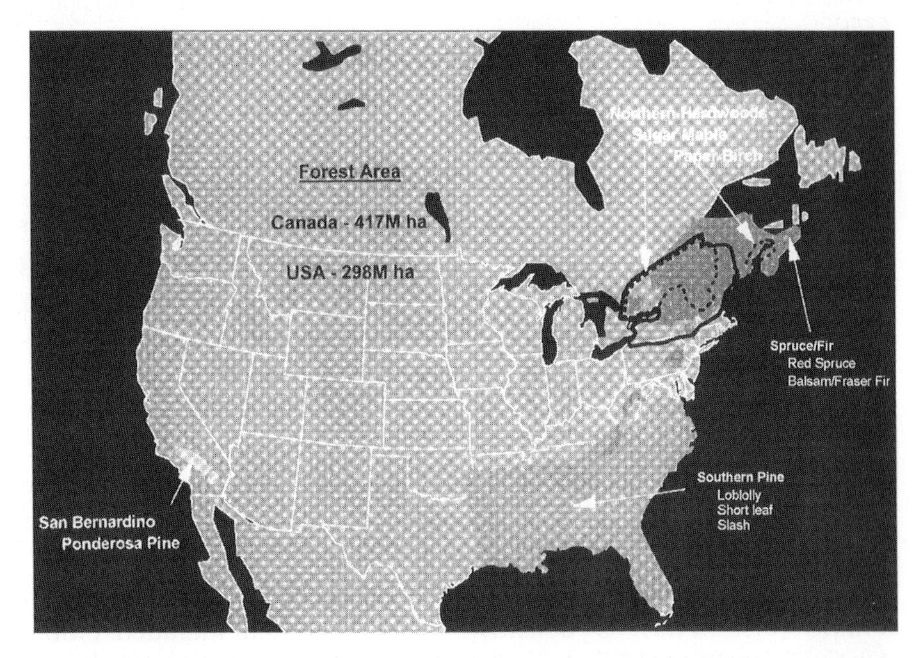

Fig. 1. Locations and area extent of four multi-disciplinary forest case studies used to evaluate process-level response to ambient levels of air pollution in regional forest types. Reproduced from McLaughlin and Percy (1999).

dieback (Arp *et al.*, 1996). Tree response has been primarily linked to soil sensitivity to acidic deposition, especially in the northern Appalachians. Accelerated loss of base cations from soils resulted in reduced soil Ca, Mg and K concentrations in sugar maple stands (Foster *et al.*, 1992). N saturation may also be partly responsible for nutrient imbalances and reduced growth on some sites (Yin *et al.*, 1994). However, O_3 cannot be excluded, although rigorous analysis of the circumstantial relationship between increased seasonal O_3 exposure and increased crown transparency has not been conducted. Interaction of cation leaching with extreme climate events may have been ultimately responsible for sugar maple decline, given the role of deep freezing of fine roots on canopy processes (Robitaille *et al.*, 1995). In eastern coastal birch forests, deterioration of white birches was first observed in the early 1980s (Magasi, 1989). Field studies subsequently documented a correlation between H^+ ion and NO_3 in fog and leaf browning (Cox *et al.*, 1996). The area also receives long-range transported O_3 from urban conurbations to the southwest and has the second highest annual mean O_3 concentration (34 ppb) in Canada (EC, 1997b).

Eastern-spruce forests occupy 10 M ha in eastern North America. In northern Appalachian spruce-fir forests, "winter injury" was first observed in the late 1950s on red spruce growing at high elevations. Recurring injury resulted in decreased radial growth and increased mortality in both cloud-/O_3-exposed, mountain (Johnson and Fernandez 1992) forests and fog-/O_3-exposed, coastal (Jagels *et al.*, 1989; Percy *et al.* 1993) red spruce forests. Growth decline of high elevation red spruce in southern Appalachian spruce-fir forests began later (1965). Canopy condition at low- and high-elevation sites declined during 1985–1989 (Peart *et al.*, 1992), while red spruce mortality increased with elevation and ranged from 3–4% yr^{-1} (Nicholas, 1992). A summary of research on decline of red spruce conducted under the Spruce-Fir Research Cooperative is available (Eagar and Adams, 1992). Dendro-climatological analysis (Cook and Johnson, 1989) has shown that growth decline was due to increased sensitivity of trees to winter (northern) and warmer late summer (southern) temperatures. Field studies have further shown that deposition of strong anions has reduced cation availability, reduced net carbohydrate production in foliage (Schaberg *et al.*, 1997), reduced photosynthesis and increased respiration (McLaughlin *et al.*, 1993). Acid deposition impacts on Ca availability (McLaughlin and Wimmer, 1999), causing reduced cold hardiness of foliage (DeHayes *et al.*, 1997) in the north, and acid deposition-induced Ca deficiency, leading to increased dark respiration, in the south are possible explanations.

Covering an area of 25 M ha in North America, significant growth declines in un-managed southern pines reported in the early 1980s (Sheffield and Cost, 1987) have been a major concern. The results of eight years of coordinated work under the Southern Commercial Forest Research Cooperative have been compiled by Fox and Mickler (1996). O_3 is known to significantly increase effects of soil moisture stress on stem growth of loblolly pine (McLaughlin and Downing, 1996). Somers *et al.* (1998) were recently able to correlate visible O_3 injury with radial growth of individual yellow-poplar trees in the Great Smokey Mountains National Park, but cautioned that this did not imply cause-effect. Acidic deposition effects in the short term are not expected to be significant (Teskey, 1995) and the main effects are expected to occur through primary impact on nutrient cycles. While 10–15% of commercial pine forest was believed to be limited by low cation supply (Binkley *et al.*, 1989), it has been shown that approximately 80% of exchangeable calcium has been lost over 30 years in a reference watershed, while base saturation has declined from > 55% in 1962 to 10% in 1990 (Richter *et al.*, 1994).

Since at least the mid-1950s, much of the 16 M ha of mixed conifer forest in Southern California has been exposed to the highest concentrations of O_3 in North America, including nighttime concentrations > 50 ppb at higher elevation sites. The detrimental role of N deposition is also well established, and is unique in that most occurs in dry form as acidic vapour, gaseous and particulate species. Deposition as high as 25–45 kg N ha^{-1} yr^{-1} has resulted in localised N saturation (Bytnerowicz and Fenn, 1996). Species most affected include ponderosa and Jeffrey pines. The complete history of the extensive, multidisciplinary case study in the San Bernardino Forest has recently been published (Miller and McBride, 1999). Effects from foliar level to successional stage have been documented. Early work by Miller (1973), when O_3 levels were highest, reported annual mortality between

2–2.5% per year. Concentrations ranging from 50–60 ppb O_3 induced foliar injury and early needle loss, decreased nutrient availability in stressed trees, reduced carbohydrate production with lessened tree vigour resulting in decreased height/diameter growth and increased susceptibility to bark beetles (Miller *et al.*, 1982). With gradually diminishing O_3 stress during 1976–1991, Miller *et al.* (1989) have reported an improvement (1974–1988) in the foliar injury index except at the most exposed plots.

2.2. *Field Evidence of Effects on Crops*

Ozone remains the most important air pollutant for crops in North America. Foliar injury from O_3 in Canada has been reported in a number of sensitive species; potato in New Brunswick, dry bean, soybean and tobacco in Quebec; dry bean, soybean, tomato, onion, tobacco, cucumber, radish, grape and peanut in Ontario; pea and potato in British Columbia (Pearson and Percy, 1997). Yield loss in some important field/horticultural crops is documented as well. In Ontario, crops considered at greatest risk include dry bean, potato, onion, hay, turnip, winter wheat, soybean, spinach, green bean, tobacco, tomato and sweet corn. Crop loss estimates range from 4% of the total U.S. $1.9 billion annual Ontario crop sales to U.S. $9 million (1986) in the Lower Fraser Valley, British Columbia.

In the U.S., foliar injury was definitively attributed to O_3 in 1959 (Heggestad and Middleton, 1959) and crop yield loss is now widespread. Adams *et al.* (1988) reported that the impacts of O_3 on crops may amount to U.S. $3 billion per year. Using annual-weighted yield reductions for four major crop species and all U.S. crops in the National Crop Loss Assessment Network (NCLAN) with estimated 1988 and 1989 O_3 exposure, Tingey *et al.* (1993) estimated yield reduction. Reductions were highly variable, depending upon species and year. Variation in corn yield loss due to O_3 was least variable, ranging from 1–20% per year while wheat was most variable, ranging from 2–80% yield reduction per year. Median yield losses across all crop growing areas in 1988, an extreme O_3 year, were 18.9% and only 3.1% in 1989, a lower O_3 year.

2.3. *Experimental Research*

There is a large body of scientific literature since the early 1970s from North American experimental research. The majority of this work was conducted with O_3 exposure of crops for the purposes of defining dose-response relationships and better defining risk at the species/varietal levels. These data were summarised during the air quality criteria review process in the U.S. (EPA, 1996) and the 1996 Canadian NOx/VOC Science Assessment (Pearson and Percy, 1997). Considerable research is also available on tree seedling exposure to wet-deposited acids (mainly simulated rain and fog) and O_3. Recent reviews of this body of work can be found in forest case study summaries above and in state of science reviews (Chappelka and Chevonne, 1992; EPA, 1996; Chappelka and Samuelson, 1998).

In North America, O_3 exposures of crops and trees have occurred in chambered laboratory and greenhouse, chambered-field, and open-air systems. Generally, methodologies became more sophisticated with time as new technology became available and new concepts forcing more "natural" exposure were developed. Portable cuvettes (Legge *et al.*, 1978) allow the controlled exposure of intact leaves/small branches while branch chambers (Teskey *et al.*, 1991) have been fitted around complete age classes of needles for controlled exposure of a combination of gases including ^{14}C tracer studies in mature pine canopies. Earlier use (Percy *et al.*, 1992) of continuous stirred tank reactor (CSTR) systems in laboratories/greenhouses designed for controlled, mass balance studies of O_3 flux, have largely gone out of use except in targeted applications, such as exposure of pine/hardwood seedlings to the extremely phytotoxic dry N species, HNO_3 (Bytnerowicz *et al.*, 1998) where excess vapour must be captured and destroyed.

Most experimental research, however, has used the chambered-field approach as open-top chambers (OTC) modified for field use with crops by Heagle *et al.* (1979). The OTCs conferred the advantage of allowing controlled gas application, statistical power through a randomised block allocation of up to five treatments (ambient-open, charcoal-filtered chamber, non-filtered chamber, $1 \times O_3$, $2 \times O_3$) with three replicate OTC's per treatment in a field setting. OTC's were used in the largest experimental O_3 fumigation project under NCLAN

(1980–1986) and data were summarised in EPA (1996). Data from NCLAN exposure-crop response regression analyses indicated that at least 50% of the species/cultivars tested were predicted to exhibit 10% yield loss at 7-hour seasonal mean O_3 concentrations of < 50 ppb (EPA, 1996). Effects were found to occur following only a few hours at > 80 ppb O_3. Ambient levels in many parts of the U.S. were considered sufficient to impair plant growth and yield. Retrospective analysis of NCLAN crop-response by Legge *et al.* (1993) used statistical techniques to compare ambient air versus non-filtered OTC treatment response and remove studies with a chamber effect. With the reduced data set, they reported a crop growth threshold as low as 35 ppb O_3. Best fit was achieved between air quality and impacts on crop yield using cumulative frequency of mid-range concentrations between 50–87 ppb O_3 (Krupa *et al.*, 1994; Legge *et al.*, 1995). While OTC's remain in use for certain applications such as relating foliar injury to local O_3 exposure (Skelly *et al.*, 1999), significant differences in micro-meteorology created within the chambers has seen their use decline. State of science at this time, while recognising the continued importance of high, short-duration O_3 concentrations, points to the increasing importance of mid-range levels in growth effects and yield loss, especially given the cumulative nature of plant response to O_3, particularly in long-lived forms such as trees.

In the case of cuvettes, branch chambers, CSTR's and OTC's, extrapolation of data to the field situation was found to be difficult due to chamber effect and age/size of plant material fumigated. Since the early 1990s, open-air fumigation systems have been favoured. Predominant are the zonal air pollution system (ZAPS) used in exposure of tree seedlings (Runeckles and Wright, 1996) and Free Air Carbon Dioxide Enrichment (FACE) (Hendrey and Kimball, 1994). The Aspen FACE in Wisconsin, U.S. is exposing hardwood species to O_3 in a randomised block design comprising 12, 30 m diameter rings including four treatments [ambient, enhanced CO_2 (+200 ppm), $1.5 \times O_3$, $CO_2 + O_3$], with three replicate rings per treatment set across a 32 ha site. While expensive to establish and operate, FACE has many advantages including: open-air exposure of plants to controlled fumigations; no chamber effects; and multi-year exposures of dynamic ecosystems under the influence of yearly climate variability. Earlier results from the aspen FACE reported highly significant effects of O_3 in aspen and birch clones such as decreased height growth, reduced radial growth, early leaf abscission, reduced below-ground growth and increased insect feeding and disease incidence (Karnosky *et al.*, 1999a). Now, after integrating four years of above-ground growth data, it is clear that O_3 offsets productivity gain in a CO_2-enriched atmosphere in rapidly growing aspen and birch, and that effects of O_3 on foliar chemistry are having a consequential effect on some insects and diseases (Isebrands *et al.*, 2001; Karnosky *et al.*, 2001; Karnosky *et al.*, 2002). Interestingly, Aspen FACE is providing confirmation in direction and magnitude of effects reported from OTC (Karnosky *et al.*, 1996) and "natural" field exposures along O_3 gradients (Karnosky *et al.*, 1999b).

3. Current Dose-Response Relationships and Risk Assessment Methods

Risk assessment involves a quantification of the likelihood of adverse effects resulting from exposure to a stress, or complex of stressors. Current approaches to defining risk to forests and crops from air pollution in North America are focused on the spatial characterisation of risk from O_3 in the U.S., and on mapping critical loads for forest soils in Canada. Methods used to assess forest risk O_3 include use of physiological models with species distribution and O_3 levels to estimate geographical extent for a range of growth responses. For mature loblolly pine and selected hardwoods in the southeast, regional application of physiological growth models with O_3 exposure data indicates an expected growth reduction of 0–35% depending upon species and O_3 level (McLaughlin and Percy, 1999). Others used experimentally-derived response data, usually from OTC's, to assess risk in the eastern U.S. (Flagler *et al.*, 1997; Taylor, 1994; Lefohn *et al.*, 1997). Outputs of these and other studies have been summarised by Chappelka and Samuelson (1998). They conclude that O_3 is affecting forest growth in the eastern U.S., but that genotype, edaphic, climatic factors, and assumptions used in the assessments, affect the assessment of risk. They also conclude that seedling stage studies are not suitable for scaling-up purposes (i.e. predictive risk assessment).

Spatial characterisation of risk using a GIS resource base is a promising tool. Hogsett *et al.* (1997) used GIS to integrate: (1) estimated O_3 exposures over forested regions; (2) measures of O_3 effects on species' and stand growth; and (3) spatially-distributed environmental, genetic and exposure influences on species' response to O_3 to characterise risk to eastern U.S. forests. Area-weighted response of annual seedling biomass loss was reported by sensitivity class: sensitive — aspen/black cherry (14–33% biomass loss over 50% of their distribution); moderately sensitive — tulip polar, loblolly pine, eastern white pine and sugar maple (5–13% biomass loss); and insensitive — Virginia pine and red maple (0–1% loss). For crops, Tingey *et al.* (1991) estimated relative yield loss for all crops using the NCLAN database across crop-growing regions in the eastern U.S., on a 20 km cell, based on estimated three-month SUM06 O_3 exposure and Weibull parameters for species' response functions. Predicted relative yield loss ranged from 0–> 30% in 1988 and 1989. Areal extent of largest loss (> 30%) was limited to portions of the mid-west, east-central and southeast in 1989; it covered the majority of crop ranges in 1988, a year of record high O_3 levels. No comparable risk assessment has been completed for Canada, although O_3 is known to affect crop yield in Ontario (Pearson, 1989) and has been circumstantially related to sugar maple crown dieback in eastern Canada (McLaughlin and Percy, 1999).

According to Nilsson and Grennfelt (1988), critical loads are quantitative estimates of an exposure to one or more pollutants below which significant harmful effects on specified sensitive elements of the environment do not occur. Regional estimates have been published, including those of Richter and Markewitz (1995) who estimated that 60% of southeastern pine soils are susceptible to accelerated cation leaching from acidic deposition. Lately, Arp *et al.* (1996) have developed critical loads for S and N deposition, by superimposing atmospheric deposition rates on soil acidification potential using a steady-state model to calculate exceedance or non-exceedance for Acid Rain National Early Warning System (ARNEWS) plots. Exceedance of > 500 eq ha^{-1} yr^{-1} is associated with an annual productivity loss of 10%. Most risk assessment has focused on the S and N deposition. However, recent compelling evidence put forward by McLaughlin and

Wimmer (1999) points to a central role for Ca physiology in terrestrial ecosystem response to air pollution.

4. Conclusions

Regional levels of O_3 are significantly impacting yield of a number of commercially significant crop species in the U.S. and in certain Canadian provinces, with important economic consequences. National SO_2 concentrations are no longer considered a threat to crops and forests, but damage continues to occur in localised areas where strong point sources exist. Acid deposition has not been shown to be a significant threat to crop yield given the annual cycle of growth. There is no conclusive evidence to date linking degree of visible, foliar injury with productivity loss in crops or trees. It is clear from process-level studies that regional levels of air pollution measured in the four diverse and widely-distributed, regionally important North American case study forests are reducing carbon reserves, increasing water stress and reducing nutrient availability (McLaughlin and Percy, 1999). Carbon reserves in particular, are critical in plant defense against a variety of biotic/abiotic stressors, which may partly explain the coincidence in frequency of insects/disease with rates of air pollution exposure in the eastern U.S. (Fig. 2). Monitoring systems established to detect change with their utilitarian assumptions, however, have not been designed to link cause and effect. Although countries do not always attribute damage to air pollutants, absence of a record does not imply that the particular cause of injury (air pollution) is not present (Innes, 1998). New, integrated monitoring and research programmes are required to link air quality and forest health-productivity (Percy *et al.*, 2000; Percy 2001). Process-level modelling should be an essential ingredient (Chappelka and Samuelson, 1998) for future risk assessment. While forest case studies in Europe and North America have documented important air pollutant impacts, patterns of decline are distinct and reflect different underlying forest processes/ disturbance histories, and, therefore, cannot be directly compared (Percy *et al.*, 1999). However, it is clear from North American forest-air pollution case histories that improvement in air quality does lead to improvement in forest health (McLaughlin and Percy, 1999).

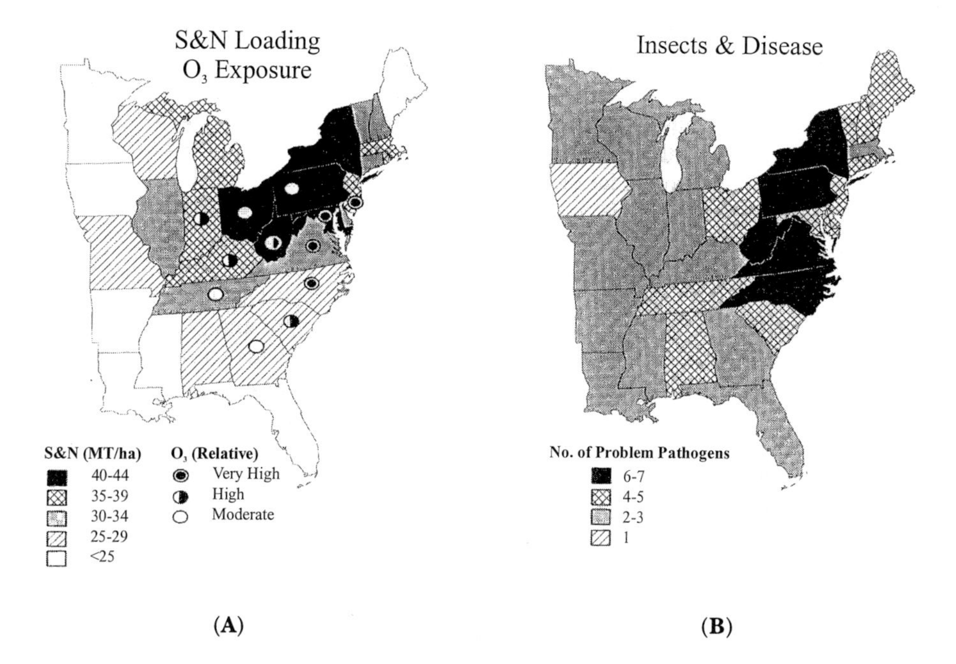

Fig. 2. Areas of the eastern U.S. having the greatest frequency of problems from forest insects and disease (**B**), related to areas receiving largest atmospheric S/N deposition and highest annual exposure to O_3 (**A**). Reproduced from McLaughlin and Percy (1999).

References

AAFC (1998) *Canadian Grains and Oilseeds Outlook, August 14, 1998: Canada, Special Crops Situation and Outlook for 1999–2000 October 14, 1999.* Agriculture and Agri-Food Canada, Policy Branch, Ottawa. http://www.agr.ca/ploicy/winn/biweekly/English/gosd/1998/aug98e.htm.

Adams R.M., Glyer D.J. and McCarl B. (1988) The NCLAN economic assessment: approach, findings and implications. In *Assessment of Crop Loss From Air Pollutants: Proceedings of An International Conference,* eds. Heck W.W., Taylor O.C. and Tingey D.T. Elsevier Applied Science, New York, pp. 473–504.

Arp P.A., Oja T. and Marsh M. (1996) Calculating critical S and N loads and current exceedances for upland forests in southern Ontario, Canada. *Can. J. Forest Res.* **26**: 696–709.

Binkley D., Driscoll C.T., Allen H.L., Schoenenberger P. and McAvoy D. (1989) *Acid Deposition and Forest Soils: Context and Case Studies in the Southeastern United States.* Springer-Verlag, New York.

Bytnerowicz A.B. and Fenn M. (1996) Nitrogen deposition in California forests: a review. *Env. Pollut.* **92**, 127–146.

Bytnerowicz A., Percy K.E., Reichers G., Padgett P. and Krywult M. (1998) Nitric acid vapour effects on forest trees — deposition and cuticular changes. *Chemosphere* **36**, 697–702.

Chappelka A.H. and Chevonne B.I. (1992) Tree responses to ozone. In *Surface Level Ozone Exposures and Their Effects,* ed. Lefohn A.S. Lewis Publishers, Chelsea, pp. 271–324.

Chappelka A.H. and Samuelson L.J. (1998) Ambient ozone effects on forest trees of the eastern United States: a review. *New Phytol.* **139**, 91–108.

Cook E.R. and Johnson A.H. (1989) Climate change and forest decline: a review of the red spruce case. *Water, Air Soil Pollut.* **48**, 127–140.

Cox R.M., Lemieux G. and Lodin M. (1996) The assessment and condition of Fundy white birches in relation to ambient exposure to acid marine fogs. *Can. J. Forest Res.* **26**, 682–688.

Dann T. and Summers P. (1997) *Ground-Level Ozone and its Precursors 1980–1993: Report of the Emissions Inventory Working Group.* Canadian 1996 NO_x/VOC Science Assessment, Environment Canada, Downsvie.

Dann T. (2001) *Trends in Ambient Levels of CWS Pollutants.* Presented at the Air and Waste Management Association Specialty Conference on Canada Wide Standards, Toronto, March 7, 2001.

DeHayes D.H., Schaberg P.G., Hawley G.J., Borer C.H., Cumming J.R and Strimbeck G.R. (1997) Physiological implications of seasonal variation in membrane-associated calcium in red spruce mesophyll cells. *Tree Physiol.* **17**, 687–695.

Eagar C. and Adams M.B. (1992) *Ecology and Decline of Red Spruce in the Eastern United States,* Ecological Studies 96. Springer-Verlag, New York.

EC (1997a) *Canadian Acid Rain Assessment Vol. 2: Atmospheric Science Assessment Report.* Environment Canada, Toronto.

EC (1997b) *National Air Pollution Surveillance (NAPS) Network Annual Summary for 1994.* Environment Canada Report EPS 7/AP/27, Ottawa.

EC (1999) *Air Quality Trends in Canadian Cities.* Environment Canada, Ottawa. http://www.ec.gc.ca/pdb/uaqt/aqfact_e.html

EPA (1996) *Air Quality Criteria for Ozone and Other Photochemical Oxidants, Vol 2.* Natural Centre for Environmental Assessment, Office of Research and Development, U.S. Environmental Protection Agency, Raleigh.

EPA (1998) *National Air Quality and Emissions Trends Report 1996.* Environmental Protection Agency, Washington.

EPA (2001) *Final Revisions to the Ozone and Particulate Matter Air Quality Standards.* http://www.epa.gov/oar/oaqps/ozpmbro/current.htm.

Flagler R.B., Brissette J.C. and Barnett J.P. (1997) In *The Productivity and Sustainability of Southern Pine Forest Ecosystems in a Changing Environment,* eds. Mickler R.A. and Fox S. EPA, 1999. Springer-Verlag, New York, pp. 73–92.

Foster N.W., Morrison I.K., Yin X. and Arp P.A. (1992) Impact of soil water deficits in a mature sugar maple forest: stand biogeochemistry. *Can. J. Forest Res.* **22**, 1753–1760.

Fox S. and Mickler R.A. (1996) *Impact of Air Pollutants on Southern Pine Forests.* Springer-Verlag, New York.

Freedman B. and Hutchinson T.C. (1980a) Long-term effects of smelter pollution at Sudbury, Ontario, on forest community composition. *Can. J. Bot.* **58**, 2123–2140.

Freedman B. and Hutchinson T.C. (1980b) Pollutant inputs from the atmosphere and accumulations in soils and vegetation near a nickel-copper smelter at Sudbury, Ontario, Canada. *Can. J. Bot.* **58**, 108–132.

Hall P., Bowers W., Hirvonen H., Hogan G., Foster N., Morrison I., Percy K., Cox R. and Arp P. (1998) *Effects of Acidic Deposition on Canada's Forests.* Natural Resources Canada Report, Canadian Forest Service, Ottawa.

Heagle A.S., Philbeck R.B., Rogers H.H. and Letchworth M.B. (1979) Dispensing and monitoring ozone in open-top field chambers for plant-effects studies. *Phytopathology* **69**, 15–20.

Heggestad H.E. and Middleton J.T. (1959) Ozone in high concentrations as cause of tobacco leaf injury. *Science* **129**, 208–210.

Hendrey G.R. and Kimball B.A. (1994) The FACE program. *Agric. Forest Meteorol.* **70**, 3–14.

Hogsett W.E., Weber J.E., Tingey D., Herstrom A., Lee E.H. and Laurence J.A. (1997) Environmental auditing: an approach for characterizing tropospheric ozone risk to forests. *Env. Manag.* **21**, 106–120.

Innes J. (1998) Role of diagnostic studies in forest monitoring programs. *Chemosphere* **36**, 1025–1030.

Isebrands J.G., McDonald E.P., Kruger E., Hendrey G., Percy K., Pregitzer K., Karnosky D.F. and Sober J. (2001) Growth responses of *Populus tremuloides* clones to interacting elevated carbon dioxide and tropospheric ozone. *Env. Pollut.*, in press.

Jagels R., Carlisle J., Cunningham R., Serrege S. and Tsai P. (1989) Impact of acid fog on coastal red spruce. *Water Air Soil Pollut.* **48**, 193–208.

Johnson D.W. and Fernandez I.J. (1992) Soil mediated effects of atmospheric deposition on eastern spruce-fir forests. In *Ecology and Decline of Red Spruce in the Eastern United States*, eds. Eagar C. and Adams M.B. Springer-Verlag, New York, pp. 235–270.

Karnosky D.F., Gagnon Z.E., Dickson R.E., Coleman M.D., Lee E.H. and Isebrands J.G. (1996) Changes in growth, leaf abscission, and biomass associated with seasonal tropospheric ozone exposures of *Populus tremuloides* clones and seedlings. *Can. J. Forest Res.* **26**, 23–37.

Karnosky D.F., Noormets A., Sober A., Percy K., Mankovska B., Sober J., Zak D., Pregitzer K., Mattson W., Dickson R., Riemenschneider D., Podila G., Hendrey G., Lewin K., Nagy J. and Isebrands J.G. (1999a) Preliminary results from the FACTS II (Aspen FACE) Experiment: Interactions of elevated CO_2 and O_3. Invited paper presented at the *International Workshop on CO_2 Issues* sponsored by the Centre for Global Environmental Research, National Institute of Environmental Study, 18 March, 1999, Tsukuba, Japan (in press).

Karnosky D., Mankovska B., Percy K., Dickson R.E., Podila G.K., Sober J., Noormets A., Hendrey G., Coleman M.D., Kubiske M., Pregitzer K.S. and Isebrands J.G. (1999b) Effects of tropospheric O_3 on trembling aspen and interaction with CO_2: results from an O_3-gradient and a FACE experiment. *Water Air Soil Pollut.* **116**, 311–322.

Karnosky D.F., Percy K.E., Xiang B., Callan B., Noormets A., Mankovska B., Hopkin A., Sober J., Jones W., Dickson R.E. and Isebrands J.G. (2001) Interacting CO_2-tropospheric O_3 and predisposition of aspen (*Populus tremuloides* Michx.) to attack by *Melampsora medusae* rust. *Global Change Biol.*, in press.

Karnosky D.F., Pregitzer K.S., Hendrey G.R., Percy K.E., Zak D.R., Lindroth R.L., Mattson W.J., Kubiske M., Podila G.K., Noormets A., McDonald E., Kruger E.L., King J., Mankovska B., Sober A., Awmack C., Callan B., Hopkin A., Xiang B., Hom J., Sober J., Host G., Riemenschneider D.E., Zasada J., Dickson R.E. and Isebrands J.G. (2002) Impacts of interacting CO_2 and O_3 on trembling Aspen: results from the Aspen FACE Experiment. *Funct. Ecol.*, submitted.

Krupa S.V., Nosal M. and Legge A.H. (1994) Ambient ozone and crop loss: establishing a cause-effect relationship. *Env. Pollut.* **83**, 269–276.

Krupa S.V., Tonneijck A.E.G. and Manning W.J. (1998) Ozone. In *Recognition of Air Pollution Injury to Vegetation: A Pictorial Atlas*, 2nd Ed. Flagler R.B. Air and Waste Management Association, Pittsburgh, pp. 2-1 to 2-28.

Lefohn A.H., Jackson W., Shadwick D.S. and Knudson H.P. (1997) Effect of surface ozone exposures on vegetation grown in the Southern Appalachian Mountains: identification of possible areas of concern. *Atmos. Env.* **31**, 1695–1708.

Legge A.H., Savage D.J. and Walker R.B. (1978) Special techniques: B. a portable gas-exchange leaf chamber. In *Methodology for the Assessment of Air Pollution Effects on Vegetation: A handbook from a Speciality Conference,* eds. Heck W.W., Krupa S.V. and Linzon S.N. Air Pollution Control Association, Pittsburgh, pp. 16–24.

Legge A.H., Nosal M. and Krupa S.V. (1993) *Development of an Ozone Air Quality Objective Based on Vegetation Effects: An Examination of Its Scientific Basis.* Phase II-Final Report, June 1993, Environment Canada, Downsview.

Legge A.H., Grunhage L., Nosal M., Jager H.-J. and Krupa S.V. (1995) Ambient ozone and adverse crop response: an evaluation of North American and European data as they relate to exposure indices and critical levels. *Angew. Bot.* **69**, 192–205.

Legge A.H., Nosal M. and Krupa S.V (1996) Modeling the numerical relationships between chronic ambient sulphur dioxide exposures and tree growth. *Can. J. Forest Res.* **26**, 689–695.

Legge A.H., Jager H.-J. and Krupa S.V. (1998) Sulfur dioxide. In *Recognition of Air Pollution Injury to Vegetation: A Pictorial Atlas*, 2nd Ed. Flagler R.B. Air and Waste Management Association, Pittsburgh, pp. 3-1 to 3-42.

Magasi L.P. (1989) *White Birch Deterioration in the Bay of Fundy Region, New Brunswick 1979–1988.* Forestry Canada Maritimes Region Report M-X-175, Fredericton.

McLaughlin S.B., Tjoelker M.G. and Roy W.K. (1993) Acid deposition alters red spruce physiology: laboratory studies support field observations. *Can. J. Forest Res.* **23**, 380–386.

McLaughlin S.B. and Downing D.J. (1996) Interactive effects of ambient ozone and climate measured on growth of mature loblolly pine trees. *Can. J. Forest Res.* **26**, 670–681.

McLaughlin S.B. and Percy K.E. (1999) Forest health in North America: some perspectives on actual and potential roles of climate and air pollution. *Water Air Soil Pollut.* **116**, 151–197.

McLaughlin S.B. and Wimmer R. (1999) Calcium physiology and terrestrial ecosystem processes. *New Phytol.* **142**, 373–417.

Miller P.R. (1973) Oxidant-induced community change in a mixed conifer forest. In *Air Pollution Damage to Vegetation.* Advances in Chemistry Series 112, ed. Naegele J.A. American Chemical Society, Washington, pp. 101–117.

Miller P.R., Taylor O.C. and Wilhour R.G. (1982) *Oxidant Air Pollution Effects on a Western Coniferous Forest Ecosystem.* Report No. EPA-600/D-82-276, Environmental Protection Agency.

Miller P.R., McBride J.R., Schilling S.L. and Gomez A.P. (1989) Trend in ozone damage to conifer forests between 1974 and 1988 in the San Bernardino Mountains in southern California. In *Effects of Air Pollution on Western Forests*, eds. Olson R.K. and Lefohn A.S. Air and Waste Management Association, Pittsburgh, pp. 309–324.

Miller P. and McBride J., eds. (1999) *Oxidant Air Pollution Impacts in the Montane Forests of Southern California: The San Bernardino Mountain Case Study.* Springer-Verlag, New York.

NAPAP (1998) *National Acid Precipitation Assessment Program Biennial Report to Congress: An Integrated Assessment.* Silver Spring.

NASS (1997) *1997 Agriculture Atlas of the United States.* USDA NAAS. http://www.nass. Usda.gov/census/census97/atlas97/index.htm.

Nicholas N.S. (1992) *Stand Structure, Growth, and Mortality in Southern Appalachian Spruce-Fir.* Ph.D. Dissertation, Virginia Polytechnic Institute and State University, Blacksburg.

Nilsson J. and Grennfelt P., eds. (1988) *Critical Loads for Sulphur and Nitrogen. Report from a Workshop, 19–24 March, 1998, Skokloster, Sweden.* Nordic Council of Ministers, Miljorapport, 1988.

NRCAN (1998) *The State of Canada's Forests 1997–1998.* Natural Resources Canada, Canadian Forest Service, Ottawa.

Pearson R.G. (1989) *Impact of Ozone Exposure on Vegetation in Ontario.* Air Resources Branch, Ontario Ministry of Environment, Toronto.

Pearson R.G. and Percy K.E. (1997) *Canadian 1996 NO_x/VOC Science Assessment: Report of the Vegetation Objective Working Group.* Environment Canada, Downsview.

Peart D.R., Nicholas N.S., Zedaker S.M., Miller-Weeks M. and Siccama T.G. (1992) Condition and recent trends in high-elevation red spruce populations. In *Ecology and Decline of Red Spruce in the Eastern United States,* eds. Eagar C. and Adams M.B. Springer-Verlag, New York, pp. 125–191.

Percy K.E., Jensen K.F. and McQuattie C.J. (1992) Effects of ozone and acid fog on red spruce needle epicuticular wax production, chemical composition, cuticular membrane ultrastructure and needle wettability. *New Phytol.* **122**, 71–80.

Percy K.E., Jagels R., Marden S., McLaughlin C.K. and Carlisle J. (1993) Quantity, chemistry and wettability of epicuticular waxes on needles of red spruce along a fog-acidity gradient. *Can. J. Forest Res.* **23**, 1472–1479.

Percy K., Bucher J. Cape J.N., Ferretti M., Heath R., Jones H.E., Karnosky D., Matyssek R., Muller-Starck G., Paoletti E., Rosengren-Brinck U., Sheppard L., Skelly J., and Weetman G. (1999) State of science and knowledge gaps with respect to air pollution impacts on forests: reports from concurrent IUFRO 7.04.00 working party sessions. *Water Air Soil Pollut.* **116**, 443–448.

Percy K.E., Karnosky D.F. and Innes J.L. (2000) Potential roles of global change in forest health during the 21st century. In *Forests and Society: The Role of Research, Sub-Plenary Sessions Volume 1, XXI IUFRO World Congress 2000, 71-12, 2000, Kuala Lumpur, Malaysia,* eds. Krishnapillay B., Soepadmo E., Arshad N.L., Wong H.H., Appanah S., Chik S.W., Manokaran N., Tong H.L. and Choon K.K., pp. 147–163.

Percy K.E. (2001) Is air pollution an important factor in international forest health? *The Effects of Air Pollution on Forest Health and Biodiversity in Forests of the Carpathian Mountains.* IOS Press, Amsterdam, in press.

Powell D.S., Faulkner J.L., Darr D.R., Zhu Z. and MacCleery D.W. (1992) *Forest Resources of the United States.* USDA Forest Service General Technical Report RM-234, Washington.

Richter D.D., Markewitz D., Wells C.G., Allen H.L., April R., Heine P.R. and Urrego B. (1994) Soil chemical change during three decades in an old-field loblolly pine (*Pinus taeda* L.) ecosystem. *Ecology* **75**, 1462–1473.

Richter D.D. and Markewitz D. (1995) Atmospheric deposition and soil resources of the southern pine forest. In *Impacts of Air Pollutants on Southern Pine,* eds. Fox S. and Mickler R.A. Springer-Verlag, New York, pp. 315–336.

Robitaille G., Boutin R. and Lachance D. (1995) Effects of soil freezing stress on sap flow and sugar content of mature sugar maples. *Can. J. Forest Res.* **25**, 577–587.

Runeckles V.C. and Wright E.F. (1996) Delayed impact of chronic ozone stress on young Douglas-fir grown under field conditions. *Can. J. Forest Res.* **26**, 629–638.

Schaberg P.G., Perkins T.D. and McNulty S.G. (1997) Effect of chronic low-level N additions on foliar element concentrations, morphology and gas exchange of mature montane red spruce. *Can. J. Forest Res.* **27**, 1622–1629.

Sheffield R.M. and Cost N.D. (1987) Behind the decline: why are natural stands in the southeast growing slower? *J. Fores.* **85**, 29–33.

Skelly J.M., Innes J., Savage J.E., Snyder K.R., Vanderheyden D., Zhang J. and Sanz M.J. (1999) Observation and confirmation of foliar ozone symptoms of native plant species of Switzerland and southern Spain. *Water Air Soil Pollut.* **116**, 227–234.

Somers G.L., Chappelka A.H., Rosseau P. and Renfro J.R. (1998) Empirical evidence of growth decline related to visible ozone injury. *Forest Ecol. Manag.* **104**, 129–137.

Stoyenhoff J., Witter J. and Leutscher B. (1998) *Forest Health in the New England States and New York.* University of Michigan, Ann Arbor.

Taylor G.E. (1994) Role of genotype in the response of loblolly pine to tropospheric ozone: effects at the whole-tree, stand, and regional level. *J. Env. Qual.* **23**, 63–82.

Teskey R.O., Dougherty P.M. and Wiselogel A.E. (1991) Design and performance of branch chambers suitable for long-term ozone fumigation of foliage in large trees. *J. Env. Qual.* **20**, 591–595.

Teskey R.O. (1995) In *Impacts of Air Pollutants on Southern Pine,* eds. Fox S. and Mickler R.A. Springer-Verlag, New York, pp. 467–490.

Tingey D.T., Hogsett W.E., Lee E.H., Herstrom A.A. and Azevedo S.H. (1991) An evaluation of various alternative ambient ozone standards based on crop yeild loss data. In *Tropospheric Ozone and the Environment: Papers From an International Conference,* eds. Berglund R.L., Lawson D.R. and McKee D.J. Air and Waste Management Association, Pittsburgh, pp. 272–288.

Tingey D.T., Olsyk D.M., Herstrom A.A. and Lee E.H. (1993) Effects of ozone on crops. In *Tropospheric Ozone: Human Health and Agricultural Impacts,* ed. McKee D.J. Lewis Publishers, Boca Raton, pp. 175–206.

Winterhalder K. (1996) Environmental degradation and rehabilitation of the landscape around Sudbury, a major mining and smelting area. *Env. Rev.* **4**, 185–224.

Yin X., Foster N.W., Morrison I.K. and Arp P.A. (1994) Tree-ring based growth analysis for a sugar maple stand: relations to local climate and transient soil properties. *Can. J. Forest Res.* **24**, 1567–1574.

CHAPTER 3

AIR POLLUTION IMPACTS ON VEGETATION IN EUROPE

M. Ashmore

1. Introduction

1.1. Background

Europe, like North America, is a diverse continent in terms of its climate, which ranges from the cold of the Arctic Circle to the heat of the Mediterranean islands, and from temperate oceanic conditions on the western seaboard to continental climates in European Russia. This climatic range inevitably leads to a wide range of natural vegetation, forests and agricultural systems. Unlike North America, the continent is also characterised by political and economic diversity; in particular, the patterns of economic development across the continent have differed significantly, which means that historically the air pollution problems in the various regions have been quite different. The importance of economic convergence, and the use of common emission and air quality standards within the European Union (EU), means that these differences are becoming less marked. This trend will continue as more countries in eastern Europe aim to meet the economic and environmental criteria to become part of the EU.

It is convenient to divide the historical and current air pollution problems of Europe into two major groups. The first of these are associated with the intensive use of coal in industry and domestically, resulting in large ground-level emissions of sulphur dioxide (SO_2) and particulates. The impacts of these emissions were often severe, but tended to be localised to areas close to the sources. The impacts of such emissions on local vegetation were already apparent in the last century. In western Europe, legislation to improve air quality, primarily driven by concerns over the effects of smog on human health, resulted in reduced concentrations of these pollutants from the 1960s, due to control of local sources and the use of tall stacks. This policy improved local concentrations of SO_2 but, as total sulphur emissions continued to increase, led to concerns over acid rain and its impacts on soils and waters. One estimate for the mid-1970s was that 5% of western Europe experienced annual mean concentrations of SO_2 above 50 $\mu g\ m^{-3}$ (Fowler and Cape, 1980), primarily in and around the industrial areas of the United Kingdom, Germany, France and Belgium. In some parts of eastern Europe, emissions of these pollutants remained high throughout the 1980s, although the situation has improved over the last decade due to economic restructuring.

The second group of air pollution problems relate to the increased emissions of nitrogen oxides (NO_x) and volatile organic compounds (VOCs), primarily from the rapid increase in the numbers of motor vehicles since the Second World War; these increases were initially greater in north-west Europe than in southern or eastern Europe. From the perspective of direct effects on vegetation, the most important consequence of these rising emissions is the increased concentrations of ozone (O_3). Measurement series have demonstrated both a steady increase in the background tropospheric O_3 concentrations over the 20th century, and the current occurrence of photochemical episodes, in which high concentrations of O_3 spread over wide areas of the continent (Jonson *et al.*, 2001).

1.2. *Emission Reductions*

A key feature of the European experience over the past two decades has been the success of international collaboration in evaluating air pollution problems, and negotiating solutions, despite the major political differences that existed prior to 1990. A key element in this success has been the establishment in 1979 of the United Nations Economic

Commission for Europe (UN/ECE) Convention on the Long-Range Transboundary Air Pollution (CLRTAP), which was established initially because of concerns over the impact of acid deposition on freshwaters. Protocols on reduction of sulphur emissions agreed under CLRTAP in 1985 and 1994 have had a major impact, and are part of a wider series of protocols covering also NO_x, VOCs, persistent organic pollutants (POPs) and heavy metals. These protocols, and the emission reductions targets agreed, are summarised in Table 1. The Gothenburg Protocol, signed in 1999, provides further reduction targets to 2010 for emissions leading to eutrophication, acidification and production of ground-level O_3. It is important to note that the emission reductions in Table 1 are not set uniformly for each country, but are based on optimising the environmental benefit of investment in emission reductions across Europe.

In addition, the 15 countries within the EU have to fulfil directives on emissions of SO_2, NO_x and VOCs, which are also listed in Table 1. Sulphur emissions in the Europe have declined by about 60% since 1980, despite growth in energy demand and GDP which would have been expected to increase emissions significantly over this period (Fig. 1). Within the 15 EU countries, emissions declined by 66% between 1980 and 1997, mainly through installation of flue gas desulphurisation and a switch to gas from coal and oil. In Eastern Europe, reductions in sulphur emissions were relatively small over the period 1980 to 1990, but large reductions have taken place since 1990, mainly due to economic restructuring, a switch from coal to natural gas, use of low-sulphur coal and construction of new power plants. In contrast to SO_2, European emissions of nitrogen oxides increased over the period 1980–1990, but have decreased subsequently, partly as a result of decreased emissions from transport in western Europe and economic restructuring in eastern Europe.

1.3. *Crops and Forests*

The wide range of climatic conditions in Europe supports a range of crop species. A recent FAO estimate is that 45% of the land area of Europe is covered by agriculture, with 57% of this being arable, 37% permanent pasture, and the remainder permanent crops such as vines

Table 1. Summary of international pollutant emission control protocols within UNECE-CLRTAP and the EU (from NEGTAP, 2001).

UNECE — CLRTAP Protocol (common name)			Objectives		
	Open for signature and [entry into force[1]]	*No. of signatures and ratifications (as of 1st Sep. 2001)[2]*	*Pollutant*	*Base year/ target year*	*Reduction (%)*
Acidification, eutrophication and ground-level O_3 (Gothenburg Protocol)	1999	31, 1	SO_2 NO_x Non-methane VOCs Ammonia[3]	1990/2010	75 50 58 12
POPs[4]	1998	36, 7	Eliminate any discharges, emissions and losses of POPs.		
Heavy metals[4]	1998	36, 10	Cadmium Lead Mercury	Reduce, control or eliminate the use of heavy metals.	
Further reduction of sulphur emissions[5] (2nd S protocol or Oslo Protocol)	1994 [1998]	28, 23	SO_2	1980/2000	62

Table 1. (*Continued*)

UNECE — CLRTAP Protocol (common name)			Objectives		
	Open for signature and [entry into force[1]]	*No. of signatures and ratifications (as of 1st Sep. 2001)[2]*	*Pollutant*	*Base year/ target year*	*Reduction (%)*
VOCs[6]	1991 [1997]	23, 21	Non-methane VOCs	1987/1999	30
NO_x[6]	1988 [1991]	25, 28	NO_x	1987/1994	Stabilisation
Thirty per cent reduction in sulphur emissions (1st S protocol or the "30% club")	1985 [1987]	19, 22	SO_2	1980/1993	30
European monitoring and evaluation programme (EMEP)	1984 [1988]	22, 38	Long-term financing of EMEP		

Table 1. (*Continued*)

UNECE — CLRTAP Protocol (common name)			Objectives		
	Open for signature and [entry into force[1]]	*No. of signatures and ratifications (as of 1st Sep. 2001)[2]*	*Pollutant*	*Base year/ target year*	*Reduction (%)*
EU	*Year initiated*		*Objectives*		
EU — 5th Environmental Action programme (5EAP)	1993		SO_2 NO_x Non-methane VOCs	1985/2000 1990/2000 1990/1999	35 30 30
National Emissions Ceilings Directive[7] (NECD)	1999		SO_2 NO_x Non-methane VOCs Ammonia	1990/2010	77 55 54 14

[1]Sixteen ratifications are needed for a protocol to enter into force.
[2]Updated status can be found at http://www.unece.org/env/lrtap/
[3]The % emission reduction target for the EU is shown, which corresponds with the overall effect of the different emission ceilings for each Member State.
[4]POPs and heavy metals will not be considered further in this report.
[5]The different emission ceilings for each Member State correspond to a 62% emission reduction for the EU.
[6]These are the same for individual Member States and for the EU.
[7]Proposed targets from the common position (June 2000) for a national emission ceilings directive (NECD).

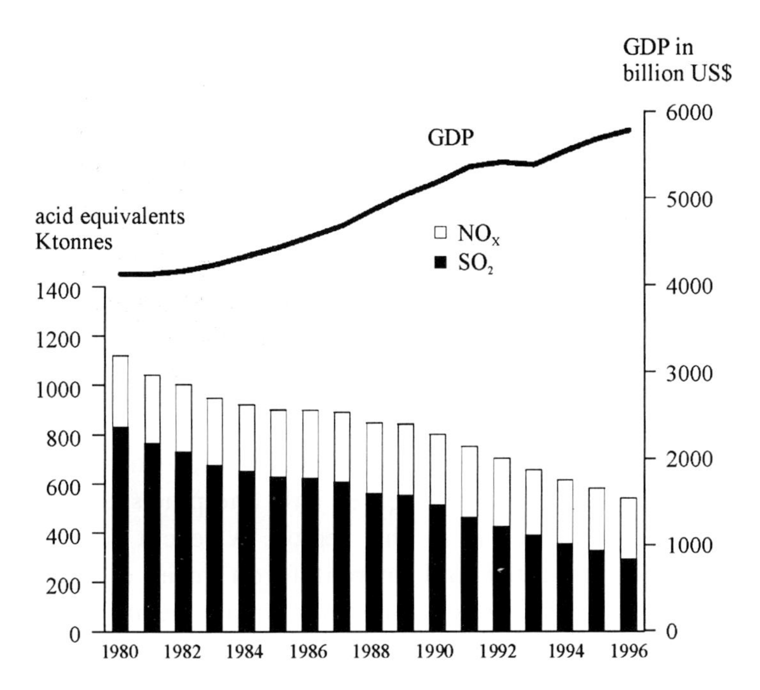

Fig. 1. Trends in European emissions of SO_2 (Mton yr^{-1}) and NO_x (Mton yr^{-1}) over the period 1980–1996, plotted alongside trends in GDP (billion $US). Redrawn from EEA (2000).

and fruit trees. The agricultural area varies widely between countries, from less than 10% in Scandinavia to 72% in the U.K. Within the EU, data exist on the contribution of different agricultural products to the total value of agricultural production. About half of this value is associated with animals (cattle, pigs, poultry, etc.) or animal products (milk, cheese, eggs). In terms of crops, the most important individual categories by value are fresh vegetables (18%), wheat (13%), grape juice and wine (12%), fresh fruits (9.5%), barley, sugar beet, potatoes and oil seeds (each 5%), maize (4%), olive oil (3%), and citrus fruit (2.5%).

Likewise, forest systems vary across the continent. A simple classification of these forests identifies four major categories (Polunin and Walters, 1985). Boreal forests are the major component in Scandinavia and northern Russia, and are dominated by evergreen coniferous species. These regions are characterised by low temperatures and nutrient-poor soils. Atlantic forests are found in the north-west of the continent, with cool summers and mild winters, where historically forests dominated by deciduous oaks, beech, and locally ash were the major land cover. Central European forests are located in regions which have a more continental climate, with colder winters and lower windspeeds; originally lowland forests were dominated by deciduous oaks and beech, and montane forests by spruce, fir and pine. However, widespread planting has modified this pattern greatly. Finally, the Mediterranean forests of southern Europe have a markedly different growth pattern and climate, with the long hot, dry summers leading to dominant species which are sclerophyllous evergreens with the main growth period in the spring. Evergreen oaks and pine species were the dominant species originally in this region. A recent study has estimated both the potential and current forest cover of Europe and deduced that 56% of Europe's forest has already been lost, primarily in historical land clearance for agriculture. Current forest cover varies greatly between countries; in the EU, for example, it varies from 5% in Ireland and 9% in the U.K. to 35% in Portugal and 45% in Greece.

2. Evidence for Direct Effects on Vegetation

2.1. *Field Studies of Crops*

Historically, field observations, or the growth of plants along field pollution gradients, were the main methods used to establish the impacts of air pollution on vegetation. While these approaches allow the effects of different pollutant mixes and concentrations to be compared, interpreting the results in terms of exposure-response relationships for specific pollutants is more difficult. One of the earliest such studies was that of Cohen and Rushton (1925), who grew a range of plant species at several sites in the city of Leeds in northern England, ranging from a highly polluted industrial location to relatively unpolluted locations at the edge of the city. The ten- to 20-fold difference in the rates of sulphur and particle deposition between these sites was associated with large reductions in the growth of the plants studied. The approach was developed in more detailed studies by Guderian and Stratmann (1962), which were designed to establish exposure-response relationships, and thresholds for adverse effects, for SO_2 on a range of species under field conditions. This work was carried out around an industrial source, the roasting furnaces of an iron-ore mine in Biersdorf, Germany, for which SO_2 was believed to be the only major pollutant.

In Europe, the first reports of visible injury to crop species from O_3 appeared in the 1970s. For example, Ashmore *et al.* (1980) reported visible injury to radish and peas following an episode of high O_3 concentrations, which was verified by experiments to demonstrate the protective effect of filtration. Acute injury to leaves of species with a market value dependent on their visible appearance, such as many horticultural crops, can cause an obvious and immediate loss of economic value. For example, an O_3 episode north of Athens in 1998 caused such severe reddening and necrosis on chicory and lettuce that local crops could not be sold (Vellisariou, 1999), with severe implications for local producers.

Visible O_3 damage to crops is most regularly reported in Mediterranean areas of Europe (Fumigalli *et al.*, 2001); for example, Velissariou *et al.* (1996) documented instances of visible foliar injury to 21 crop species in Italy, Spain and Greece. Agricultural and horticultural crops for which visible damage to commercial fields have been reported in Europe in O_3 episodes include: bean, clover, corn, grape, peanut, potato, soybean, tobacco, wheat, courgette, chicory, grapevine lettuce, muskmelon, onion, parsley, peach, peas, pepper, radish, beetroot, spinach, tomato and watermelon. Plate 1 illustrates O_3 injury to four of these species.

In the case of O_3, because of its regional distribution, it is difficult to use spatial variation in plant growth or crop yield as a basis for assessing the direct impacts on vegetation. Field treatments with protectant chemicals have played a role in assessing the effects of ambient O_3. Using EDU, one such protectant chemical, co-ordinated experiments using a common protocol have been carried out with white clover across Europe. Pooling the results of such experiments has provided a powerful method of developing exposure-response relationships or demonstrating the extent of adverse effects across the continent (Ball *et al.*, 1998).

Plate 1. Examples of O_3 injury to crop species in Europe: **(A)** tobacco (source: G. Lorenzini); **(B)** bean (source: G. Mills) **(C)** grapevine (source: D. Vellisariou); and **(D)** wheat (source: G. Mills). From UN/ECE (2002).

2.2. *Field Studies of Forests*

The most dramatic effects of air pollution on forest ecosystems in Europe have occurred in the so-called Black Triangle, covering parts of southern Poland, eastern Germany and northern Czechoslovakia (now the Czech Republic and Slovakia). The extensive and severe forest damage in this region was associated primarily with emissions from coal-fired power stations burning lignite, a fuel with high sulphur and low energy content. An assessment of the scale of the problem in the 1970s (Godzik, 1984) suggested that at least one million hectares of forests, primarily of Norway spruce, were severely injured, with considerably greater areas subject to some impact of SO_2. The degree of forest damage in both Poland and Czechoslovakia increased dramatically through the 1960s and 1970s, and in both of these countries, the greatest degree of forest damage was associated with the areas of highest SO_2 emissions. While soil acidification may have been a significant factor in this major problem, there is little doubt that direct impacts of atmospheric SO_2, complemented in specific locations by emissions of fluorides and metals, were the major cause of the decline. Studies in western Europe in the 1970s also indicated negative correlations between the performance of conifers such as Scots pine and atmospheric concentrations of SO_2 in industrial regions of the U.K. and West Germany (Farrar *et al.*, 1977; Knabe, 1970).

Concern over impacts of air pollution on forests became intense in western Europe in the late 1970s and 1980s, when evidence emerged of declines in the health of beech, fir and spruce in rural areas which, unlike in central and eastern Europe, had low SO_2 concentrations. In many cases, the unhealthy trees were characterised by deficiencies in foliar nutrient concentrations. The concern over forest decline in the 1980s led to many field and experimental studies, most of which focussed on the roles of soil acidification and eutrophication. However, field studies also provided evidence of the possible role of ozone as a further stress factor (Sandermann *et al.*, 1997).

Detailed monitoring programmes are now in place across Europe to assess the condition of forests through visual assessments of defoliation and discolouration at nearly 20,000 sites in 31 countries. More detailed physical, chemical and physiological measurements are made at almost 1000 sites to contribute to understanding of the causes of change. The fact that these programmes use a co-ordinated assessment and standardised methodology means that long-term trends in tree health can be identified.

Figure 2 shows trends in mean crown density for major European tree species from this survey. In the past decade, there has been little change in the crown condition of Scots pine, Norway spruce or beech. However, there has been a consistent decline in the crown condition of the Mediterranean species holm oak and maritime pine and, to a lesser extent, of European oak. These mean trends disguise large spatial variation across Europe in crown condition both in terms of absolute values and trends over the last decade. Statistical analysis of the more detailed stand data suggests that climate, soils and stand age are the most important factors affecting tree vitality, but that higher sulphur and nitrogen deposition and higher O_3 exposures are associated

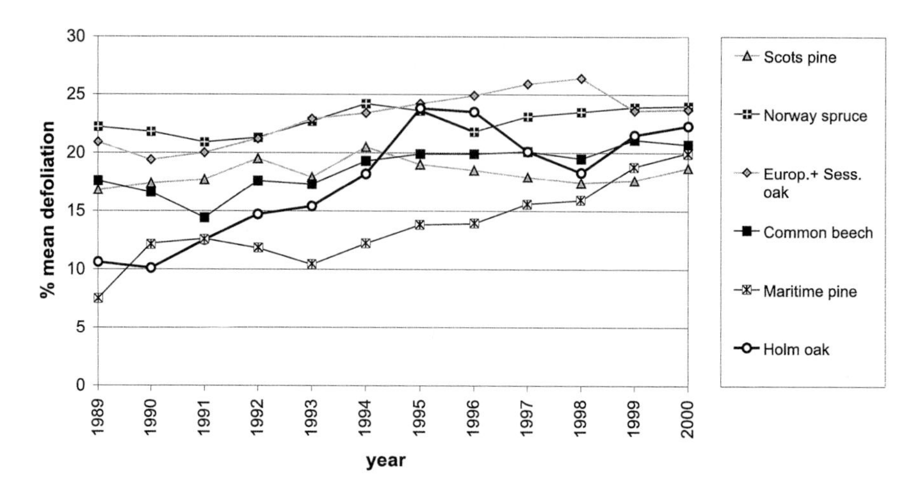

Fig. 2. Trend in crown condition of six species, expressed as mean percentage defoliation, averaged across 5900 monitoring plots across Europe. Redrawn from Fischer *et al.* (2001).

with poorer crown condition in deciduous species, especially beech (Klap *et al.*, 1997).

Data on trends in forest production provide a contrast, with most long-term studies suggesting that yield has increased over recent decades (Spieker *et al.*, 1996). This trend to increased yield undoubtedly partially reflects changes in silvicultural practice, but both increased carbon dioxide (CO_2) concentrations and increased nitrogen deposition may play a role.

The evidence of visible damage to tree species caused by O_3 is limited in Europe, although this may reflect the lack of systematic surveys of injury. For example, recent field surveys in southern Switzerland and in Spain, regions with relatively high O_3 exposures, have revealed O_3-like symptoms on a large number of shrub and deciduous tree species, with experimental studies confirming the diagnosis in several species (Skelly *et al.*, 1999). Baumgarten *et al.* (2000) reported characteristic visible symptoms of O_3 injury both on young seedlings and mature beech trees, and showed that this was related to a critical cumulative uptake by the tree. In southern Europe, there have been a number of records of visible injury which

Plate 2. Examples of O_3 injury to tree species in Europe: **(A)**. willow (source: M. Schaub); **(B)** beech (source: M. Schaub); and **(C)** ash (source: M. Sanz). From UN/ECE (2002).

resemble O_3 symptoms, and in the case of at least one species, Aleppo pine, this has been confirmed by experimental fumigation (Gimeno *et al.*, 1992). Plate 2 illustrates recent examples of O_3 injury on tree species in Europe.

However, it remains unclear whether O_3 has had a significant long-term effect on the growth and vitality of European forests. Very few studies have attempted to relate O_3 exposure to forest growth. In one such study, Braun *et al.* (1999) examined the diameter growth of beech trees at 57 permanent plots in Switzerland over a four-year period in relation to the estimated O_3 exposures at these sites, and a range of other site and climatic variables. A significant negative association between radial growth and O_3 exposure was found, with a slope which was greater than that determined in earlier experimental studies with beech seedlings (Braun and Fluckiger, 1995).

Critical levels of O_3 have been set to prevent significant growth reductions in trees due to O_3 (see Sec. 2.3 below). Assessments suggest that these critical levels are exceeded in about 70% of the deciduous forest stock and 40% of the coniferous forest stock across Europe (Matyssek and Innes, 1999). This critical level is set at an exposure which caused significant effects in experimental seedlings (a 10% reduction in annual growth) and would therefore be expected to have large cumulative effects on the growth and yield of mature trees in the field. However, there is little evidence from field studies that O_3 is affecting forest growth and vitality across Europe. It is not clear how far this reflects a lack of well-designed field studies or the fact that experiments with seedlings have little value in predicting impacts on mature forest systems.

2.3. *Experimental Studies with Crops*

It is useful to divide experimental studies carried out in Europe into two categories: studies designed to assess the effect of the ambient air pollution at a particular location, primarily through examining responses to filtering the air, and fumigation studies designed to test the effect of adding controlled concentrations of particular pollutants.

One of the earliest studies of the effects of filtration was carried out by Bleasdale (1952), in central Manchester; this study clearly demonstrated the adverse effects of the urban pollutant mix on the growth of grass species. Subsequent studies in urban areas of the U.K. in the 1970s similarly found large significant effects of filtration, which it was initially assumed were primarily caused by SO_2. However, when the results of these experiments were compared with those of fumigating the same grasses with controlled concentrations of SO_2, large discrepancies in the dose-response relationship became apparent. For example, Figure 3 shows an analysis of data for controlled fumigation studies of perennial ryegrass (closed symbols), with a dose-response relationship, plotted alongside urban filtration experiments (open symbols) with the same species (Bell, 1985). It is clear from such data analyses that SO_2 alone could not cause adverse effects of the magnitude observed in the filtration experiments. This apparent contradiction was resolved in a series of experiments which demonstrated that,

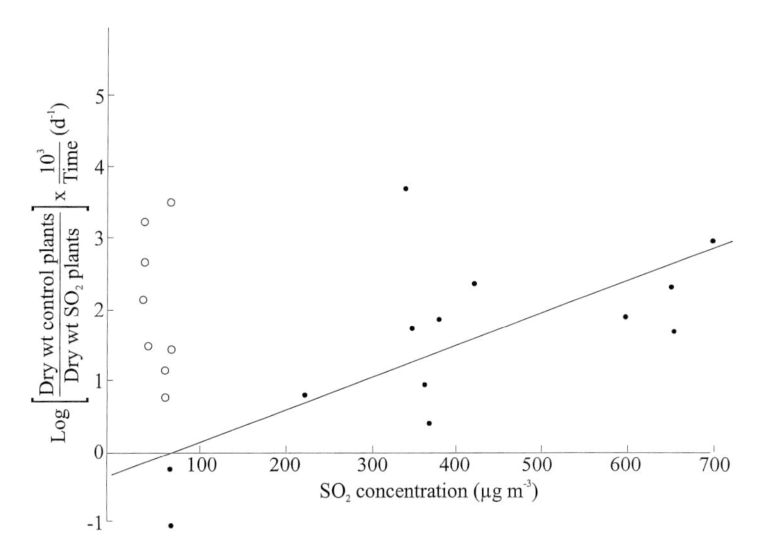

Fig. 3. Data from fumigation experiments with SO_2 (closed circles) and filtration experiments (open circles) with perennial ryegrass. The plant response is expressed as the log of the ratio of biomass in control and pollution-treated plants divided by experimental duration. The regression line is fitted to the fumigation experiments only. From Bell (1985).

at the concentrations then prevalent in urban areas, SO_2 and NO_2 in combination had a much greater effect than the individual gases in isolation (Ashenden and Mansfield, 1978); subsequent work demonstrated a biochemical mechanism for this synergistic response.

These studies of urban air pollution and SO_2 were primarily conducted using closed chambers. In the 1970s and 1980s, open-top chambers began to replace these as the standard experimental tool since they provided more realistic experimental conditions. Such chambers have been used to investigate the responses of a range of species in rural areas where O_3 is the main pollutant, and have clearly demonstrated the potential for significant growth and yield reductions due to O_3 in many parts of Europe (e.g. Ashmore 1984; Schenone and Lorenzini, 1992).

Considerable attention has been paid in recent years to the development of exposure-response relationships for O_3, primarily using open-top chambers (Jager *et al.*, 1993). Unlike in the US NCLAN

(National Crop Loss Assessment Network) programme (see Percy, this volume), standard protocols have not been employed at all the study sites. However, combining the data from different experimental studies has proved a valuable method of testing whether common relationships can explain the effects of O_3 across a number of sites and years. The exposure index initially used in North America was the 7-hour seasonal mean O_3 concentration (see Chap. 2). In Europe, a new index was developed to summarise the cumulative seasonal O_3 exposure, termed the AOT40 (Accumulated exposure Over a Threshold concentration of 40 ppb) index (Fuhrer *et al.*, 1997). The use of this index allowed a linear relationship to be established with wheat yield reductions from a pooled data-set covering 17 experiments in six different countries (Fig. 4).

While the derivation of this relationship for wheat, as major arable crop, is of value, further analysis is needed to compare the responses

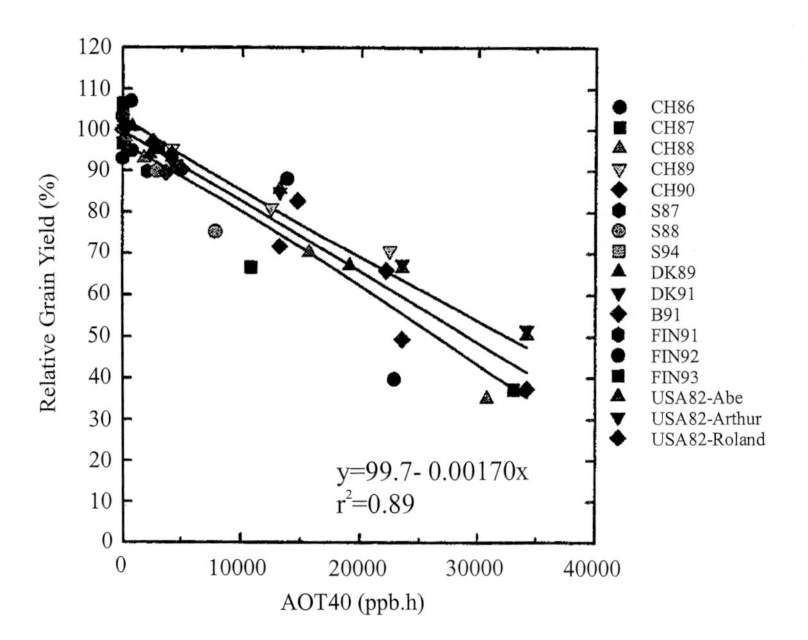

Fig. 4. Relationship between wheat grain yield, expressed relative to a control treatment, and seasonal AOT40 exposure, derived from 17 experiments in six countries; the codes refer to the country and the year of the experiment. From Fuhrer (1996).

of other crop species. Figure 5 shows a recent compilation of experimental data from the EU and the United States for six crops, using the 7-hour mean O_3 concentration, which indicates considerable consistency in exposure-response relationships from the two continents (Mills *et al.*, 2000). Soybean and wheat are the most sensitive crops in terms of yield, followed by potato, maize and rape, with barley showing no significant effect even at high exposures.

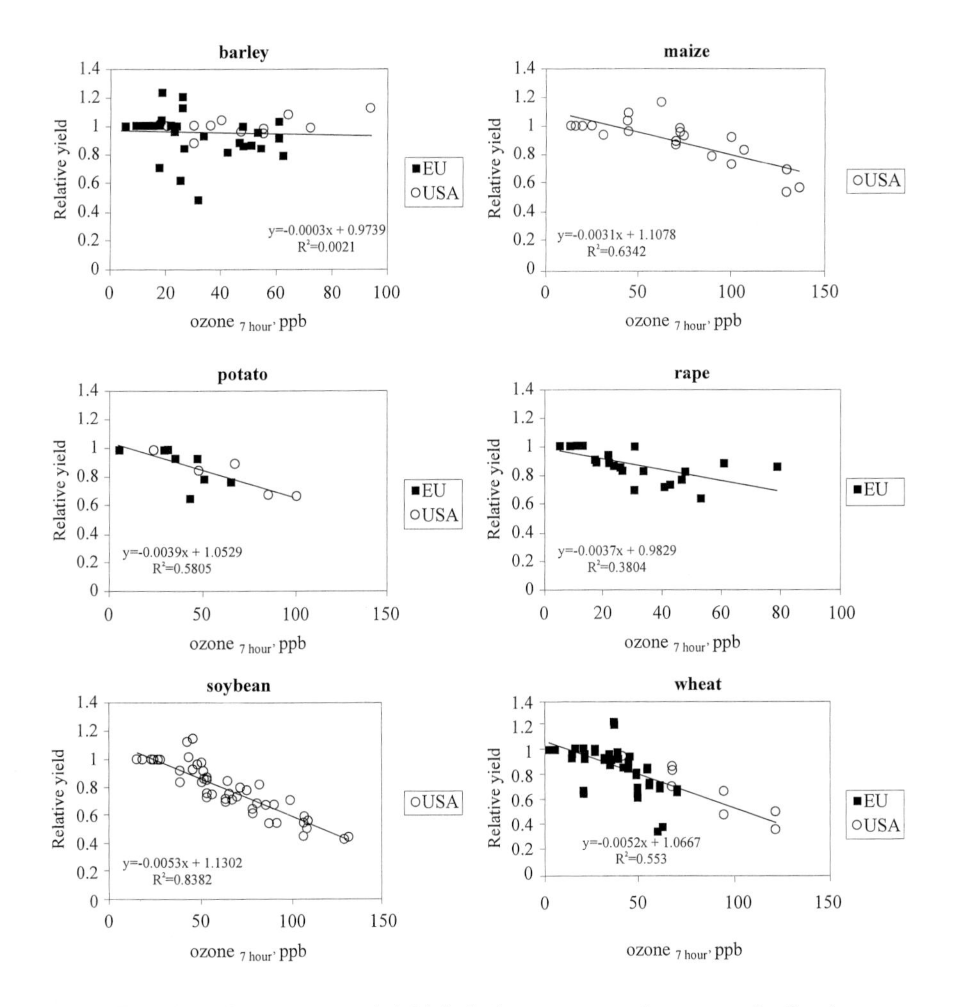

Fig. 5. Functions for response of yield (relative to a control treatment) of major crops to O_3 developed from literature on experimental studies in the EU (closed squares) and the U.S. (open circles). From Mills *et al.* (2000).

3. Risk Assessment

3.1. *Air Quality Criteria and Critical Levels*

The major focus of risk assessment in Europe over the last two decades has been on the derivation and application of air quality criteria for effects on vegetation, soils and freshwaters. In particular, the use of critical loads to describe the deposition rates above which long-term acidification of soils and freshwaters may occur has been closely linked to international negotiations on emission control. There are three organisations which currently play a major role in developing and applying air quality criteria in Europe:

(1) The World Health Organisation's Regional Office for Europe produced a set of "Air Quality Guidelines for Europe" in the mid-1980s, which were revised in the late 1990s (WHO, 2000). Although these documents concentrate on air quality guidelines to protect human health, guidelines have also been developed for vegetation. The WHO guidelines have no direct link to the development of emission control policy. However, they have strongly influenced the development of national air quality guidelines in Europe, and the process of scientific review leading to the revised guidelines has been influential in the development of air quality standards within the EU.
(2) The EU has the power to set air quality limit values that are binding on all member states; these states are bound to report exceedances of these values and take abatement measures to meet them. Until recently, the EU has only set limit values to protect human health. However, it has now issued a Directive that sets limit values for certain pollutants to protect vegetation or ecosystems; these limit values apply only in rural background areas, and not in urban or industrial areas where the focus of policy measures will be effects on human health.
(3) The United Nations Economic Commission for Europe (UN/ECE) has played an important role in the international negotiation of emission reductions based on targets for long-term environmental protection, through the Convention on Long-Range Transboundary Air Pollution. As well as its work on the use of critical loads

Table 2. Summary of air quality guidelines, limit values and critical levels applied in Europe.

	Ozone	Sulphur dioxide	Nitrogen oxides
Nature of index	Cumulative exposure over growing season (ppb.h)	Annual or winter mean (μg m^{-3})	Annual mean of sum of NO and NO_2 concentrations (ppb), expressed as μg m^{-3} based on conversion for NO_2
UNECE	3000 ppb.h in daylight hours over three months (agricultural crops and natural vegetation) 10,000 ppb.h in daylight hours over six months (forests)	30 (agricultural crops) 20 (forests) 15 (forests in harsh climates) 10 (lichens)	30 (All vegetation)
WHO	5300 ppb.h in daylight hours over three months (agricultural crops) 10,000 ppb.h in all hours over six months (forests)	As for UNECE	As for UNECE
EU		20 (All ecosystems away from vicinity of sources)	30 (All vegetation)

for acidification, UN/ECE has provided the framework for the development of critical levels, the gaseous concentrations above which direct adverse effects on vegetation may occur.

The values recommended for the three air pollutants of greatest concern in terms of direct effects — O_3, SO_2 and NO_x — by the three organisations listed above are summarised in Table 2.

Ozone

As mentioned in the previous section, European approaches to devising air quality criteria for O_3 have differed significantly from those in the U.S., which have been based on seasonal mean concentrations. The aim in Europe was to develop an exposure index which reflected the cumulative nature of O_3 impacts and which incorporated the capacity of vegetation to detoxify O_3 entering the leaf. An appropriate index, termed AOT40, was identified from experimental data from open-top chamber studies across Europe to establish the effects of O_3 on wheat, carried out as part of a wider EU programme (Jager *et al.*, 1993). This AOT40 index for crops is calculated by summing the differences between the actual concentration and 40 ppb for each daylight hour during a growing season of three months when the concentration exceeds 40 ppb. Using the relationship illustrated in Fig. 4, a critical level, or air quality guideline, could be set for any chosen value of yield loss; in practice, a yield loss of 5% averaged over several years has been adopted, leading to a critical level of 3000 ppb.h (Karenlampi and Skarby, 1996).

For forests, the difficulty in setting critical levels for O_3 relates to the lack of long-term field evidence, such as that for SO_2, on mature forest systems. Therefore, use has been made of short-term studies with young trees in open-top chambers. Data for beech have been employed to estimate a critical level of 10,000 ppb.h over a growing season of six months, based on an annual change in above-ground growth of 10%, the minimum which can normally be demonstrated to be statistically significant in this type of experiment. However, this empirical approach cannot be used to assess longer-term effects of O_3, which will require modelling-based approaches. Furthermore, the responses of mature trees may be quite different from these young experimental trees, a factor which requires much further investigation.

Sulphur Dioxide

In contrast to the standards for O_3, which have been derived from experimental studies, those for SO_2 are primarily based on field observation. Standards for SO_2 were first proposed by International Union of Forest Research Organisations (IUFRO) in 1978 (Wentzel, 1983). Standards of 50 $\mu g\ m^{-3}$ to protect trees in good growing conditions and 25 $\mu g\ m^{-3}$ to protect trees growing on poor sites were established. The latter value was based on field evidence from Finland and mountain situations in central Europe that conifer species are much more sensitive to SO_2 under cold temperatures.

More recent analysis of the available data under the auspices of UN/ECE and of WHO has led to the development of a series of critical levels, or air quality guidelines, for SO_2 for different types of vegetation (cf. Table 2). The lowest value is set to protect lichen species, many of which were lost from urban areas of Europe due to the effects of air pollution. The value of 10 $\mu g\ m^{-3}$ was based on a field study around a newly established rural point source of pollution which demonstrated community change at this concentration (Will-Wolf, 1981). The values for trees and natural vegetation are based on an analysis of the rate of breakdown of Norway spruce stands in Czechoslovakia as a function of SO_2 concentration and altitude (Makela *et al.*, 1987). The lower value of 15 $\mu g\ m^{-3}$ in colder areas, and the application of the guideline as a winter, as well as an annual, mean concentration reflects the greater impact of SO_2 at low temperatures. For crops, the critical level of 30 $\mu g\ m^{-3}$ is based on an analysis of experimental data to identify the lowest concentration producing an adverse effect — a study of perennial ryegrass (Bell, 1985).

Nitrogen Oxides

The amount of evidence on which to base an air quality guideline is much lower for NO_x than for SO_2 or O_3. The values adopted by

WHO (2000), following the critical levels proposed by the UN/ECE, are based on analysis of the existing data using standard ecotoxicological approaches to estimate the lowest observed effect concentration. It is important to note that this analysis treats stimulations and reductions of growth together; this is because of the evidence that physiological responses to increased nitrogen deposition tend to increase sensitivity to stresses such as cold and drought, and increase the performance of insect herbivores. Because of recent evidence that nitric oxide (NO) can be as phytotoxic as nitrogen dioxide (NO_2), the values are set for the sum of both pollutants, expressed as NO_2 equivalents.

3.2. *Quantification of Pollutant Impacts*

As has been emphasised throughout this section, evaluation of air quality criteria and impacts, and development of policy for emission control for regional air pollution problems, has largely taken place through international bodies in Europe. For sulphur and nitrogen deposition, critical loads set to prevent long-term acidification or eutrophication have been the tool used for risk assessment, but in terms of direct effects on forests and crops, O_3 is now the major regional pollutant of concern.

Plate 3 provides model estimates of seasonal AOT40 values across Europe for crops in 1995, and as predicted in 2010 after the implementation of the Gothenburg Protocol (NEGTAP, 2001). The precursor emission reductions under this protocol are predicted to lead to significant reductions in AOT40 values across Europe, although these may be offset to some extent by rising global background concentrations (Jonson *et al.*, 2001). The critical level of 3000 ppb.h (3 ppm.h) is still predicted to be exceeded over all of central and southern Europe in 2010. Therefore, a simple switch from exceedance to non-exceedance of the critical level cannot capture the full benefits of this modelled reduction in O_3 exposures; instead a quantification of yield losses, and their economic consequences, is needed.

In theory, exposure-response relationships, such as those shown in Figs. 4 and 5, could be integrated with databases on crop distributions and O_3 concentrations to estimate the regional impacts of O_3, or

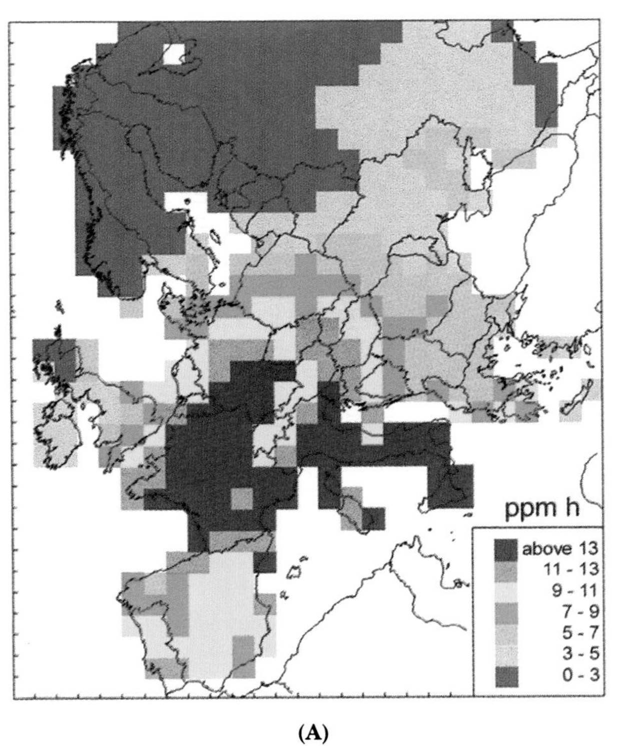

(A)

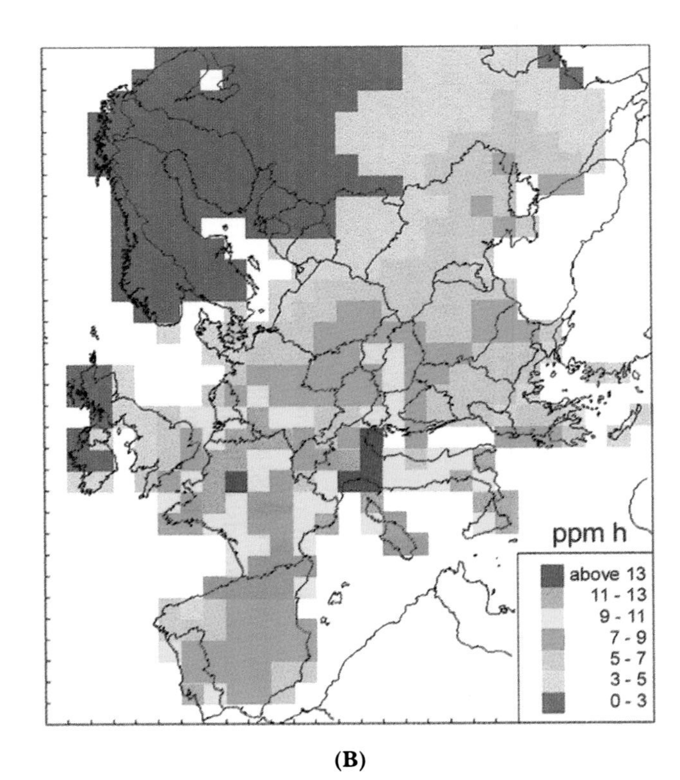

(B)

Plate 3. Modelled seasonal AOT40 exposures of crops across Europe as estimated by the EMEP model using **(A)** 1995 precursor emissions and **(B)** those predicted under the Gothenburg Protocol in 2010. From NEGTAP (2001).

rather to assess the benefits of measures to reduce O_3 concentrations. A recent evaluation, in the context of the development of EU emission control policy, estimated the annual benefit of reducing the gap between current O_3 exposure and the critical level by about 30% across the Member States as about 2000 million Euro (Holland *et al.*, 1999). Interestingly, this value is consistent in scale with estimates for the U.S. made a decade earlier. The benefits in the EU study in terms of forestry were an order of magnitude lower, but only considered the effects of timber production, and not the wider societal value of healthy forests. Furthermore, the estimates do not include important effects, such as increased sensitivity to insect pests or plant pathogens and altered grain or forage quality.

Considerable caution needs to be exercised in using such experimental exposure-response relationships to assess the regional impacts of O_3. In typical exposure-response experiments, O_3 is added simultaneously in all treatments, i.e. under the same atmospheric and micrometeorological conditions. When comparing the impacts of O_3 in different locations, or in different years, it is important to appreciate that the different climates may lead to variable impacts of O_3. For example, Grünhage *et al.* (1997) showed that in the field the highest O_3 concentrations tend to occur under meteorological conditions which limit the dose of O_3 absorbed by the plant. This is because of the high resistance to O_3 flux across the atmospheric boundary layer to the vegetation and because such concentrations often occur with high vapour pressure deficits, which lead to low values of stomatal conductance. These high resistances can cause large vertical gradients in O_3 concentrations over crops which lead to large differences in AOT40 values between normal measurement height and the crop surface, a factor which needs to be considered when assessing the likely impact of the O_3 exposures mapped in Plate 3.

The value of using O_3 flux, as opposed to seasonal mean concentrations or AOT40 exposures, is well illustrated by a recent analysis of experimental data for wheat collected over several seasons in southern Sweden by Pleijel *et al.* (2000). The yield data were first related to the O_3 exposure over the period of grain filling (Fig. 6). This showed a wide variation in the slope of the relationships in different years, with a yield loss of 10% being predicted at AOT40 values that differ by an

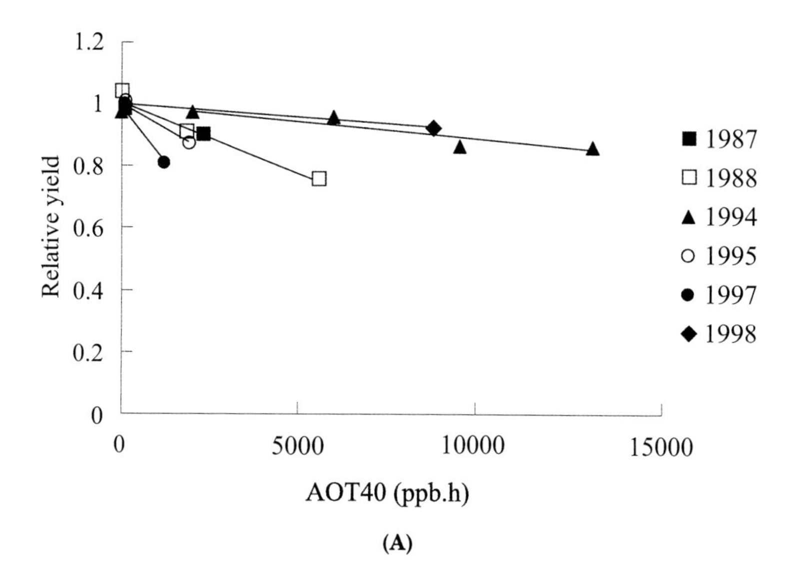

(A)

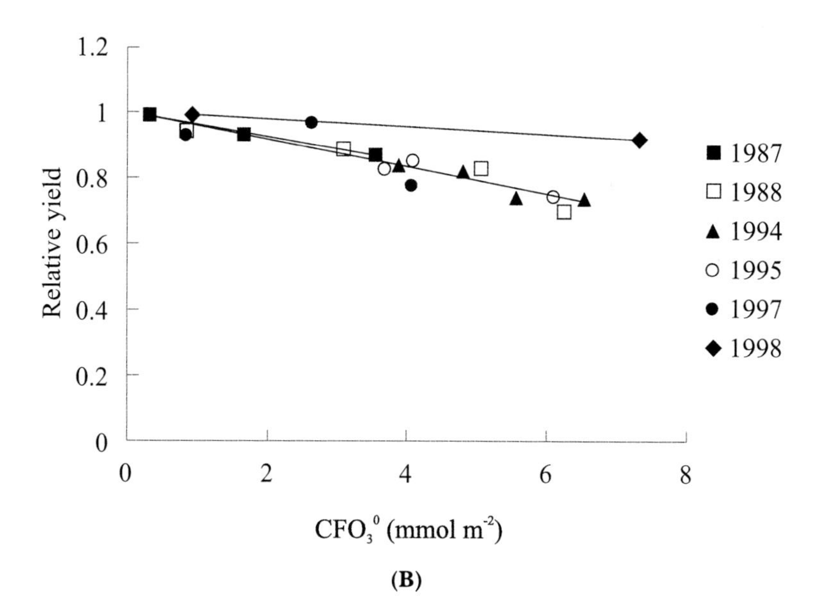

(B)

Fig. 6. Linear regression of grain yield of wheat, relative to control treatments in a series of open-top chamber exposure-response studies in southern Sweden, plotted against **(A)** AOT40 and **(B)** cumulative O_3 uptake (CFO_3) during the period of grain filling. From Pleijel *et al.* (2000).

order of magnitude over the different years. Figure 6 shows the same data plotted against the modelled dose of O_3 absorbed through the stomata over the same period. Five of the six experiments now fall on a common line, indicating that by incorporating the effects of irradiance, temperature and vapour pressure deficit in modifying stomatal O_3 uptake, a stronger mechanistic relationship to yield is obtained. The outlier in Fig. 6(b) may be explained by the fact that winter wheat, rather than spring wheat, was used in that one year.

The implications of using O_3 flux, rather than O_3 exposure, in assessing risks of O_3 impacts across Europe were demonstrated by Emberson *et al.* (2000). While AOT40 values are higher in central than in northern Europe (cf. Plate 3), the modelled fluxes were similar in the two areas. This was largely because the higher O_3 concentrations tended to be associated with higher temperatures and vapour pressure deficits, and hence lower stomatal conductance.

4. Conclusions

Air pollution problems in Europe have changed dramatically over the last 30 to 40 years. While local impacts of urban or industrial sources on crops and forests still occur, successful national and international policies to reduce sulphur emissions has had a dramatic effect on local and regional SO_2 concentrations. The major impacts on vegetation are now perceived primarily as transboundary issues, which require collaboration between countries in monitoring and evaluating problems, and in assessing the benefits of further emission control. In terms of direct impacts on vegetation, O_3 is seen to be the most important future problem, especially given the evidence of long-term increases in global background concentrations.

In terms of impacts on forests, acidification, eutrophication and ground-level O_3 may all affect crown condition. O_3 has been shown to cause visible injury on some sensitive species, but most data on growth and physiological responses derive from short-term studies on young trees, and there is a lack of good field and experimental data to assess its longer-term impact on forest growth and vitality. Empirical exposure-response relationships for crops derived from open-top chamber studies have been used to assess impacts of O_3 across the continent. However, the difficulty of extrapolating exposure-response relationships from chambers to field conditions with any confidence places considerable uncertainty on current estimates of the benefits of policies to reduce O_3 exposures at different locations. For both crops and forests, greater use of mechanistic models would improve the basis for future risk assessment.

References

Ashenden T.W. and Mansfield T.A. (1978) Extreme pollution sensitivity of grasses when SO_2 and NO_2 are present in the atmosphere together. *Nature* **273**, 142–143.

Ashmore M.R., Bell J.N.B. and Reily C.L. (1980) Visible injury to crop species in the United Kingdom. *Env. Pollut.* (*Series A*) **21**, 209–215.

Ashmore M.R. (1984) Effects of ozone on vegetation in the UK. In *Ozone*, ed. Grennfelt P. IVL, Goteborg, pp. 92–104.

Ball G.R., Benton J.M., Palmer-Brown D., Fuhrer J., Skarby K., Gimeno B.S. and Mills G.E. (1998) Identifying factors which modify the effects of ambient ozone on white clover (*Trifolium repens* L.) in Europe. *Env. Pollut.* **103**, 7–16.

Baumgarten M., Werner H., Haberle K.-H., Emberson L.D., Fabian P. and Matyssek R. (2000) Seasonal ozone response of mature beech trees (*Fagus sylvatica*) at high altitude in the Bavarian forest (Germany) in comparison with young beech grown in the field and in phytotrons. *Env. Pollut.* **109**, 431–442.

Bell J.N.B. (1985) SO_2 effects on the productivity of grass species. In *The Effects of SO_2 on Plant Productivity*, eds. Winner W.E., Mooney H.A. and Goldstein R.A. Stanford University Press, Stanford, pp. 209–216.

Bleasdale J. (1952) Atmospheric pollution and plant growth. *Nature* **169**, 376–377.

Braun S. and Fluckiger W. (1995) Effects of ambient ozone on seedlings of *Fagus sylvatica* L and *Picea abies* (L.) Karst. *New Phytol.* **129**, 33–44.

Braun S., Rihm B., Schindler C. and Fluckiger W. (1999) Growth of mature beech in relation to ozone and nitrogen deposition: an epidemiological approach. *Water Air Soil Pollut.* **116**, 357–364.

Cohen J.B. and Rushton A.C. (1925) *Smoke: A Study of Town Air.* Edward Arnold, London.

EEA (2000) *Environmental Signals.* European Environment Agency, Copenhagen.

Emberson L.D., Ashmore M.R., Cambridge H., Tuovinen J.-P. and Simpson D. (2000) Modelling stomatal flux across Europe. *Env. Pollut.* **109**, 403–413.

Farrar J.F., Relton J. and Rutter A.J. (1977). Sulphur dioxide and the scarcity of *Pinus sylvestris* in the industrial Pennines. *Env. Pollut.* **14**, 63–68.

Fischer R., de Vries W., Dobbertin T., Haußmann G., Landmann M., Lorenz M., Mayer P., Mues V., Nakos G., Rademacher P., Seidling W., van Tol G. (2001) *Forest Condition in Europe.* Executive Report, UN/ECE and EC, Geneva and Brussels.

Fowler D. and Cape J.N. (1980) Air pollutants in agriculture and horticulture. In *Effects of Gaseous Air Pollution in Agriculture and Horticulture,* eds. Unsworth M.H. and Ormrod D.P. Butterworth Scientific, London, pp. 3–26.

Fuhrer J. (1996) The critical level for effects of ozone on crops, and the transfer to mapping. In *Critical Levels for Ozone in Europe: Testing and Finalizing the Concepts,* eds. Karenlampi L. and Skärby L. UN/ECE Workshop Report, University of Kuopio, pp. 27–43.

Fuhrer J., Skärby L. and Ashmore M.R. (1997) Critical levels for ozone effects on vegetation in Europe. *Env. Pollut.* **97**, 91–106.

Fumagalli I., Gimeno B.S., Vellissariou D., de Temmerman L. and Mills G. (2001) Evidence of ozone- induced adverse effects on crops in the Mediterranean region. *Atmos. Env.* **35**, 2583–2587.

Gimeno B.S., Velissariou D., Barnes J.D., Inclan R., Pena J.M. and Davison A.W. (1992) Danos visibles por ozono en aciculas de *Pinus halepenis* Mill. en Grecia y Espana. *Ecologia* **6**, 131–134.

Godzik S. (1984) Air pollution problems in some central Euopean countries — Czechoslovakia, the German Democratic Republic and Poland. In *Gaseous Pollutants and Plant Metabolism,* eds. Koziol M.J. and Whatley F.R. Butterworths, London, pp. 25–34.

Grünhage L., Jager H.J., Haenel H.D., Hanewald K. and Krupa S. (1997) PLATIN (Plant Atmosphere Interaction) II Co-occurrence of high ambient ozone concentrations and factors limited plant absorbed dose. *Env. Pollut.* **98**, 51–60.

Guderian R. and Stratmann H. (1962). *Frielandversuche zur Ermittlung von Scwefeldioxydwirkungen auf die Vegetation.* Forschunsberichte des Landes Nordrhein — Westfalen, Part 1, No. 1118.

Holland M.R., Forster D. and King K. (1999) *Cost-Benefit Analysis.* Section 5.2. of EU Ozone Position Paper, European Commission, Brussels.

Jager H.J., Unsworth M., de Temmerman L. and Mathy P. (1993) *Effects of Air Pollution on Agricultural Crops in Europe.* Air Pollution Research Report 45, CEC Brussels.

Jonson J.E., Sundet J.K. and Tarrason L. (2001) Model calculations of present and future levels of ozone and ozone precursors with a global and a regional model. *Atmos. Env.* **35**, 525–537.

Karenlampi L. and Skärby L. (1996) *Critical Levels for Ozone in Europe: Testing and Finalising the Concepts.* UN/ECE Workshop Report, University of Kuopio, Finland.

Klap J., Voshaar J.O., de Vries W. and Erisman J.W. (1997) Relationships between crown condition and stress factors. In *Ten years of Monitoring Forest Condition in Europe* eds. Muller-Edzards C., de Vries W. and Erisman J.W. UNECE and EU, Geneva and Brussels.

Knabe W. (1970) Kiefenwaldverbreiting und Schwefeldioxide — Immissionen in Ruhrgebeit. Staub. *Reinhaltung der Luft* **39**, 32–35.

Makela A., Materna J. and Schopp W. (1987) *Direct Effects of Sulfur on Forests in Europe: A Regional Model of Risk.* Working paper 87-57, International Institute for Applied Systems Research, Laxenburg, Austria.

Matyssek R. and Innes J.L. (1999) Ozone — a risk factor for trees and forests in Europe? *Water Air Soil Pollut.* **11**, 199–226.

Mills G., Hayes F., Buse A. and Reynolds B. (2000) *Air Pollution and Vegetation.* UNECE ICP Vegetation Annual Report 1999/2000, NERC (ISBN 1-870393-55-4).

NEGTAP (2001) *Transboundary Air Pollution: Acidification, Eutrophication and Ground-Level Ozone in the UK.* Report of the National Expert Group on Transboundary Air Pollution, UK Department for Environment, Food and Rural Affairs, London.

Pleijel H., Danielsson H., Karlsson G.P., Gelang J., Karlsson P.E. and Sellden G. (2000) An ozone flux-response relationship for wheat. *Env. Pollut.* **109**, 453–462.

Polunin O. and Walters M. (1985) A *Guide to the Vegetation of Britain and Europe.* Oxford University Press.

Sandermann H., Wellburn A.R. and Heath R.L. (1997) *Forest Decline and Ozone.* Ecological Studies Vol. 127, Springer-Verlag, Berlin.

Schenone G. and Lorenzini G. (1992) Effects of regional air pollution on crops in Italy. *Agric. Ecosyst. Env.* **38**, 51–59.

Skelly J.M., Innes J.L., Savage J.E., Snyder K.R., Vanderheyden D., Zhang J. and Sanz M.J. (1999) Observation and confirmation of foliar ozone symptoms on native plant species of Switzerland and southern Spain. *Water Air Soil Pollut.* **116**, 227–234.

Spieker H., Melikainen K., Kohl M. and Skovegaard J. (1996) *Growth Trends in European Forests.* Springer-Verlag, Berlin.

UN/ECE (2002) *Damage to Vegetation by Ozone Pollution.* Brochure produced by the ICP Vegetation and the ICP Forests, eds. Mills G., Sanz M. J. and Fischer R.

Velissariou D., Gimeno B.S., Badiani M., Fumigalli I. and Davison A.W. (1996) Records of O_3 visible injury in the ECE Mediterranean region. In *Critical Levels for Ozone in Europe: Testing and Finalizing the Concepts.* eds. Karenlampi L. and Skärby L. UN/ECE Workshop Report, University of Kuopio, pp. 343–350.

Vellissariou D. (1999) Toxic effects and losses of commercial value of lettuce and other vegetables due to photochemical air pollution in agricultural areas of Attica, Greece. In *Critical Levels for Ozone — Level II*, eds. Fuhrer J. and Achermann B. Swiss Agency for Environment, Forest and Landscape, Bern, pp. 253–256.

Wentzel K.F. (1983) IUFRO studies on maximal SO_2 emissions standards to protect forests. In *Effects of Accumulation on Air Pollutants in Forest Ecosystems*, ed. Reidel D. Dordrecht, pp. 295–302.

WHO (2000) *Air Quality Guidelines for Europe*, 2nd Ed. WHO, Copenhagen.

Will-Wolf S. (1981) Structure of corticolous lichen communities before and after exposure to emissions from a "clean" coal-fired power station. *Bryologist* **83**, 281–295.

CHAPTER 4

AIR POLLUTION IMPACTS ON VEGETATION IN JAPAN

T. Izuta

1. Introduction

In Japan, air pollution is one of the most serious environmental problems not only for people, but also for vegetation. However, the environmental quality standards of air pollutants such as sulphur dioxide (SO_2), nitrogen dioxide (NO_2), suspended particle matter (SPM) and photochemical oxidants have been established for human health, not for the protection of plants.

In the 1960s and 1970s, air pollution induced by SO_2 and NO_2 became one of the most serious environmental problems especially in the big cities such as Tokyo and Osaka, and visible foliar injuries and abnormal defoliation could be observed in crops and street trees, respectively. Since the early 1970s, we could frequently observe visible foliar injuries induced by photochemical oxidants such as ozone (O_3) and peroxyacetyl nitrate (PAN) in many plant species. The areas where the risk of air pollution is highest in Japan are considered to be the Kantoh district including the Tokyo Metropolis and Kinki district including Osaka Prefecture, because the concentrations of air pollutants are relatively high in these districts. Recently, there has been a considerable interest in the relationships between air pollutants and decline of forest tree species in mountain areas.

In this chapter, the present situation of air pollution in Japan, and field evidence and experimental research on the impacts of air pollutants on Japanese crop and woody plants is discussed. A risk assessment of the effects of O_3 on the yield of Japanese rice and dose-response relationship between O_3 and dry matter production of Japanese forest tree species are described.

2. Air Pollution Emissions and Concentrations

In the 1960s, the atmospheric concentration of SO_2 was very high in many areas of Japan, and 60 ppb was recorded as the annual average concentration in 1967. Since then, there has been a gradual decrease in the concentration of SO_2 as a result of the implementation of various measures, such as the increased use of scrubbers to remove sulphur from gases from stationary sources and fuels. The environmental quality standard for SO_2 states that the daily average of hourly values should not exceed 40 ppb with individual hourly values not exceeding 100 ppb. Recently, the atmospheric concentration of SO_2 is relatively low all over Japan, and the annual average value was 9 ppb in 1996 (Japan Environment Agency, 1998).

In the metropolitan area of Japan, total emissions of nitrogen oxides (NO_x) are approximately 222,500 tonnes per year, of which more than 50% are emitted from motor vehicles. Recently, atmospheric concentrations of NO_2 have remained stable in the big cities such as Tokyo and Osaka. There are many sites where the concentrations of NO_2 exceed the air quality standard for this pollutant, which restricts daily averages of hourly values to maximum concentrations of 40–60 ppb. In 1996, approximately 4% of general monitoring stations and 35% of roadside monitoring stations reported NO_2 concentrations that exceeded the air quality standard. In 1996, annual average concentrations of NO_2 monitored at general and roadside monitoring stations were approximately 28 ppb and 41 ppb, respectively (Japan Environment Agency, 1998).

Since 1974, the annual average concentration of SPM has remained stable at approximately 40 $\mu g\ m^{-3}$ in Japan. The air quality standard for SPM with a diameter of 10 μm or less (PM_{10}) states that the daily average of hourly values should not exceed 100 μg m^{-3} with hourly values not exceeding 200 μg m^{-3}. In 1996, approximately 32% of general monitoring stations and 58% of roadside monitoring stations reported PM_{10} concentrations that exceeded the air quality standard (Japan Environment Agency, 1998).

At the present time, photochemical oxidants are considered by Japanese environmental scientists to be the most dangerous air pollutant not only for people, but also for crops and trees. The environmental quality standard for photochemical oxidants in Japan states that the daily average of hourly values should not exceed 60 ppb. Warnings are issued when the hourly concentration of photochemical oxidants reaches 120 ppb and alarms when it reaches or exceeds 240 ppb. In 1996, the number of warnings was 95, but no alarms were issued (Japan Environment Agency, 1998).

3. Impacts of Air Pollution

3.1. *Field Evidence*

Since 1973, many researchers belonging to the local self-governing body in the Kantoh district have been evaluating air pollution induced by photochemical oxidants, and its effects on plants such as morning glory, taro and peanut cultivated under field conditions (The Union of Air Pollution Protection Association in Kantoh District, 1996). From 1988 to 1992, visible foliar injuries induced by photochemical oxidants could be observed in morning glory and taro in all the prefectures of Kantoh district (Fig. 1). Especially in Tokyo Metropolis, Kanagawa Prefecture and Saitama Prefecture, the degree of visible foliar injuries was greater than that in the other areas such as Chiba and Nagano Prefectures.

Izuta *et al.* (1993) investigated relationships between the dose of ambient O_3 and growth parameters of radish plants cultivated as an indicator plant in small open-top chambers (OTC) at Fuchu, Tokyo in 1987–1989. Visible foliar injuries such as white fleck and chlorosis could be observed on the cotyledons and 1st true leaf of the plants cultivated in the non-filtered treatment during periods of 7 days with relatively high concentrations of O_3 above 100 ppb. As shown

Fig. 1. Visible foliar injuries induced by photochemical oxidants in taro (source: Kuno, 1994).

in Fig. 2, the extent of reduction in the area of cotyledons and total dry weight of the plants grown in the non-filtered treatment (NF) compared to those in the charcoal-filtered treatment (CF) increased with increasing dose of O_3. These results suggest that the growth of sensitive crop plants cultivated in the suburban areas of Tokyo Metropolis suffered from negative effects of ambient O_3 in the late 1980s.

At the present time, declines in health of species such as Japanese fir, Veitch's silver fir, maries fir, Siebold's beech and birch can be observed in the mountain areas of Japan (Fig. 3). However, the causes and mechanisms of these phenomena are still not clarified because field surveys in the forest-decline areas are insufficient. Atmospheric concentrations of O_3 above 100 ppb have been frequently recorded, with a peak concentration of 242 ppb in August of 1995 at approximately 1500 m above sea level in the Tanzawa mountainous district of Kanagawa Prefecture, where decline of mature Siebold's beech can be observed (Totsuka *et al.*, 1997). Therefore, O_3 is considered to be an important factor relating to tree dieback and/or forest decline observed in Japan (Kanagawa Prefecture, 1994).

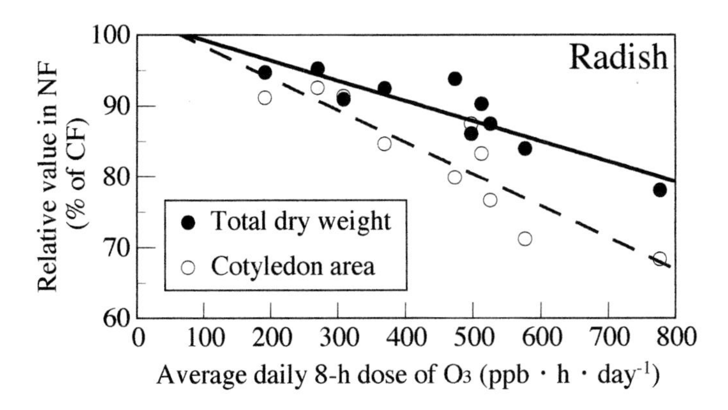

Fig. 2. Relationships between the average daily 8-hour (8:00–16:00) dose of O_3, and cotyledon area and total dry weight of radish plants cultivated in the non-filtered treatment (NF) relative to those in the charcoal-filtered treatment (CF) of small open-top chamber (source: Izuta *et al.*, 1993).

Fig. 3. Declines of Veitch's silver fir and maries fir in the Oku-Nikko mountainous district of Tochigi Prefecture.

In Japan, there is no monitoring network for PAN. Nouchi *et al.* (1984) evaluated air pollution injury induced by PAN using petunia plants as an indicator plant in Tokyo Metropolis in 1982 and 1983. They could observe typical symptoms of PAN-induced foliar injuries in plants grown at five sites from May to October. This result indicates that air pollution by PAN had already spread all over Tokyo Metropolis in the 1980s. In the 1990s, PAN-induced visible foliar injuries were also observed on peanut, watermelon, soybean and petunia in the Kanto district.

3.2. *Experimental Research*

Over the past 30 years, several research institutes and universities in Japan have been conducting experimental studies on the effects of gaseous air pollutants such as SO_2, NO_2, O_3 and PAN on visible foliar injury, growth and physiological functions of Japanese crop plants. However, there are few data concerning the relationships between the dose of gaseous air pollutants and the yield or quality of Japanese crop plants (Kobayashi, 1999). Little is known about the effects of SPM on plants native to Japan. Hirano *et al.* (1995) reported that exposure to dust reduced net photosynthetic rates of cucumber and kidney bean plants by shading the leaf surface, and that the additional absorption of incident radiation by the dust increased the leaf temperature.

In the 1990s, several experimental studies have been reported on the effects of O_3 on the growth and physiological functions of Japanese forest tree species (Izuta, 1998). The dry weight growth of Japanese cedar, the most representative Japanese conifer, seems to be relatively tolerant to exposure to O_3 below 100 ppb for several months (Miwa *et al.*, 1993; Shimizu *et al.*, 1993; Matsumura *et al.*, 1996 and 1998). However, the exposure of the seedlings to concentrations of O_3 above 150 ppb for several months caused a reduction in the root dry mass, which induced an increase in the shoot/root ratio (Miwa *et al.*, 1993; Matsumura *et al.*, 1996 and 1998). The relative sensitivity to O_3 based on the appearance and/or degree of visible foliar injuries and dry weight growth are quite different among tree species native to Japan. As shown in Fig. 4, Matsumura *et al.* (1996) reported that Japanese zelkova was relatively sensitive to ambient levels of O_3 below 100 ppb compared with Japanese cedar and cypress.

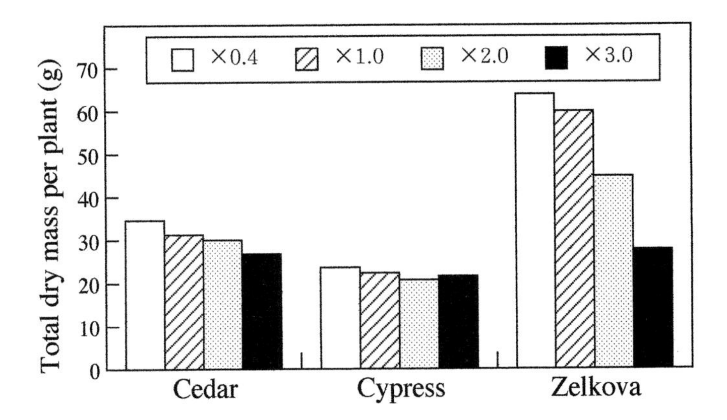

Fig. 4. Total dry mass per plant of Japanese cedar, cypress and zelkova seedlings exposed to four levels of O_3 for 24 weeks. The seedlings were exposed to four simulated profiles with diurnal fluctuations of O_3: 0.4, 1.0, 2.0 and 3.0 times the ambient O_3 concentration at Abiko, Chiba Prefecture (source: Matsumura *et al.*, 1996).

Limited information is available on the physiological responses of Japanese forest tree species to gaseous air pollutants. Shimizu *et al.* (1993) reported that the transpiration activity of Japanese cedar seedlings was reduced by exposure to 100 ppb O_3 for 3 weeks, which might be caused by stomatal closure. Izuta *et al.* (1996) reported that the net photosynthetic rate of three-year-old Siebold's beech seedlings was reduced by exposure to 75 or 150 ppb O_3 for 18 weeks without significant changes in the gaseous phase diffusive conductance to CO_2 and dark respiration rate of the leaves. This result indicates that O_3-induced inhibition of net photosynthesis in the seedlings was not due to stomatal closure. As shown in Fig. 5, the CO_2-response curves of net photosynthetic rate in the O_3-exposed seedlings suggest that O_3 induced reductions in the regeneration rate of ribulose-1, 5-bisphosphate (RuBP), and the activity and/or amount of RuBP carboxylase/oxygenase, which is the key enzyme in photosynthesis. Furthermore, Matsumura *et al.* (1996) reported that exposure to ambient levels of O_3 for 24 weeks reduced net photosynthetic rate and dark respiration rate in Japanese zelkova and cedar seedlings, but not in Japanese cypress seedlings. This result suggests that there is a great

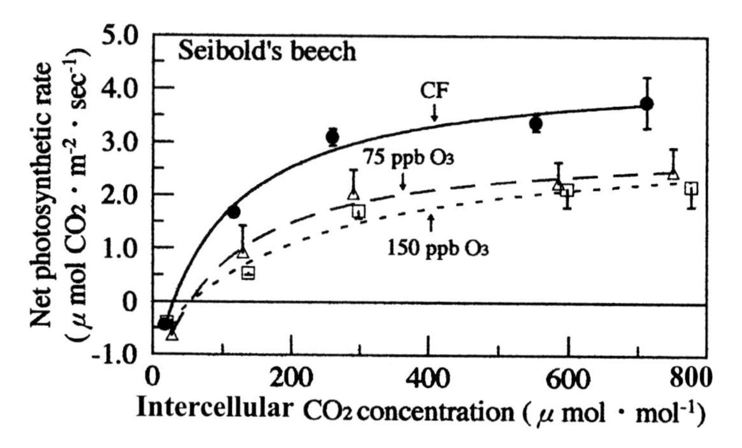

Fig. 5. The intercellular CO_2-response curves of net photosynthetic rate in Seibold's beech seedlings exposed to charcoal-filtered air (CF), and 75 or 150 ppb O_3 for 18 weeks (source: Izuta *et al.*, 1996).

variation in the sensitivity of physiological functions to O_3 among Japanese forest tree species.

4. Dose-Response Relationship and Risk Assessment

In Japan, a dose-response relationship between O_3 and crop yield has been obtained in rice. Kobayashi (1992) evaluated the impacts of O_3 on rice production in the Kantoh district from 1981 to 1985, using a model based on simulation of plant growth process. The changes in total dry matter, leaf number and leaf area of rice plants were simulated on a one-day time step with daily average concentration of O_3, daily integral of insolation and average air temperature. The rice yield loss due to O_3 (RYL) was expressed as the relative difference between the yield simulated with actual O_3 concentration (Y) and that simulated with a base O_3 concentration of 20 ppb (Yb), that is, $RYL = (Yb - Y)/Y$. The simulated rice yield loss correlated much better with the average concentrations of O_3 for the reproductive growth stage than those for the whole growing season. As shown in Fig. 6, the simulated rice yield loss due to O_3 was greater in the central part of Kantoh district (Tokyo Metropolis and Saitama Prefecture) than in the Pacific coast or the Eastern part (Chiba and Ibaraki

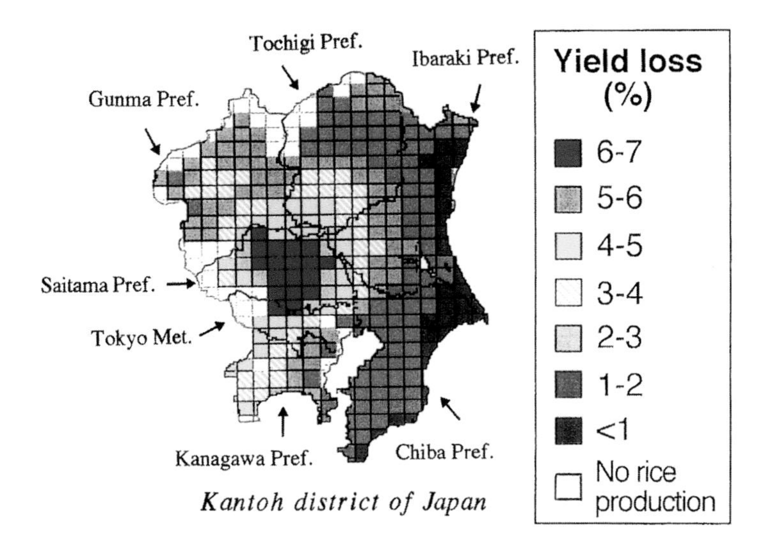

Fig. 6. The average simulated rice yield loss due to O_3 in the Kantoh district of Japan in 1981–1985 (source: Kobayashi, 1999).

Prefectures). The average rice yield loss by O_3 during the five years ranged from 0 to 7% over the district. Furthermore, the estimated total rice production loss due to O_3 in the Kantoh district ranged from 16,000 metric tons in 1981 to 78,500 metric tons in 1985, which corresponded to 1.1% and 4.6% of the total rice production in the district, respectively (Kobayashi, 1992; Kobayashi, 1999).

At the present time, we have no original method to evaluate critical levels of gaseous air pollutants such as O_3 for protecting Japanese forest tree species. Therefore, it is very difficult to conduct risk assessment of the effects of air pollutants on Japanese forest ecosystem. However, based on the data of Izuta and Matsumura (1997), and Matsumura and Kohno (1999), the relationships between the accumulated exposure over a threshold of 40 ppb (AOT40, Fuhrer *et al.*, 1997) and relative increment of total dry weight per seedling can be evaluated for 17 tree species. As shown in Fig. 7, there are great variations in the sensitivity to O_3 among the 17 tree species. In the eight coniferous tree species, the sensitivity to O_3 is relatively high in Japanese red pine and Eastern white pine, but is low in Japanese

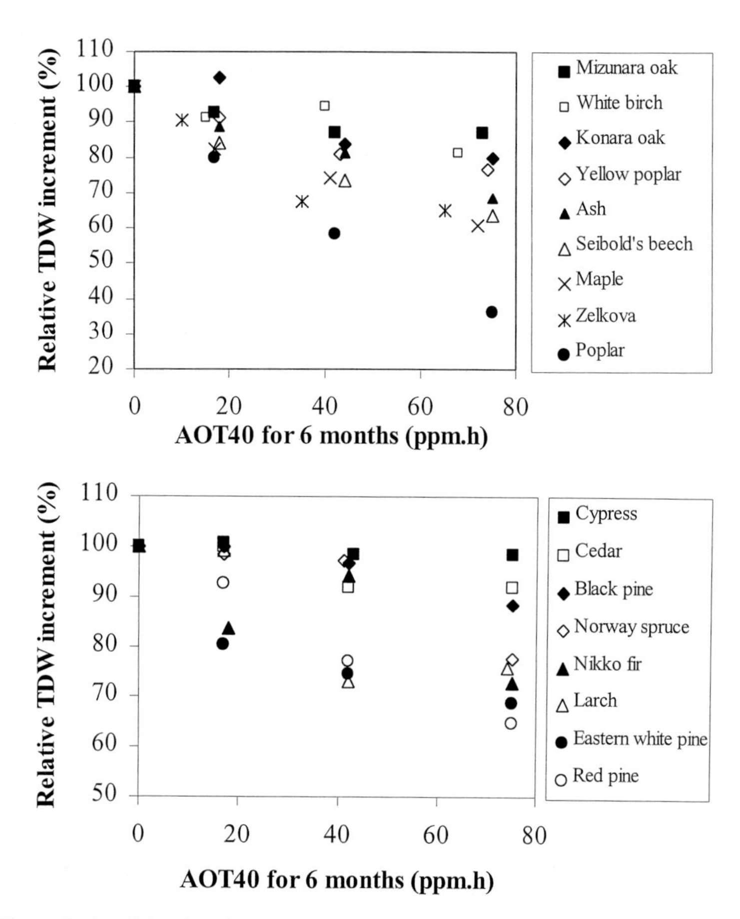

Fig. 7. The relationships between AOT40 for 6 months and relative increment of total dry weight per plant (TDW) of 17 tree species (source: Izuta and Matsumura, 1997; Matsumura and Kohno, 1999).

cypress and cedar. In the nine deciduous broad-leaved tree species, poplar, Japanese zelkova, maple and Seibold's beech are sensitive, but Mizunara oak and Japanese white birch are not sensitive to O_3. The AOT40 values for a 10% loss in the relative increment of total dry weight per seedling are approximately 8000 ppb.h in poplar, and 12,000–21,000 ppb.h in Japanese red pine, Eastern white pine, Japanese zelkova, maple, Siebold's beech and Japanese ash.

When we apply risk assessment methods developed in Japan to the simulation of air pollution impacts on vegetation in the other Asian countries, many factors need to be taken into account. In general, the sensitivity of crop plants cultivated in the Asian countries to air pollutants such as O_3 are different among species and cultivars (Nouchi *et al.*, 1988; Izuta *et al.*, 1999). Meteorological conditions such as light intensity, air temperature and humidity during the crop-growing seasons are different among the Asian countries. These factors affect the degree of negative effects of gaseous air pollutants such as O_3 on growth and net photosynthetic rate of crop plants (Izuta *et al.*, 1988; Izuta *et al.*, 1991). Therefore, until the effects of air pollutants on crop plants cultivated in each Asian country, and the interactions between the effects of air pollutants and meteorological factors are clarified, it will not be possible to evaluate the dose-response relationships and critical levels of air pollutants for protecting crop production in the Asian countries. On the other hand, because the most important air pollutant in the Asian developing countries is considered to be SO_2, further studies are needed on the effects of ambient levels of SO_2 on vegetation native to Asian countries.

5. Conclusions

At the present time, the gaseous air pollutants which pose the greatest threat to vegetation in Japan are probably photochemical oxidants such as O_3. Therefore, we must evaluate critical levels of O_3 for protecting agricultural productivity and forest ecosystems. Furthermore, it is necessary to establish the environmental quality standards of air pollutants for protecting vegetation.

Unfortunately, tree dieback and forest decline can be observed, and high concentrations of O_3 have been detected in several forest-decline areas of Japan. Based on the results obtained from experimental studies, ambient levels of O_3 are considered to be sufficient to reduce dry matter production and physiological functions such as photosynthesis of Japanese forest tree species. Therefore, it is necessary to make a monitoring network of air pollutants in the mountainous areas of Japan.

References

Fuhrer J., Skärby L. and Ashmore M.R. (1997) Critical levels for ozone effects on vegetation in Europe. *Env. Pollut.* **97**, 91–106.

Hirano T., Kiyota M. and Aiga I. (1995) Physical effects of dust on leaf physiology of cucumber and kidney bean plants. *Env. Pollut.* **89**, 255–261.

Izuta T., Funada S., Ohashi T., Miyake H. and Totsuka T. (1988) Effects of ozone on the growth of radish plants at different temperatures. *J. Japan Soc. Air Pollut.* **23**, 209–217 (In Japanese with English summary).

Izuta T., Funada S., Ohashi T., Miyake H. and Totsuka T. (1991) Effects of low concentrations of ozone on the growth of radish plants under different light intensities. *Env. Sci.* **1**, 21–33.

Izuta T., Miyake H. and Totsuka T. (1993) Evaluation of air-polluted environment based on the growth of radish plants cultivated in small-sized open-top chambers. *Env. Sci.* **2**, 25–37.

Izuta T., Umemoto M., Horie K., Aoki M. and Totsuka T. (1996) Effects of ambient levels of ozone on growth, gas exchange rates and chlorophyll contents of *Fagus crenata* seedlings. *J. Japan Soc. Atmos. Env.* **31**, 91–105.

Izuta T. and Matsumura H. (1997) Critical levels of tropospheric ozone for protecting plants. *J. Japan Soc. Atmos. Env.* **32**, A73–A81 (in Japanese).

Izuta T. (1998) Ecophysiological responses of Japanese forest tree species to ozone, simulated acid rain and soil acidification. *J. Plant Res.* **111**, 471–480.

Izuta T., Takahashi K., Matsumura H. and Totsuka T. (1999) Cultivar difference of *Brassica campestris* L. in the sensitivity to O_3 based on the dry weight. *J. Japan Soc. Atmos. Env.* **34**, 137–146.

Japan Environment Agency (1998) Present situation of air pollution. In *White Paper on Environment for 1998.* Tokyo, Japan, pp. 393–430 (in Japanese).

Kanagawa Prefecture (1994) *Research Report of Acid Rain Survey* (in Japanese).

Kobayashi K. (1992) Modeling and assessing the impact of ozone on rice growth and yield. In *Tropospheric Ozone and the Environment II*, ed. Berglund R. L. Air and Waste Management Association, Pittsburgh, USA, pp. 537–551.

Kobayashi K. (1999) Assessing the impacts of tropospheric ozone on agricultural production. *J. Japan Soc. Atmos. Env.* **34**, 162–175 (in Japanese with English summary).

Kuno H. (1994) Indicator plants for monitoring air pollution. In *The Simple Measuring and Evaluation Method on Air Pollution*, ed. Japan Society of Air Pollution, pp. 1–10.

Matsumura H., Aoki H., Kohno Y., Izuta T. and Totsuka T. (1996) Effects of ozone on dry weight growth and gas exchange rate of Japanese cedar, Japanese cypress and Japanese zelkova seedlings. *J. Japan Soc. Atmos. Env.* **31**, 247–261 (in Japanese with English summary).

Matsumura H. and Kohno. Y. (1999) Impact of O_3 and/or SO_2 on the growth of young trees of 17 species: an open-top chamber study conducted in Japan. In *Critical Levels for Ozone — Level II. UN-ECE Workshop Report,* eds. Fuhrer J. and Achermann B.

Matsumura H., Kobayashi T. and Kohno Y. (1998) Effects of ozone and/or simulated acid rain on dry weight and gas exchange rates of Japanese cedar, Nikko fir, Japanese white birch and Japanese zelkova seedlings. *J. Japan Soc. Atmos. Env.* **33**, 16–35 (in Japanese with English summary).

Miwa M., Izuta T. and Totsuka T. (1993) Effects of simulated acid rain and/ or ozone on the growth of Japanese cedar seedlings. *J. Japan Soc. Air Pollut.* **28**, 279–287 (in Japanese with English summary).

Nouchi I., Takasaki T. and Totsuka T. (1988) Relative photochemical oxidant sensitivity of agricultural and horticultural plants. *J. Japan Soc. Air Pollut.* **23**, 355–370 (in Japanese with English summary).

Nouchi I., Ohashi T. and Sofuku M. (1984) Atmospheric PAN concentrations and foliar injury to petunia indicator plants in Tokyo. *J. Japan Soc. Air Pollut.* **19**, 392–402 (in Japanese with English summary).

Shimizu H., Fujinuma Y., Kubota K., Totsuka T. and Omasa K. (1993) Effects of low concentrations of ozone (O_3) on the growth of several woody plants. *J. Agric. Meteorol.* **48**, 723–726.

The Union of Air Pollution Protection Association in Kantoh District (1996) *Actual Situation of Photochemical Smog-Induced Plant Injury in Kantoh District.,* pp. 1–57 (in Japanese).

Totsuka T., Aoki M., Izuta T., Horie K. and Shima K. (1997) Comparison of the air pollutant concentration and soil conditions on south-slope of damaged beech stands and north-slope of healthy beech stands. *Tanzawa Range Environment Studies, Kanagawa Prefecture* pp. 93–96 (in Japanese).

CHAPTER 5

AIR POLLUTION IMPACTS ON VEGETATION IN AUSTRALIA

F. Murray

1. Introduction

Australia has a large land mass (7.69 million km^2), and a relatively small population (18.75 million in 1998) largely clustered around a few large cities at coastal locations in the southern half of Australia (ABS, 2000). About 60% of the Australian population lives in five cities, and 88% lives in cities and large towns (Department of Environment, Sport and Territories, 1996). The Australian economy has been dominated by agriculture, mining and mineral processing. In recent decades, the relative importance of manufacturing, tourism, and other tertiary sector activities has grown considerably.

Australia is an old weathered land mass with mostly nutrient-poor soils. A third of Australia is arid, receiving less than 250 mm of rainfall a year, and another third is semi-arid, receiving 250–500 mm per year of rainfall. Less than 10% of Australia has soils able to sustain productive agriculture or dense vegetation. Only 6% of the land is arable. The main vegetation is woodland and shrubland. Before the arrival of Europeans, 9% of Australia was covered by forest (Department of Environment, Sport and Territories, 1996). By 1965, only 2% was forested (Commonwealth Bureau of Census and Statistics, 1966). Most of the clearing was for agriculture. The area of all crops, excluding pasture, in 1998 was 21.6 million ha, including 10.4 million ha of wheat,

3.5 million ha of barley, 0.94 million ha of oats, 0.7 million ha of cotton and 0.5 million ha of grain sorghum (ABS, 2000).

With a large land mass, a small population, and a small manufacturing sector, it is commonly stated that Australia does not have a significant problem with air pollution outside of the large cities and industrial areas (Department of Arts, Heritage and Environment, 1985; Department of Environment, Sport and Territories, 1996). However, ambient concentrations of sulphur dioxide (SO_2) in some mineral processing towns can exceed 0.5 ppm, and with large cities such as Sydney, Melbourne, Brisbane and Perth located in sunny climates, ozone (O_3) levels in these cities in summer frequently exceed WHO guidelines (NEPC, 1997).

Problems with air pollution in the rapidly growing cities and mining areas were apparent from the early days of the industrialisation of Australia. Air pollution, largely associated with SO_2, in the mineral processing towns was extreme. The forested hills around Queenstown in Tasmania were denuded of vegetation, and almost all plant life killed within many kilometres of the smelter, by the roasting of metal sulphide ores in the town. Holiday processions and men setting out for work become lost in the small town as visibility was sometimes only a metre or two as a result of the formation of SO_2 aerosols during smelting of pyrite in the damp climate (Blainey, 1967).

Damage was also caused to native vegetation, including *Eucalyptus* and *Acacia* species, in the arid climates at Mount Isa in Queensland, and Kalgoorlie in Western Australia, due to SO_2 emissions from roasting of ores. Vegetation up to several kilometres around these point sources of emissions was damaged by SO_2. Similarly, localized damage by SO_2 to vegetation has occurred around other major industrial and mineral processing areas in most states in Australia. With improved pollution control, process changes, and fuel changes, these events are less common now than ten or 20 years ago.

Changes in the world economy, due to falling commodity prices, globalisation of trade, changes in technology, and other factors, have been reflected in the Australian economy, and the emissions profile. Tighter regulation, increased enforcement, increased financial and other business penalties for non-compliance or offences, have also contributed to changes in the characteristics of emissions, and damage to vegetation. The Australian population continues to grow (increasing by more than 50% since 1968). The large cities have grown largely by immigration from other countries and rural areas, including a growth of 50% in Sydney's population since 1969 to 4 million in 1998, and Perth's population has doubled since 1969 to 1.2 million in 1998 (ABS, 2000). The growth in the vehicle population is even faster than the growth in the human population. There were about 4.6 million registered vehicles in 1969, and 12.1 million vehicles in 1998 (ABS, 2000).

There is a changing profile of air pollutants. In many of the cities and industrial areas, reductions in emissions and ambient concentrations of particulate matter, SO_2, carbon monoxide and lead have been achieved. However, the rapidly increasing vehicle population fleet has meant that, despite measures to control vehicle emissions such as the mandatory use of catalytic converters on all new vehicles since 1986, ambient levels of nitrogen oxides (NO_x), O_3, and volatile organic compounds in the large cities remain high.

2. Emissions and Current Concentrations in Cities and Industrial Areas

There is limited information available about national emissions of air pollutants in Australia. While emissions of some common pollutants have been estimated for the capital cities and some major industrial sources, many of these estimates are more than 15 years old, and many substantial sources have not been included in the following estimates (see NEPC, 1997 and 1998). Ambient air quality monitoring information is available for the cities, and some industrial areas, and the following data are from NEPC (1997 and 1998). However, typical concentrations outside of these areas are not available, as 95% of Australia is not monitored for air pollution (Department of Environment, Sport and Territories, 1996)

Emissions of NO_x from the five largest cities, and the regions around them, total about 480 kt per year. Ambient levels of NO_x in the cities occasionally exceed 160 nl l^{-1} for 1 hour. Total emissions of volatile organic compounds have not been estimated reliably.

Biogenic emissions are important in the airsheds of most Australian cities, with guess estimates ranging from about a quarter to about two-thirds of total volatile organic compound emissions. However, the quantification of biogenic emissions is inadequate. Typically, ambient concentrations of O_3 in the largest cities exceed 100 nl l^{-1} for 1 hour on a few occasions each year. For the five largest cities, between 1993–1995, the three-year highest levels ranged from 74–113 nl l^{-1} for 8 hours.

Emissions of SO_2 in Australia are associated with mineral processing and electricity generation, often far from the largest cities. Although national estimates are not yet available, annual emissions of SO_2 from the major industrial centres in Australia are estimated to be 1303 kt. Annual emissions from Mount Isa are estimated to be 500 kt, and Kalgoorlie emissions are estimated to be 372 kt. Australia's national emissions of SO_2 exceed those of many of the large European nations. Ambient concentrations of SO_2 in the large cities are normally low, below about 50 nl l^{-1} for 1 hour. However, concentrations in towns in some mineral processing regions have sometimes exceeded 700 nl l^{-1} for 1 hour.

Estimates of total emissions of particles in Australia are not available. Measurements of PM_{10} in urban areas indicate annual average levels of 25–40 μg m^{-3}, with peak 24-hour levels normally up to 90–110 μg m^{-3}. Occasionally, levels reach up to 150 μg m^{-3}.

3. Impacts Of Air Pollution

Most of the published research on effects of air pollution on vegetation in Australia has been conducted using open-top chambers (Heagle *et al.*, 1973), closed chambers (Doley, 1986) and field studies. More research has been conducted on impacts on vegetation of SO_2 and fluoride than on the other pollutants.

3.1. *Sulphur Dioxide (SO_2)*

There is a limited knowledge about the ambient concentrations of SO_2 that produce adverse responses in the native vegetation of Australia, and in the crop varieties bred for, and growing, under Australian climatic conditions.

Many eucalypts appear to be very sensitive to SO_2. O'Connor *et al.* (1974) screened 131 Australian native species in temperature controlled cabinets for acute visible injury to foliage. The exposure concentrations of SO_2 were 1000, 2000 and 3000 nl 1^{-1}. They reported that *Eucalyptus* was the most sensitive genus, showing visible injury after three hours exposure to 1000 nl 1^{-1}. Norby and Kozlowski (1981) found *Eucalyptus camaldulensis* and *E. globulus* were injured after less than 24 hours exposure to 350 nl 1^{-1}, and they were the most sensitive of five species tested. Howe and Woltz (1981) screened 43 mixed ornamental species from around the world for SO_2. They reported that all of the Eucalyptus species tested (*E. amphifolia, E. robusta, E. torreliana* and *E. viminalis*) showed high sensitivity to acute SO_2 injury.

Extensive studies of the effects of SO_2 on Australian crops and native trees have been conducted using open top chambers, with continuous and intermittent exposures of about six months duration. The results are summarised in Figs. 1 and 2. Tables 1 and 2 summarise the data used in these figures.

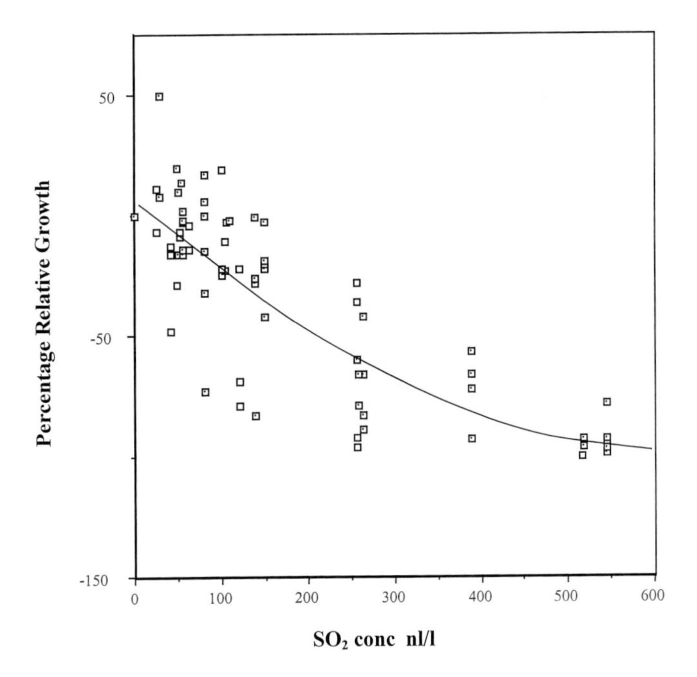

Fig. 1. Yield response to SO_2 exposure of various Australian agricultural crops.

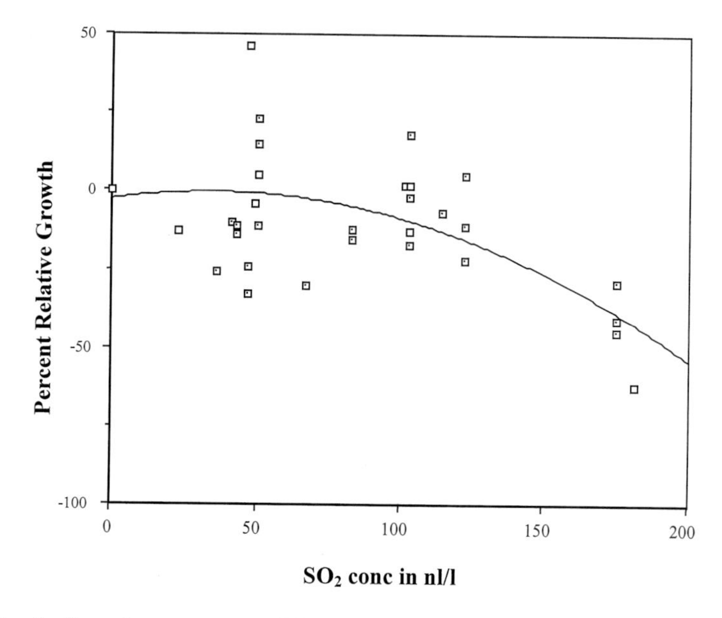

Fig. 2. Growth response to SO_2 exposure of various Eucalyptus species.

Agricultural crops grown in Australia demonstrate a wide range of yield response to exposure to SO_2. This ranges from stimulation of growth and production, especially at low concentrations, to growth reduction of 82% for the particularly sensitive species burr medic (*Medicago polymorpha cv. Santiago*) due to exposure to approximately 150 nl l^{-1} SO_2 for 4 hours/day (Fig. 1). Different cultivars within the same species can show different responses, and this is consistent with studies conducted in other parts of the world. The results collated in Fig. 1 are generally comparable with the results collated from U.K. experiments by Roberts (1984).

When the response of *Eucalyptus* species or crops to SO_2 was plotted according to daily duration of exposure, no trends were evident. Exposure for 4 hours/day produced a similar response to exposure for 8 hours/day (Murray *et al.*, 1992c). Concentration of SO_2 was found to be a more important determinant of plant response than daily duration of exposure. Similarly, total length of exposure has little effect on the magnitude of vegetation response to SO_2 within the range 90–130 days (Murray *et al.*, 1992c).

Table 1. Data used to establish a response model to assess the impacts of SO_2 on agricultural crops.

Common name and cultivar	Species	Concentration of SO_2 in nl/l	Reference
Barley cv. Clipper	*Hordeum vulgare*	49, 100	Murray and Wilson (1988b)
Barley cv. Schooner	*Hordeum vulgare*	41, 121, 257, 517	Murray and Wilson (1990b)
Barley cv. Schooner	*Hordeum vulgare*	55, 149, 262, 544	Murray *et al.* (1992)
Barrel medic cv. Parragio	*Medicago truncatula*	41, 120, 255, 515	Murray and Wilson (1991)
Barrel medic cv. Parragio	*Medicago truncatula*	55, 149, 262, 544	Murray *et al.* (1994b)
Burr medic cv. Santiago	*Medicago polymorpha*	80, 139, 255, 388	Murray *et al.* (1994a)
Ladino clover cv. Regal	*Trifolium repens*	29, 81	Murray (1985b)
Lucerne cv. CUF 101	*Medicago sativa*	28, 81	Murray (1985a)
Lucerne cv. Siriver	*Medicago sativa*	55, 149, 262, 544	Murray *et al.* (1994b)
Maize cv. QK 958	*Zea mays*	51, 106	Murray and Wilson (1990a)
Navy bean cv. Gallaroy	*Phaseolus vulgaris*	53, 109	Murray and Wilson (1990a)
Oats cv. Echidna	*Avena sativa*	80, 139, 255, 388	Murray *et al.* (1991)
Peanut cv. Virginia bunch	*Arachis hypogaea*	52, 105	Murray and Wilson (1990a)
Ryegrass cv. Tetralite	*Lolium perenne*	26, 62	Murray (1984b)
Soybean cv. Dragon	*Glycine max*	52, 104	Murray and Wilson (1990a)
Sub clover cv. Trikkala	*Trifolium subterraneum*	41, 120, 255, 515	Wilson and Murray (1990a)

Table 1. (*Continued*)

Common name and cultivar	Species	Concentration of SO_2 in nl/1	Reference
Sub clover cv. Trikkala	*Trifolium subterraneum*	55, 149, 262, 544	Murray *et al.* (1992b)
Sub clover cv. Woogenellup	*Trifolium subterraneum*	26, 62	Murray (1984b)
Triticale cv. Currency	*X. triticosecale*	80, 139, 255, 388	Murray *et al.* (1991)
Wheat cv. Banks	*Triticum aestivum*	42, 121, 257, 517	Wilson and Murray (1990b)
Wheat cv. Banks	*Triticum aestivum*	55, 149, 262, 544	Murray *et al.* (1994c)
Wheat cv. Eradu	*Triticum aestivum*	49, 100	Davieson *et al.* (1990)
Wheat cv. Halberd	*Triticum aestivum*	49, 100	Murray and Wilson (1988b)
White clover cv. Haifa	*Trifolium repens*	80, 139, 255, 388	Murray *et al.* (1994a)

Concentration was the best descriptor of exposure for these exposure types, and the regression equation below gives an initial guideline for the threshold of SO_2 effects on growth and yield due to long term and frequent exposures.

Pooling the data for all crop species gave a generalised response curve that could be used to predict the SO_2 concentration at which a given yield or growth loss occurred (Fig. 1). This response relationship is described by the equation:

$$y = 1.0366 - (5.973.10^{-3}) \times 0.8213 \quad (r^2 = 0.714)$$

where $y = \%$ change in yield or growth

$x = SO_2$ concentration $(\text{nl}\,\text{l}^{-1})$

This equation relates to a mean SO_2 concentration predicted from continuous exposure, exposure for 8 hours/day, 4 hours/day, and 2.5 hours/day, 3 days/week. The bulk of the data came from

Table 2. Data used to establish a response model to assess the impacts of SO_2 on *Eucalyptus* species.

Eucalyptus species	Concentration of SO_2 in nl/l	Reference
E. pilularis	122, 175, 332	Wilson and Murray (1994)
E. rudis	50, 103	Clarke and Murray (1990)
E. tereticornis	49, 101	Murray and Wilson (1988a)
E. moluccana	43, 83	Murray (1984a)
E. marginata	47, 103	Murray and Wilson (1988c)
E. calophylla	47, 103	Murray and Wilson (1988c)
E. calophylla	50, 103	Clarke (1987)
E. regnans	50, 122, 175, 332	Wilson and Murray (1994)
E. crebra	43, 83	Murray (1984a)
E. microcorys	50, 122, 175, 332	Wilson and Murray (1994)
E. gomphocephala	47, 103	Murray and Wilson (1988c)
E. gomphocephala	41, 114	Fulford and Murray (1990)
E. wandoo	23, 36, 67, 181	Murray *et al.* (1991)

experiments conducted with an exposure regime of 4 hours/day each day for four to five months.

There was enormous variability of response of different species of Eucalyptus to the same SO_2 concentration (Fig. 2). For example, exposure to 50 nl l^{-1} SO_2 produced a range of growth responses from −36% for *E. gomphocephela* to +54% for *E. marginata*. This was not due to differences in daily duration of exposure, total length of exposure or environmental conditions. The difference in response is solely attributable to differences in the responses to SO_2 among different species. The high level of species variation suggests that generalised SO_2 response curves (Fig. 2) for these species, while providing an indication of the average response and range of responses, cannot be used to predict the response of an individual species which has not been investigated.

Like *Eucalyptus* species, other native species demonstrate a wide range of response to SO_2. In these experiments, injury was assessed primarily in terms of growth or yield changes. However, four out of five native species (*Acacia saligna, Banksia attenuata, Banksia menziesii* and *Eucalyptus wandoo*) exposed to 376 nl l^{-1} for one hour in experiments developed acute visible injury (Murray *et al.*, 1991). Tissue death was evident immediately after exposure, so the threshold SO_2 concentration for visible injury in sensitive native species is lower than 376 nl l^{-1} for one hour. Frequent exposure to 181 nl/l SO_2 for 2.5 hours, 3 days/week resulted in the development of severe visible injury in these four species. Although the exact number of exposures that occurred before onset of visible injury is not known, it was well developed after eight weeks of fumigation.

3.2. *Fluoride*

Fluoride emissions caused damage to native vegetation and crops, especially grapevines, in the Hunter Valley of New South Wales and the Swan Valley near Perth (Cox and Jones, 1980; Murray, 1982; Leece *et al.*, 1986). Research has attempted to characterise the relationship between ambient concentrations of fluoride, leaf concentrations, and effects on visible injury, reduced growth and yield of grapevines and a few other species, but the quantitative relationships are still not well understood. Plate 1 illustrates typical symptoms of visible injury caused by hydrogen fluoride (HF).

Studies using open-top chambers located in the field around established grapevines in a vineyard investigated the relationship between ambient concentrations of fluoride, leaf concentrations, and effects on visible injury, reduced growth and yield of grapevines fumigated for a full growing season. The studies found that foliar necrosis occurred at fluoride concentrations of 0.17 μg m^{-3} or more after 99 days, but at levels up to 0.28 μg m^{-3} there were few significant effects on growth or yield, and little fluoride accumulation in berries (Murray, 1983 and 1984c). Leece *et al.* (1986) reported on the relationship between fluoride accumulation by leaves of grapevines, and visible injury. They reported that elevated leaf fluoride concentrations up to 42 μg g^{-1} did not produce leaf necrosis in well-managed vineyards. This contrasts with the findings of Cox and Jones (1980) and Horne *et al.* (1981). Leece *et al.* (1986) hypothesised that leaf fluoride levels above 26 μg g^{-1} can be associated with fluoride injury when grapevines are subject to soil water or osmotic stress after fumigation.

Plate 1. Examples of foliar injury caused by exposure to HF: **(A)** anthocyanins on the leaf margins of *Eucalyptus europhylla*; **(B)** distorted growth to leaflets of *Ailanthus* species; and **(C)** necrotic legions on leaves of *Persoonia laevis* (source: A. Davison).

Australian native vegetation shows a considerable range of variation in response to fluoride fumigation, as studies elsewhere have shown. Experimental data based on potted plants in closed chamber studies are summarised by Doley (1986). The results of ten years of field studies around fluoride sources show that sensitive Australian native species demonstrate unacceptable damage at atmospheric concentrations of gaseous fluoride above 0.5 $\mu g\ m^{-3}$ as a three-month average (Mitchell *et al.*, 1981). Horne *et al.* (1981) studied effects of fluoride emissions in vineyards for a number of years and found foliar damage at concentrations of 0.18 and 0.27 $\mu g\ m^{-3}$ as 24-hour seasonal averages in two seasons. They found a decline in vigour of grapevines in the field after several years exposure to fluoride concentrations of about 0.2 $\mu g\ m^{-3}$. They recommended an air quality standard to protect grapevines of 0.1 $\mu g\ m^{-3}$ as a growing season 24-hour average.

3.3. *Other Pollutants*

Exposure of Eucalyptus species to NO_x in open-top chambers for 169 days demonstrated that the exposure had a bimodal effect on growth, with some differences between species. There was growth stimulation at small exposures (25 or 50 nl l^{-1} for one per month or once per week) and a progressive growth decrease as exposure increased (91 or 187 nl l^{-1} and frequent exposures) (Murray *et al.*, 1994d).

Most studies of the effects of O_3 on Australian native species have been short-term studies using high concentrations to assess acute visible injury (O'Connor *et al.*, 1975). In a longer term study, eight *Eucalyptus* species were exposed to a diurnally varied concentration of 26 or 172 nl l^{-1} O_3 (as a 7-hour mean) for 7 hours/day, 5 days in every 14 days, for 18 weeks (Monk and Murray, 1995). The results showed considerable differences in the responses of the different species. Some species showed no visible injury or growth changes, but others showed up to 90% leaf injury and 30% growth reductions. The findings were generally consistent with those found for North American woody perennials and northern European herbaceous species (Monk and Murray, 1995). Exposure of pasture grasses in Australia to 90 nl l^{-1} O_3 for 4 hours/day for 5 days per week for five weeks demonstrated growth reductions of 14–21% (Horsman *et al.* 1980).

Based upon dose-response data from the U.S. and European studies and a limited amount of Australian data, including field pollution gradient studies using O_3-sensitive varieties of plants, current ambient concentrations of O_3 in the regions around cities are likely to reduce growth and yield of sensitive crops, and damage some areas of biodiversity importance.

A number of studies have been conducted on the effects on Australian vegetation of mixtures of pollutants, including mixtures of SO_2 and HF, SO_2 and nitrogen dioxide (NO_2) and others, but insufficient studies have been conducted to reach any clear conclusions on impacts of interactions.

Studies of acid deposition in Australia show that there are some substantial industrial sources with local effects and some regional influences. However, acid deposition over most of Australia is at lower rates than in Europe, North America and parts of Asia (Murray, 1989).

4. Conclusions

Very little research has been conducted on the impacts of air pollution on Australian vegetation. The highest priority should be given to investigating the impacts of photochemical oxidants, and to the impacts of deposition of nitrogen and sulphur compounds on nutrient poor ecosystems of biodiversity importance. Fumigation studies of species of biodiversity or agricultural importance in areas at risk are needed. The concentrations and distribution of peaks should be realistic, and the conditions of fumigation should be as similar to field conditions as possible.

The regions most at risk from the adverse impacts of air pollution are the areas in and around the major cities, where photochemical oxidants pose risks to areas of biodiversity importance such as national parks and nature reserves, and damage to the growth and yield of sensitive crops and landscape vegetation. In particular, this applies to the regions around Sydney, Melbourne, Brisbane, Perth and Adelaide.

SO_2, HF and acid deposition pose risks to sensitive crops and landscape vegetation, and areas of biodiversity importance around some mineral processing and fuel combustion facilities. These effects are locally significant, but generally not regionally pervasive. Areas in or around the Hunter Valley, Latrobe Valley, Gladstone, Mount Isa, Kalgoorlie, Port Pirie, Wollongong, and other areas are at risk.

References

ABS (2000) *2000 Yearbook Australia.* Australian Bureau of Statistics, Canberra.

Blainey G. (1967) *The Peaks of Lyall*, 3rd Ed. Melbourne University Press, Melbourne.

Clarke K. (1987) Growth and anatomical effects of low concentrations of sulfur dioxide on *Eucalyptus rudis* and *Eucalyptus calophylla.* Murdoch University, Perth.

Clarke K. and Murray F. (1990) Stimulatory effects of SO_2 on growth of *Eucalyptus rudis* Endl. *New Phytol.* **115**, 633–637.

Commonwealth Bureau of Census and Statistics (1966) *Official Yearbook of the Commonwealth of Australia 1966.* Commonwealth Bureau of Census and Statistics, Canberra.

Cox W.J. and Jones L.T. (1980) *Fluorine Toxicity in Grapevines. A Case Study.* Technical Bulletin No. 61, Western Australian Department of Agriculture, Perth.

Davieson G., Murray F. and Wilson S. (1990) Effects of sulfur dioxide, hydrogen fluoride singly and in combination on growth and yield of wheat in open-top chambers. *Agric. Ecosys. Env.* **30**, 317–325.

Department of Arts, Heritage and Environment (1985) *State of the Environment in Australia, 1985.* Australian Government Publishing Service, Canberra.

Department of Environment, Sport and Territories (1996) *Australia: State of the Environment 1996.* CSIRO Publishing, Melbourne.

Doley D. (1986) *Plant-Fluoride Relationships.* Inkata Press, Melbourne.

Fulford G. and Murray F. (1990) Morphogenic changes in *Eucalyptus gomphocephala* exposed to SO_2. *Env. Experimental Bot.* **30**, 343–347.

Heagle A.S., Body D.E. and Heck W.W. (1973) An open-top field chamber to assess the impact of air pollution on plants. *J. Env. Qual.* **2**, 365–368.

Horne R.W., Cox W.J. and Stokes M.C. (1981) Atmospheric fluoride pollution in vineyards. In *Proceedings of the Seventh International Clean Air Conference*, eds. Webb K.A. and Smith A.J. Ann Arbor Science, Ann Arbor, pp. 495–507.

Horsman D.C, Nicholls A.O. and Calder D.M. (1980) Growth responses of *Dactylis glomerata, Lolium perenne* and *Phalaris aquatica* to chronic ozone exposure. *Australian J. Plant Physiology* **7**, 511–517.

Howe T.K. and Woltz, S.S. (1981) Symptomatology and relative susceptibility of various ornamental plants to acute airborne sulfur dioxide exposure. *Proc. Florida State Horti. Soc.* **94**, 121–123.

Leece D.R., Scheltema J.H., Anttonen T. and Weir R.G. (1986) Fluoride accumulation and toxicity in grapevines *Vitis vinifera* L. in New South Wales. *Env. Pollut.* (*Series A*) **40**, 145–172.

Mitchell A.A., Dowling B.J. and Scheltema J.H. (1981) Effects of fluoride on Australian vegetation. In: *Proceedings of the Seventh International Clean Air Conference.* eds. Webb K.A. and Smith, A.J. pp. 479–493. Ann Arbor Science, Ann Arbor.

Monk R.J. and Murray F. (1995) The relative tolerance of some *Eucalyptus* species to ozone exposure. *Water Air Soil Pollut.* **85**, 1405–1411.

Murray F. (1982) *Fluoride Emissions. Their Monitoring, and Effects on Vegetation and Ecosystems.* Academic Press, Sydney.

Murray F. (1983) Response of grapevines to fluoride under field conditions. *J. Am. Soc. Hort. Sci.* **108**, 526–529.

Murray F. (1984a) Effects of sulfur dioxide on three *Eucalyptus* species. *Australian J. Bot.* **32**, 139–145.

Murray F. (1984b) Responses of subterranean clover and ryegrass to sulphur dioxide under field conditions. *Env. Pollut.* (*Series A*) **36**, 239–249.

Murray F. (1984c) Effects of long-term exposure to hydrogen fluoride on grapevines. *Env. Pollut.* (*Series A*) **36**, 337–349.

Murray F. (1985a) Changes in growth characteristics of lucerne (*Medicago sativa* L.) in response to sulphur dioxide exposure under field conditions. *J. Exp. Bot.* **36**, 449–457.

Murray F. (1985b) Some responses of ladino clover (*Trifolium repens* L. cv. Regal) to low concentrations of sulphur dioxide. *New Phytol.* **100**, 57–62.
Murray F. and Wilson S. (1988a) Joint action of sulfur dioxide and hydrogen fluoride on growth on *Eucalyptus tereticornis. Env. Exp. Bot.* **28**, 343–349.
Murray F. and Wilson S. (1988b) Joint action of sulphur dioxide and hydrogen fluoride on the yield and quality of wheat and barley. *Env. Pollut.* **55**, 239–249.
Murray F. and Wilson S. (1988c) Effects of sulphur dioxide, hydrogen fluoride and their combination on three *Eucalyptus* species. *Env. Pollut.* **52**, 265–279.
Murray F. (1989) Acid rain and acid gases in Australia. *Arch. Env. Contam. Toxicol.* **18**, 131–136.
Murray F. and Wilson S. (1990a) Yield responses of soybean, maize, peanuts and navy beans exposed to SO_2, HF and their combination. *Env. Exp. Bot.* **30**, 215–223.
Murray F. and Wilson S. (1990b) Growth responses of barley exposed to SO_2. *New Phytol.* **114**, 537–541.
Murray F., Clarke K. Wilson S. and Monk R. (1991) *Development of Australian Secondary Ambient Air Quality Criteria for Sulphur Dioxide and Nitrogen Oxides.* Final Report to the Energy Research and Development Corporation, Murdoch University, Perth.
Murray F. and Wilson S. (1991) The effects of SO_2 on the final growth of *Medicago truncatula. Env. Exp. Bot.* **31**, 319–325.
Murray F. Wilson S. and Monk R. (1992a) NO_2 and SO_2 mixtures stimulate barley grain production but depress clover growth. *Env. Exp. Bot.* **32**, 185–192.
Murray F., Clarke K. Wilson S. and Monk R. (1992b) The relationship between exposure to sulphur dioxide and yield or growth of crops and trees in Australia. In *Proceedings of the 11th International Clean Air Conference, Brisbane,* eds. Best P., Bofinger N. and Cliff D. Clean Air Society of Australia and New Zealand, Vol. 1, pp. 273–286.
Murray F., Monk R., Clarke K. and Qifu M. (1994a) Growth response of N. and S-deficient white clover and burr medic to SO_2, NO and NO_2. *Agric. Ecosyst. Env.* **50**, 113–121.
Murray F. Wilson S. and Qifu M. (1994b) Effects of SO_2 and NO_2 on growth and nitrogen concentrations in lucerne and barrel medic. *Env. Exp. Bot.* **34**, 319–328.
Murray F. Wilson S. and Samaraweera S. (1994c) NO_2 increases wheat grain yield even when SO_2 is present at low concentrations. *Agric. Ecosyst. Env.* **48**, 115–123.
Murray F., Monk R. and Walker C.D. (1994d) The response of shoot growth of Eucalyptus species to concentration and frequency of exposure to nitrogen oxides. *Forest Ecol. Manag.* **64**, 83–95.

NEPC (1997) *Draft National Environmental Protection Measure and Impact Statement for Ambient Air Quality*. National Environmental Protection Council, Adelaide.

NEPC (1998) *National Environment Protection Measure for Ambient Air Quality*. National Environment Protection Council, Adelaide.

Norby R.J. and Kozlowski T.T. (1981) Interactions of SO_2 concentrations and postfumigation temperature on growth of five species of woody plants. *Env. Pollut.* (*Series A*) **25**, 27–39.

O'Connor J.A., Parbery D.G. and Strauss W. (1974) The effects of phytotoxic gases on native Australian plant species: Part 1. Acute injury due to ozone. *Env. Pollut.* **7**, 7–23.

O'Connor J.A., Parbery D.G and Strauss W. (1975) The effects of phytotoxic gases on native Australian plant species: Part 2. Acute effects of sulphur dioxide. *Env. Pollut.* **9**, 181–192.

Roberts T.M. (1984) Long-term effects of sulphur dioxide on crops: an analysis of dose-response relations. *Philo. Trans. Roy. Soc. London* **B05**, 299–316.

Wilson S.A. and Murray F. (1990a) Response of subterranean clover in open top chambers to chronic exposure to sulphur dioxide. *Agric. Ecosyst. Env.* **32**, 283–293.

Wilson S.A. and Murray F. (1990b) SO_2-induced growth reductions and sulphur accumulation in wheat. *Env. Pollut.* **66**, 179–191.

Wilson S.A. and Murray F. (1994) The growth response of sclerophyllous *Eucalyptus* species to SO_2 exposure compared to *Pinus radiata*. *Forest Ecol. Manag.* **68**, 161–172.

Air Pollution Impacts on Vegetation in Developing Countries

As described in Chap. 1, increases in gaseous air pollution over recent decades have been experienced in many industrial and urban centres of Asia, Africa and Latin America primarily as a result of rapid economic growth, industrialisation and urbanisation with associated increases in energy demands. To date, most attention has focused on the impact of these industrial and urban emissions on human health whilst little is known about pollutant concentrations and exposure patterns in many suburban and rural areas and associated impacts on the local vegetation. The information collated in Chaps. 2 to 5 clearly shows that a large body of evidence exists documenting the adverse impacts of air pollution on forest trees and agricultural crops for certain industrialised countries and regions. This chapter aims to assess the extent of air pollution impacts on vegetation in rapidly industrialising countries with a view to investigating the suitability of tools designed to assess the extent of air pollution damage in Europe and North America, to other continents. The information needed to enable such risk assessments to be performed is discussed further in Chap. 14.

This chapter collates evidence obtained from reviews commissioned from experts from China, Taiwan, India, Pakistan, Egypt, South Africa, Brazil and Mexico describing regional emissions, vegetation distributions and pollutant concentrations. This information is related to observations of injury to crops and forests in the field in an attempt to define levels of ambient pollutant concentrations that cause vegetation damage. In addition, key observations of damage including transect studies along pollution gradients and controlled experimental investigations of impacts on selected crop and forest species are described for each of the pollutants SO_2, O_3, NO_x and SPM.

Establishing the extent of our existing knowledge of air pollution effects is crucial to the process of evaluating progress made in our understanding of pollution impacts to vegetation ▶

in the developing world and to ensure that knowledge gaps that do exist can be identified and addressed. Ultimately, a primary aim of collecting such information is to define dose-response relationships or exposure limits for damage to enable assessments of the probable extent of vegetation damage across the developing world, both for the present day and in the future.

CHAPTER 6

AIR POLLUTION IMPACTS ON VEGETATION IN CHINA

Y. Zheng and H. Shimizu

1. Introduction

China has a population of 1.22 billion people making up one-fifth of the world's population; over the last 15 to 20 years, China has undergone a period of rapid economic development. This rapid expansion in the economy and in the population has inevitably resulted in a growing demand for energy. In comparison with most of the other developing countries in Asia, China has a higher rate of increase in energy consumption (mainly coal) and is short of efficient air pollution control measures. This has resulted in greater air pollutant emissions of sulphur dioxide (SO_2), nitrogen oxides (NO_x) and carbon dioxide (CO_2) than the average rate of increase in developed countries (Kato, 1996). The rise in ambient air pollution levels has already caused an alarming deterioration in the air quality in many Chinese cities. Indeed, a recent report indicates that five of the world's ten worst cities for air pollution are located in China (Yunus *et al.*, 1996) and the air pollution problem is so severe that it is believed to be responsible for more than one million deaths per year across the country (i.e. about one in every eight deaths nationwide; Florig, 1997). Moreover, it has been estimated that in 1995, the total economic losses resulting from acid rain and SO_2 amounted to 110×10^9 Chinese yuan (13×10^9 US dollar) — equivalent to 2% of the gross national product (Editing Committee of China Environmental Yearbook (ECCEY), 1998).

1.1. *Emissions*

In China, more than 75% of primary energy utilisation is supplied by domestic coal. The total coal consumption in 1990 was 1052 million tons and in 1995 reached 1280 million tons. It is estimated that by the year 2000, total coal consumption will possibly reach 1500 million tons per year (ECCEY, 1998). Most of the coal that is used to supply energy has a high sulphur and dust content. The national average sulphur content is 1.35% and in some provinces, like Sichuan, the provincial average attains levels as high as 3.19%, Yunan (3.09%), Guizhou (2.95%), and Guangxi (2.22%). This reliance on coal as the principle source of energy has resulted in an air pollution climate dominated by SO_2, particulates, NO_x and acid rain.

Systematic monitoring of air quality was only introduced in China in the late 1970s and today a network of monitors is maintained and provides reliable measurements of SO_2, NO_x, total suspended particulate matters (SPM), carbon monoxide and acid rain. This monitoring network is chiefly concerned with assessing the likely impacts on human health so the vast majority of monitoring stations are located in urban areas. Much less information exists about ground-level air pollutant concentrations in rural regions.

Sulphur dioxide

In 1995, annual SO_2 emissions amounted to c. 23.7 million tons. Based on previous increases in the rate of emissions, it is forecast that total SO_2 emissions will reach 27.3 million tons by the year 2000 and will reach 33 million tons by the year 2010. Recent estimates show that Chinese SO_2 emissions account for c. 69% of the total SO_2 emissions from Asia (Kato, 1996), and at present China is the biggest SO_2 emitter in the world. In 1997, 79% of these SO_2 emissions were associated with industrial sources and the remainder were injected into the atmosphere by a wide variety of domestic sources (ECCEY, 1998). Most SO_2 emissions are associated with the eastern part of China, where there is a high population density and much industrial activity. For example, the five provinces along the coast of Bo Hai and the Yellow Sea account for only 9% of the total land area of the country, but are responsible for c. 40% of total SO_2 emissions (Wang *et al.*, 1993).

In 1997, the ground-level annual daily mean SO_2 concentration was 66 μg m^{-3}, ranging from 3–248 μg m^{-3} with 52% of the northern cities exceeding the Chinese secondary air quality standard (SAQ) for SO_2 (60 μg m^{-3}), in comparison with 38% of the southern cities. Annual mean SO_2 concentration in the northern cities was 72 μg m^{-3}, while it was 60 μg m^{-3} in the southern cities. In the northern region, the highest levels of SO_2 were monitored in Taiyuan and Jinan; annual mean SO_2 concentrations reached 248 μg m^{-3} and 173 μg m^{-3}, respectively, in 1997. In the southern region, the highest levels of SO_2 were monitored in Yibin and Chongqing where the annual mean SO_2 concentrations reached 216 and 208 μg m^{-3} respectively in 1997, although levels were even higher in 1996. (ECCEY, 1997 and 1998).

Nitrogen oxides

From the early 1980s to the present day, vehicle numbers in Beijing city have increased dramatically from 0.3 million to over 1 million. Because of the growing industrial activity and the rapid increase in vehicle numbers, NO_x emissions in China increased 33-fold between 1950 and 1990 (Wang *et al.*, 1997). It is estimated that NO_x emissions in China amount to 4.9 million tons accounting for 48% of the total emissions in Asia in 1987 (Kato, 1996).

In 1997, the national annual daily mean NO_x concentration was 45 μg m^{-3}; ranging from 4–140 μg m^{-3}. The annual daily mean NO_x concentrations were 49 μg m^{-3} and 41 μg m^{-3} in the northern and southern cities, respectively. In more than 36% of all the cities, the NO_x concentrations exceeded the Chinese SAQ for NO_x (50 μg m^{-3}). The bigger cities, with high vehicle numbers, normally exhibit higher NO_x concentrations. For example, in 1997, the monitoring programme in Beijing shows that in the northern part of China, highest NO_x levels reached 133 μg m^{-3} while in the southern part of the country, Guangzhou and Shanghai exhibit the worst problem with NO_x levels reaching 140 and 105 μg m^{-3}, respectively (ECCEY, 1998).

Suspended particulate matter

The burning of coal with a high dust content by industry and domestic sectors plus the large number of vehicles running on a limited number of roads of poor quality, has resulted in huge emissions of particulate matter into the atmosphere. It is estimated that of the total SPM concentration of 227 μg m^{-3} in Chongqing, 43% result from the combustion of coal, c. 36% from heavy industry (mostly steel manufacturing plants) and c. 9% originates from roads (Chen *et al.*, 1996). In 1997, the national annual daily mean (TSPM) is 291 μg m^{-3}, ranging from 32–741 μg m^{-3}. The national annual mean of dust deposition is 15.3 tons km^{-2} $month^{-1}$. The annual mean in northern cities is 21.48 tons km^{-2} $month^{-1}$, the highest levels recorded in Anshan (53.17 tons km^{-2} $month^{-1}$) while the annual mean in southern cities is 9.29 tons km^{-2} $month^{-1}$, with the highest levels recorded in Wuhan (19.16 tons km^{-2} $month^{-1}$) (ECCEY, 1998).

Fluorides

The most serious fluoride problems are associated within the local vicinity of brickyards, ceramic factories, aluminium plants, phosphate fertiliser plants and cement factories. A large amount of fluoride is emitted from the burning of coal, clay and other materials of high fluoride content. It is estimated that in China the average fluoride emission rate from brickyards is 663 kg fluorides per million bricks produced (Liang *et al.*, 1992). Assuming the average fluoride content of coal to be 157 mg kg^{-1}, the total national fluoride emissions associated with coal burning are calculated to be c. 0.2 million tons. Monitoring data show that between July and December 1991 the average fluoride concentration at ground-level in Changchun was 0.77 μg F m^{-3} (Inoue *et al.*, 1995). In Zhongguanchun (Beijing), the average fluoride concentration in early summer is 0.218 μg F m^{-3} and this rises in winter to an average of 1.198 μg F m^{-3} (Feng, personal communication). Fluoride levels can reach values of 22.6 μg F m^{-3} or higher around some brickyards in local areas (Sun *et al.*, 1998).

Ozone

The ground-level air pollutant composition in China is quite different from that in many industrialised countries in Europe, North America and Southeast Asia. In most of these situations, NO_x and ozone (O_3) are the major air pollutants and SO_2 concentrations have declined dramatically in recent years (Yunus *et al.*, 1996). In China, NO_x pollution is not yet as serious as SO_2. To date, little attention has been paid to O_3 and little information exists on ground-level O_3 concentrations in China; measurements are restricted to a few inner-city locations or industrialised areas in Lanzhou, Beijing (Tang *et al.*, 1988) and Shanghai (Xu and Zhu, 1994). Only recently have data become available for O_3 concentrations in rural areas and this is restricted to individual studies conducted in and around the city of Chongqing (Zheng *et al.*, 1998), and four other rural sites (Chameides *et al.*, 1999). At this stage, because of the need for nationwide monitoring data, it is not possible to summarise the O_3 pollution climate in China. However, the limited data that are available shows that in some places, O_3 is potentially high enough to cause adverse effects on vegetation. For example, records show the maximum hourly mean O_3 concentration attained more than 400 ppb in Xigu, Lanzhou, in 1981, and 332 and 254 ppb in 1982 and 1983, respectively (Wang *et al.*, 1997). In Shanghai and Chongqing, hourly mean O_3 concentrations reached c. 100 ppb (Xu and Zhu, 1994; Zheng *et al.*, 1998), while in Linan (a rural area 60 km from Hangzhou city) hourly mean O_3 concentrations attained 120 ppb, and in Waliguan (a rural area in Qinghai) levels of 130 ppb were recorded (Wang, personal communication). Research in the USA has established that crop yields were reduced by 10% or more when SUM06 (three-month sum of all daytime one-hour-averaged O_3 > 60 ppb) is ~15–25 ppm.h. Monitoring data in Linan indicate that the SUM06 reached 15–31 ppm.h between the end of 1994 and the beginning of 1995 (Chameides *et al.*, 1999). Most of the time O_3 levels at the majority of monitoring sites are below levels considered to be phytotoxic to vegetation. However, the rapid growth of industry and increases in the number of motor vehicles is forecast to result in a growing photochemical oxidant problem in the foreseeable future (Zheng *et al.*, 1998; Chameides *et al.*, 1999).

1.2. *Vegetation*

Air pollution in China is mainly associated with anthropogenic activities. More than 94% of the population is concentrated in the southeast part of the country. In this region, the main land cover/use types are forest (including woodland and shrubs) and farmland (Sun *et al.*, 1994). In terms of vegetation, this region can be sub-divided into five sectors:

(1) The northeast which includes the whole area east of Daxinanlin and north of Liaodong Peninsula. Here, the main agricultural species are soybean, sugar beet, wheat and corn. Forests are typified by cold temperate conifers (including Xingan deciduous pine, *Picea sylvestris* var. *mongolica*, etc. and several broadleaf tree species *Betula platyphyll*, *Quercus mongolica*, *Populus davidiana*, etc.) and mid-temperate conifers and deciduous broadleaf trees (including the main conifer species *Pinus koraiensis*, *Picea jezoensii* var. *microsperma*, *Abies nephrolepis*, Xingan deciduous pine, *Larix olgensis* var. changbaiensis, *Picea korainsis*, etc. and main broadleaf tree species *Quercus mongolica*, *Betula ermanii*, *Populus davidiana*, *Populus ussuriensis*, etc.).
(2) The Huang-Huai-Hai region, including the plains formed around the Huanghe River, Huaihe River and Haihe River. Here, the main agricultural species are wheat, cotton, maize, peanut, soybean, sweet potato, potato and tobacco. Forests are typified by warm temperate deciduous broadleaf trees (dominated by *Quercus*, *Pinus tabulae formis*, *Betula*, *Populus davidiana* and *Acer*, etc.). In this region, forests are sparse and are restricted to rural mountain areas far away from industrial sources of air pollution.
(3) The middle and lower reaches of the Yangtze and Hanjiang Rivers which include the area south of Qinglin mountain and the Huaihe River. Here, the main agricultural species are rice, oilseed rape, wheat, tobacco, mulberry tree, tea, fruit trees (citrus, peach, etc.) and sweet potato. Forests are typified by northerly sub-tropical evergreen broadleaf trees and deciduous broadleaf trees. Due to the long history of human occupation in this region, the dwindling native evergreen broadleaf trees and deciduous broadleaf trees are restricted to high elevation mountain areas. The main forests

in this region comprise *Pinus massoniana* Lamb. and *Cunninghamia lanceolata* (Lamb.).

(4) The southwest which includes Sichuan, Chongqing, Yunan and Guizhou. Here the main agricultural species are rice, oilseed rape, wheat, tobacco, mulberry tree, tea, fruit trees (e.g. citrus and peach) and sweet potato. The main tree species are *Fagaceae, Lauraceae, Pinaceae* and *Taxodiaceae.* Economically important tree species are *Fraxinus chinensis* Roxb., *Trachycarpus fortunei* (Hook. f.) H. Wendle., *Sapium sebiferum* Roxb., *Vernicia fordii* and Walnuts.

(5) South China, including Guangdong, Guangxi, Fujian and Hainan Island. Here the main agricultural species are rice, sugar cane, peanut, mulberry tree, tea, rubber tree, coconut, coffee, pepper, banana, pineapple and several other tropical fruit trees. Forest type is typified by tropical evergreen broadleaf forest and some tropical rainforest but the majority of forests comprise *Pinus massoniana* Lamb.

Generally, agricultural land is located around cities and industrial areas at low elevation and forests are distributed in mountain areas. Due to the huge demand for fresh fruit and vegetables in the cities and the poor transportation system, most fruit and vegetable producing areas are located around the densely populated areas. Consequently, fruit trees and vegetables are considered to be at the greatest risk from air pollutants originating from industrial sources, followed by other crop plants then forests. The available evidence suggests that air quality has not deteriorated sufficiently to cause deterioration in the health of forests in China. Carmichael *et al.* (1995) conducted air quality measurement in five sites which covered areas including the northern, Yangqing (Beijing), the southern, Huitong and Yueyang (Hunan) and the southwestern, Mt. Simian (Sichuan) and Luchongguan (Guiyang) regions. Their results show that at most of these sites, the annual SO_2 concentration is lower than $20\,\mu g\ m^{-3}$ which is below the critical level set to protect forests in Europe (Bell, 1992). At only one site, Luchongguan, was the annual daily mean SO_2 concentration greater than $20\,\mu gm^{-3}$. To date, there have been no reports of a widespread deterioration in forest condition in China, although concerns have been raised about localised declines in tree condition around some large cities or industrialised areas in southwest of China (Liao, 1988; Yu *et al.*, 1988). The best documented case is Nanshan in Chongqing (Zheng, 1991) as described in Sec. 2.1. Due to the alarming pollution situation in many urban areas, the Chinese government is currently putting effort into relocating some of the most heavily polluting industries and introducing legislation to build taller stacks to disperse the air pollutants over greater distances. These actions may reduce the present levels of pollution in the inner city areas and relieve concerns over effects on human health. However, this policy may at the same time result in higher levels of air pollutants in rural areas where forest and natural vegetation are situated.

2. Impacts of Air Pollution

2.1. *Field Evidence*

Air pollution impacts on vegetation have been under-researched in China. To date, field evidence is limited to the impacts of SO_2, fluorides and particulate matter and/or the combined effects of these pollutants on vegetation. Below, we focus on several key studies that have been conducted to date.

Chongqing is the biggest industrial city in southwest China with a population ca 2.5 million living in the inner city. In 1995, the annual daily mean SO_2 concentration was 340 μg m^{-3}, with daily mean concentrations reaching 940 μg m^{-3}. The annual daily mean SPM concentration was 320 μg m^{-3}, with daily mean concentrations reaching 2240 μg m^{-3}. Mean dust deposition was 15.9 tons km^{-2} (Jiang and Zhang, 1996) and the annual daily mean NO_x concentration was 67 μg m^{-3} (ECCEY, 1998). O_3 was generally found to be below the limit of phytotoxicity. However, during summer months, peak hourly mean concentrations reached values of 93 ppb, and exposures during intermittent O_3 episodes commonly approached (or exceeded) United Nations Economic Commission for Europe (UN/ECE) and World Health Organization (WHO) short-term guidelines for the protection of the most sensitive vegetation (Zheng *et al.*, 1998). Due to the heavy air pollution load, the trees lining the streets of this city have had to be replaced three times since the 1960s. At present, the main urban tree species is the native *Ficus lacor Buch.* Ham. After several days without precipitation, red and brown necrotic lesions and a thick layer of black dust can often be seen on the surface of the leaves. Compared with "clean areas", the date of the sprouting of new shoots is delayed in the spring and leaves fall earlier in the autumn (Zheng and Chen, 1991). It is also reported that air pollutants adversely affect peach, cherry, citrus and some other fruit tree species in the parks in and around the city. These effects appear as either a lack of, or a shortened period of flowering, early leaf abscission and premature dropping of fruit (Zheng and Chen, 1991). Zheng and Chen (1991) conducted a field survey in which the growth and yield of vegetable species was compared between a place with relatively clean air (Beibei, 50 km north of Chongqing City) and a polluted intermediate suburb of Chongqing. The results showed that the leaves of the vegetables grown around the city often accumulate a layer of black dust and growth is not as vigorous as in Beibei — resulting on average in c. 24% reduction in final yield.

Nanshan is a mountain adjacent to the southeast of Chongqing City and is impacted by pollutants from the city. Nanshan has c. 2000 ha of forest, mainly *Pinus massoniana* Lamb. It has been reported that since the beginning of the 1980s, more than half of these trees have died and 85% of the remaining pine trees show some degree of injury (Zheng, 1991). The symptoms of injury on these trees include needle tip necrosis, thin crown, reduced needle length, premature abscission, branch dieback, and reduced radial growth. Broadleaf trees (including *Robinia pseudoacacia, Quercus dentata, Eucalyptus robusta* and *Erythrina variegata*) also display necrotic lesions (Yu *et al.*, 1990; Zheng, 1991). The problems have attracted public attention and researchers have investigated the reasons for this tree dieback. Most of the research results indicate that the problems in this area are related to ambient levels of air pollution, with most studies focusing on the impacts of SO_2 (Yu *et al.*, 1990; Zheng, 1991; Shen *et al.*, 1995). Similar concerns were raised regarding the impacts of SO_2 on 6000 ha of *Pinus armentii* Franch in Maocaoba, Fengjie in Sichuan Province in 1980s (Shan and Totsuka, 1997). Here, more than 90% of the trees were killed by pollutants emitted from sulphur-processing factories. SO_2 concentrations monitored within the immediate vicinity of one of these factories were recorded at 134–1462 μg m^{-3} and at a distance of 2 km concentrations were 2.8 μg m^{-3}. Severe forest damage (96% of trees died) was observed in areas where concentrations were 0.35 μg m^{-3} and less severe damage (90% of trees died) occurred where concentrations were 0.162 μg m^{-3}.

Impacts of air pollutants on crop plants and tree species around many power stations, steel plants and cement factories have been reported in China (Bao and Zhu, 1997; Fu *et al.*, 1996; Wu *et al.*, 1990). Bao and Zhu (1997) conducted an investigation around a coal-burning power station in Hunan Province. The air pollutants from this power station resulted in severe impacts on local agricultural productivity and sometimes caused 100% losses in yield. Investigators selected four sites close to the power station and one site away from the power station as a control (located in a relatively clean area). Results showed that the average leaf CO_2 assimilation rate of citrus trees was c. 55% lower in the polluted area than in the clean area; flower numbers per tree were c. 15% lower and fruit number per tree was reduced by 50%. Dust not only reduced the yield of vegetables and fruit, but also increased heavy metal content and reduced quality (Fu *et al.*, 1996; Wu *et al.*, 1990). Wu *et al.* (1990) conducted an investigation and results showed that dust around a steel plant in Chongqing significantly increased heavy metal contents in leaves of lettuce (Mn by 8.7 times, Cu by 6.5 times, Zn by 3.8 times, and Pb by 11.4 times). The Pb content of lettuce leaves exceeded the Chinese National Standard for food by c. six times.

A ceramic factory in Jiangbei, Chongqing, started production in the early 1990s. Two months later the local vegetation (bamboo, vegetables and other agricultural crops) downwind of the factory started to show typical fluoride injury with necrosis around the margins and tips of leaves. One year later almost all the bamboo had been destroyed within 400 m of the factory (Zheng, 1993 unpublished). Although largely circumstantial, observational evidence of this kind is common in China (Sun *et al.*, 1998; Fluorides Project Research Team, 1994). However, Sun *et al.* (1998) conducted a field study around an aluminium factory in Zhengzhou to assess the effects of fluoride emitted from the factory on the growth of winter wheat. Their results showed that there

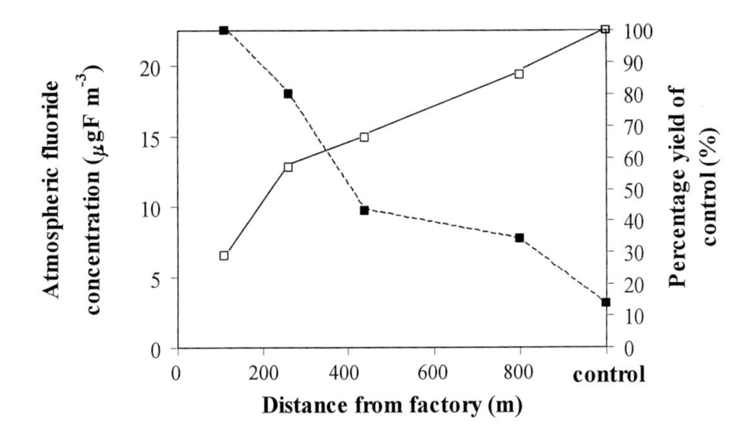

Fig. 1. Atmospheric fluoride concentration (filled squares) and the impact of fluorides on winter wheat yield (open squares) around an aluminium factory in Zhengzhou, China. Control is located 800 m away from the factory, in the prevailing wind direction. Percentage yield of control (%) = yield of polluted area/yield of control (drawn according to Sun *et al.*, 1998).

were significant reductions in the growth and yield of winter wheat within at least 800 m of the factory (see Fig. 1).

2.2. *Experimental Research*

Since the most common gaseous air pollutant in China is SO_2 and the pollutant has caused widespread deposition of acid rain (more than 40% of the land in China is covered by acid rain), the majority of experimental research has focused on the impacts of this pollutant on vegetation in China. Few controlled studies have been directed at quantifying the impacts of SO_2 and even fewer have attempted to investigate the combined effects of SO_2 and other air pollutants (e.g. NO_x, O_3, fluoride and SPM). Most of the research has been conducted using unrealistically high pollutant concentrations for a short time period and little work has been directed at investigating the mechanisms underlying effects. The following sections focus on those few studies where environmentally relevant pollutant concentrations have been employed and where attempts have been made to study dose-response relationships.

Sulphur dioxide

A substantial project, "The Ecological and Environmental Effects and Economic Losses Caused by Acid Rain", was supported by the central government from 1985 to 1990 to study the effects of acid rain on soil and vegetation in China. From 1991 to 1995, the central government financed another similar project called "The Study of the Effects of Acid Deposition on Ecology and Environment in China". This project aimed to examine the individual effects of acid rain and SO_2 on vegetation and to study the combination effects of these two air pollutants (Feng *et al.*, 1999). In the latter project, field-based open-top chambers (OTCs) were employed to assess the effects of a range of SO_2 concentrations (clean air, 132, 264, 396 and 666 $\mu g\ m^{-3}$; seven-hour d^{-1}) on the growth and yield of several crop plants (oilseed rape, bean, tomato, carrot, soybean, cotton, wheat, barley and rice) from seedling to the time of final harvest. Based on this study, yield-response relationships were established (see Fig. 2). The threshold concentration of SO_2 resulting in a 5% reduction in biomass (Table 1) and yield (Table 2) was calculated using linear regression analysis. In addition, yield losses caused by SO_2 pollution in seven provinces

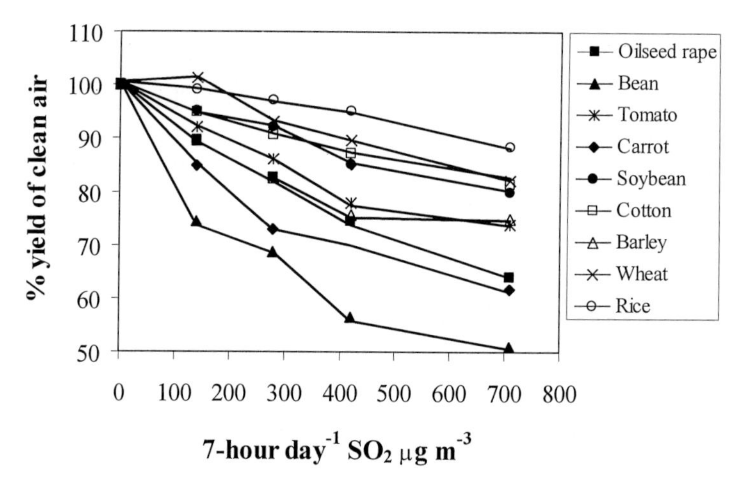

Fig. 2. Percentage yields of crop plants exposed to different concentrations of SO_2 in comparison with plants in clean air. % yield of clean air = yield of plant exposed to SO_2/yield of plant in clean air. Drawn according to data from Feng *et al.* (1999).

Table 1. The thresholds of SO_2 at which plants show a 5% biomass reduction. Thresholds were calculated from linear regression analysis between SO_2 concentrations and biomass (Feng *et al.*, 1999).

Species	SO_2 ($\mu g\ m^{-3}$)	Species	SO_2 ($\mu g\ m^{-3}$)
Bean	72	Soybean	174
Carrot	93	Cotton	199
Oilseed rape	98	Barley	140
Tomato	133	Wheat	186

Table 2. The thresholds of SO_2 at which plants show a 5% yield reduction. Thresholds were calculated from linear regression analysis between SO_2 concentrations and yields (from Feng *et al.*, 1999).

Species	SO_2 ($\mu g\ m^{-3}$)	Species	SO_2 ($\mu g\ m^{-3}$)
Oilseed rape	98	Cotton	199
Bean	72	Barley	140
Tomato	134	Wheat	186
Carrot	93	Rice	456
Soybean	174		

(Jiangsu, Zhejiang, Anhui, Fujian, Hunan, Hubei and Jiangxi) in southern China were estimated from the yield-response relationships established. Results indicated that SO_2, rather than acid rain, is the main factor responsible for losses in yield in these provinces. The average yield reduction due to the combined effects of SO_2 and acid rain was 4.34% in the seven provinces in the middle of the 1990s. Vegetable yield was reduced by 7.8%, wheat by 5.41%, soybean by 5.73% and cotton by 4.99%. The area impacted by phytotoxic pollutant concentrations was calculated to be 19.05% of the total agricultural land in the seven provinces.

Ozone

There is little experimental data available concerning the effects of environmentally relevant O_3 concentrations on vegetation in China.

However, the study by Zheng *et al.* (1998) draws attention to the potential for adverse effects. In this experiment, 11 cultivars of Chinese crops (aubergine, cauliflower, Chinese leaves, tomato, lettuce, wheat, maize, radish, courgette, green pepper and rice) commonly grown in the Chongqing region were screened for their relative ozone sensitivity under controlled conditions. Over a four-week period, plants were exposed to a low ozone concentration during the night of 15 ppb rising to a mid-day maximum of 75 ppb. The results indicated that in terms of the effects on growth rate, several Chinese cultivars (green pepper, rice, aubergine, cauliflower, etc.) were as sensitive to ozone as some of the most commonly used bioindicators (e.g. "Cherry Belle" radish and plantain).

Fluoride

Many experiments have been conducted in China to study the effects of fluoride on the mulberry tree because of its importance in silk production, especially in Zhejiang and Jiansu provinces where both silk and the manufacture of bricks form the cornerstone of the economy. It was found that to protect silkworm from the effects of fluoride, to which it could be exposed by feeding on mulberry tree leaf, the ground-level concentration of fluoride had to be lower than 0.36 $\mu g\ F\ m^{-3}$ (Shentu, 1993). Using OTCs, Zeng *et al.* (1990) exposed several Chinese species to high HF for a short period (one to 12 days) or low HF over a protracted period (one to 32 days) and assessed visible leaf damage. By regression analysis, a table of thresholds for protecting plants from visible leaf injury was produced (Table 3).

Table 3. The thresholds of atmospheric fluorides for protecting plants from leaf visible injury (adapted from Zeng *et al.*, 1990).

$\mu g\ F\ m^{-3}$	Rice	Broad bean	Yunan Pine	Huashan Pine	Pear
Hourly mean	238	128	81	85	915
Daily mean	40.2	16.9	25.2	36.7	224.7
Monthly mean	3.6	1.8	5.8	9.9	13
Seasonal mean	1.1	0.7	3.0	5.8	

Suspended particulate matter (SPM)

Ding and Lei (1991) conducted an interesting experiment in which different amounts of dust, collected from coal burning factory stacks, were applied to a range of fruit and vegetable plants at different growth stages. In addition, they exposed the same species grown in pots to dust in the field. Results showed that the sensitivities of leaf photosynthesis to dust pollution varied between species and even the same species showed different sensitivities dependent on growth stage, with higher sensitivity during the early part of the growing season. Plants in the greenhouse were more sensitive than plants grown in the field. Figure 3 shows the dose-response relationships generated from this experiment. Most of the vegetables started to show reductions in leaf photosynthesis when the dust deposition was 20 g m^{-2} $month^{-1}$, and all showed reduced rates of photosynthesis when dust deposition reached 50 g m^{-2} $month^{-1}$. Monitoring data indicates that in many localities in and around urban areas, dust deposition is higher than

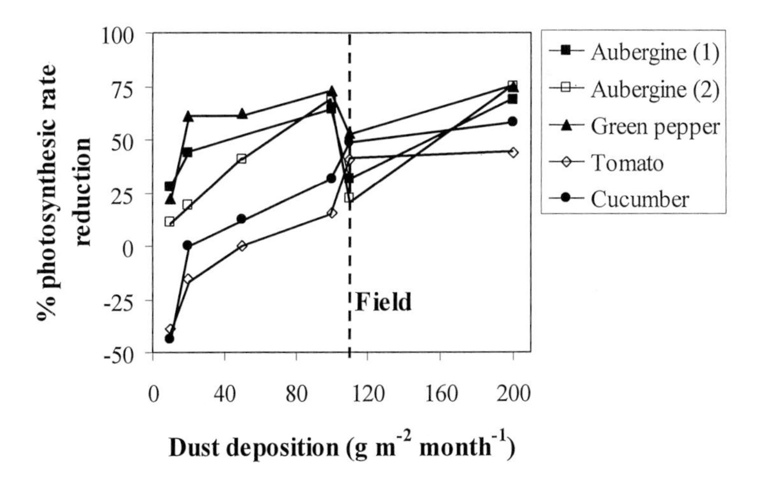

Fig. 3. Effects of coal burning dust on leaf photosynthesis of fruit vegetables at the early growth stage. Data on the dotted line are collected in the field and the rest are from the greenhouse experiments. % photosynthesis rate reduction = [(Control − Treatment)/Control] × 100, where Control is without applying dust on the leaves. Aubergine (1), ball-shaped; Aubergine (2), long aubergine. Drawn according to data from Ding and Lei (1991).

20 g m^{-2} $month^{-1}$ in China, and in some places exceeds 50 g m^{-2} $month^{-1}$ (ECCEY, 1998).

Combined effects of air pollutants on vegetation

In most parts of China, as elsewhere, vegetation is exposed to a complex cocktail of air pollutants (SO_2, NO_x, SPM, O_3, etc.). Technically, it is difficult to investigate the combined effects of this pollutant load in the same experiment under controlled conditions, especially in developing countries where there is a shortage of exposure facilities. Consequently, although some controlled studies on air pollution combinations have been carried out in China, most employed unrealistically high concentrations of air pollutants for short periods of time to study physiological and biochemical effects (Dai *et al.*, 1994; Zhou *et al.*, 1993). There are a few field fumigation experiments but these lack continuous air pollutant monitoring data. For example, Zheng *et al.* (1996) grew four vegetable species in pots in different locations, 5 km, 10 km and > 20 km downwind of the city of Chongqing. Results showed that in comparison with the yield at the "clean site" (Jieshi, > 20 km), yields of all four vegetable species at the two sites closest to the city were significantly decreased by ambient air pollution (see Table 4). It was also found that the numbers of aphids on some of the vegetable leaves were significantly higher at the sites closer to the city.

Table 4. Yields (g $plant^{-1}$) of four vegetable species located at different distances from Chongqing city (data following the names of the sites are distances from Chongqing city). For individual species, data identified by the same superscript letter are not significantly different at the 5% level (from Zheng *et al.*, 1996).

	Bagongli, (5 km)	Chalukou (10 km)	Jieshi (20 km)
Brassica juncea involuta	40.3^a	92.5^b	337.5^c
Brassica oleracea capitata	101.8^a	255.5^b	255.3^b
Lactuca sativa ungustana	28.0^a	163.2^c	116.7^b
Raphanus sativus	45.7^a	82.9^b	235.1^c

3. Applicability of European and North American Air Quality Standards

Based on our present knowledge, a cautious approach should be taken to applying European and North American critical levels and secondary air quality standards (SAQs) to assess the risk from air pollution to vegetation in China. At the very least, the following three aspects should be taken into consideration before applying such standards:

(1) Plants vary in their response to air pollution under different environmental conditions (e.g. light, temperature, soil type, nutrient status and water conditions) (Ashmore and Marshall, 1998). Climate, soils and some other environmental factors differ quite significantly in China compared to Europe and North America.
(2) Plants also vary in their response to different air pollutant combinations (Barnes and Wellburn, 1998). Air pollutant composition in China and developed countries is quite different from that in Europe and North America. For example, in most regions of China, SO_2 and particulate matter are the two most prevalent air pollutants and maintain higher concentrations for longer time periods compared to developing countries (see Sec. 1.1). In developed countries NO_x and O_3 tend to be the two most important air pollutants.
(3) It is well known that different plant species and cultivars vary in their response to air pollution. The plant species and cultivars in China will vary from those found in other countries. In addition, there is evidence that some plant species are able to evolve a resistance under a changing air pollution environment (Barnes *et al.*, 1999). After so many years exposure to the specific air pollutant combination in China, some vegetation may have obtained certain air pollution resistance properties. Dose-response studies in China showed that for most of the plant species, the thresholds for SO_2 appear to be much higher than SAQs of North America (see Sec. 2.2). However, this may be a consequence of the regression analysis being performed on experimental data collected under high SO_2 fumigations (i.e. exposures were always greater than 120 $\mu g\ m^{-3}$).

4. Future Research Directions

To be able to quantitatively assess the risks posed by air pollutants to vegetation in China the following knowledge gaps remain to be filled:

(1) There is a need to expand the air pollutant monitoring networks into agricultural and forest areas. To date, most of the air pollution monitoring stations have been restricted to sites located in inner city or industrial areas.
(2) Monitoring data relating to ground-level ozone and fluoride concentrations is lacking in most areas. From the available data, it is clear that the concentrations of both these pollutants approach levels in some regions that are generally considered to be phytotoxic (see Sec. 1.1).
(3) Almost all of the air pollution impact studies conducted to date have focused on agricultural crops and forest trees — virtually nothing is known about the impacts of air pollutants on natural (or semi-natural) herbaceous vegetation.
(4) There is a severe shortage of reliable yield-response data applicable to Chinese conditions, or the way in which the presence of other air pollutants may influence these relationships. Acquisition of these data will require the application of state-of-the-art technology for the field-based assessment of effects in order that critical levels/SAQs based on sound scientific principles can be derived for Chinese vegetation types. The studies will require the injection of funding to permit employment of OTCs and/or the development of free air exposure systems in the South and North of China — where the climates are very different. Moreover, studies should focus not only on agricultural crops and forest trees, but also consider effects on natural vegetation, especially endangered species.

5. Conclusions

SO_2, NO_x, particulate matter and fluorides are considered to be the main gaseous air pollutants in present day China. Concerns regarding O_3 pollution are increasing with the rapid growth of motor vehicle numbers and industry. SO_2, NO_x and particulate matter are high enough to cause adverse effects on vegetation in and around most of the city and industrial areas in the southeast part of China. In these areas, vegetables, fruit trees and some other agricultural crops are likely to be the most affected vegetation. Some forests close to cities and industrial areas are also affected, but most of the forests are far from these air pollutant sources. There is likelihood that the expansion of industry and increase of NO_x and VOC emissions will put more forest and agricultural areas at risk from the combination effects of SO_2, NO_x and O_3 in the near future. Adverse effects of fluorides on vegetation are mostly recorded around industrial areas, especially aluminium industrial complexes and brickyards. It is likely that fluoride pollution is playing a not insignificant role in the widespread combination effects of air pollutants on vegetation. To be able to assess the risk of air pollutants posed to vegetation and to establish realistic air quality guidelines in China, more experimental studies are needed.

Acknowledgements

The authors wish to thank Dr. Jeremy Barnes for his comments on an earlier draft of this manuscript and for undertaking language corrections, Dr. Lisa Emberson for her support and Mr. C. Wang and Ms Y. Feng for providing access to unpublished information. This review was written during Youbin Zheng's tenure of a Japanese Science and Technology Agency Fellowship.

References

Ashmore M.R. and Marshall F.M. (1998) Direct impacts of pollutant gases on crops and forests. In *Regional Air Pollution in Developing Countries*, eds. Kuylenstierna J. and Hicks K. Stockholm Environment Institute, York, pp. 21–33.

Bao W. and Zhu Z. (1997) Study of the combination effects of dust and SO_2 on citrus tree. *Agro-Env. Protect.* **16,** 16–19 (in Chinese).

Barnes J., Bender J., Lyons T. and Borland A. (1999) Natural and man-made selection for air pollution resistance. *J. Exp. Bot.*

Barnes J. and Wellburn A. (1998) Air pollutant combinations. In *Responses of Plant Metabolism to Air Pollution and Global Change*, eds. De Kok L.J. and Stulen I. Backhuys Publishers, Leiden, pp. 147–164.

Bell J.N.B. (1992) A reassessment of critical levels for SO_2. In *Critical Levels of Air Pollutants for Europe.* Egham, UK.

Carmichael G.R., Ferm M., Adikary S., Ahmad J., Mohan M., Hong M.-S., Chen L., Fook L., Liu C.M., Soedomo M., Tran G., Suksomsank K., Zhao D., Arndt R. and Chen L.L. (1995) Observed regional distribution of sulphur dioxide in Asia. *Water Air Soil Pollut.* **85**, 2289–2294.

Chameides W., Li X., Tang X., Zhou X., Luo C., Kiang C.S., John J., St. Saylor R.D., Liu S.C., Lam K.S., Wang T. and Giorgi F. (1999) Is ozone pollution affecting crop yields in China? *Geophys. Res. Lett.* **26,** 867–870.

Chen S., Zheng Y., Zhao Q. and Meng M. (1996) Identify the sources of particulate matter in ground-level atmosphere in Choingqing by CMB. *Chongqing Env. Sci.* **18,** 24–28 (in Chinese).

Dai Z., Zhou J., Wu Z. and Xu Y. (1994) The effects of SO_2 + NO_2 on plant free proline concentrations. *China Env. Sci.* **14,** 190–195 (in Chinese).

Ding Q. and Lei H. (1991) Effects of coal burning dust on photosynthesis of fruit vegetable. *Agro-Env. Protect.* **10,** 208–211 (in Chinese).

Editing Committee of China Environmental Yearbook (1997) *China Evironmental Yearbook (1997).* China Evironmental Yearbook Press, Beijing (in Chinese).

Editing Committee of China Environmental Yearbook (1998) *China Evironmental Yearbook (1998).* China Environmental Yearbook Press, Beijing (in Chinese).

Feng Z., Cao H. and Zhou S. (1999) *Effects of Acid Deposition on Ecosystems and Recovery Study of Acid Deposition Damaged Forest.* China Environmental Science Press, Beijing (in Chinese).

Florig H.K. (1997) China's air pollution risks. *Env. Sci. Technol./News* **31,** 274–279.

Fluorides Project Research Team (1994) Impacts of fluorides on rice yield and the countermeasure. *Agro-Env. Protect.* **13,** 127–131 (in Chinese).

Fu L., Meng F., Liu C., Chen Q. and Ban X. (1996) The effects of cement dust on soil and crops. *Agro-Env. Protect.* **15,** 221–224 (in Chinese).

Inoue K., Zhang Y., Sakai K., Kakuta F. and Zhao J. (1995) Influence of airborne particulate matters transported from the Asian continent on water-insoluble, soluble and gaseous fluoride concentrations of aerosols in Japan. *Japan J. Soil Sci. Plant Nutrit.* **66,** 223–232 (in Japanese).

Jiang L. and Zhang D. (1996) Present situation and forecasting of environmental quality in Chongqing. *Chongqing Env. Sci.* **18,** 33–37 (in Chinese).

Kato N. (1996) Analysis of structure of energy consumption and dynamics of emission of atmospheric species related to the global environmental change (SO_x, NO_x and CO_2) in Asia. *Atmos. Env.* **30,** 757–785.

Liang W., Wu Q., Zhang J. and Gu J. (1992) Study of the fluoride emission rates from brickyards and ceramic factories in China. *Agro-Env. Protect.* **11,** 220–223 (in Chinese).

Liao Z.Q. (1988) An investigation on damage of tungoil tree (*Aleurites fordii*) caused by air pollution in the Zhongba region, Jiangyou County Sichuan Province. *Chinese J. Env. Sci.* **9**, 86–89 (in Chinese).

Shan Y. and Totsuka T. (1997) Forest damage in air pollution area. In *China Environment Handbook,* ed. Sadakata M. Science Forum, Tokyo, pp. 133–139 (in Japanese).

Shen J., Zhao Q., Tang H., Zhang F., Feng Z., Okita T., Ogura N. and Totsuka T. (1995) Concentrations and deposition of SO_2, SO_4^{2-} etc. in a Chongqing suburban forested area. *Water Air Soil Pollut.* **85**, 1299–1304.

Shentu M. (1993) The progress and problems in the studies of atmospheric fluoride pollution. *Agro-Env. Protect.* **12**, 137–139 (in Chinese).

Sun H., Shen Y., Shi Y., Zhang Z. and Zhang Q. (1994) *Agricultural Natural Resources and Regional Development of China.* Jiangsu Science and Technology Press, Nanjing (in Chinese).

Sun Z., Wei Y., Zhang T., Yun Z. and Jia X. (1998) Effects of fluorides on the growth of winter wheat around an aluminium factory. *Agro-Env. Protect.* **17**, 22–25 (in Chinese).

Tang X., Li I., Chen D., Bai Y., Li X., Wu X. and Chen J. (1988) The study of photochemical smog pollution in China. *Proceedings of the 3rd Joint Conference of Air Pollution Studies in Asia, Tokyo,* Japan Society of Air Pollution.

Wang W., Zhang W., Hong X. and Shi Q. (1993) Study on factors related to acidity of rain water in China. *J. China Env. Sci.* **13**, 401–407 (in Chinese).

Wang W., Gao S. and Sakamota K. (1997) Photochemical smog (mainly NO_x, HC). In *China Environment Handbook,* ed. Sadakata M. Science Forum, Tokyo, pp. 79–81 (in Japanese).

Wu D., Qing C. and Gao S. (1990) Study of the effects of dust emitted from a steel plant on soil-vegetable system. *Agro-Env. Protect.* **9**, 13–16 (in Chinese).

Xu J. and Zhu Y. (1994) Some characteristics of ozone concentrations and their relations with meteorological factors in Shanghai. *Atmos. Env.* **28**, 3387–3392.

Yu S.W., Yu Z.W., Ma G.J., Zhu C.L., Bian Y.M., Chen S.Y. and Zhu X.Y. (1988) Preliminary investigation on the causes of masson pine forest decline in Nanshan region of Chongqing, Sichuan Province. *Chinese J. Env. Sci.* **9**, 77–81 (in Chinese).

Yu S.W., Bian Y.M., Ma G.J., Luo J.J. (1990) Decline of masson pine forest in Nanshan, Chongqing and air pollution. *Acta Scientiae Circumstantiae* **10**(3), 378–383.

Yunus M., Singh N. and Iqbal M. (1996) Global status of air pollution: an overview In: *Plant Response to Air Pollution,* eds. Yunus M. and Iqbal M. John Wiley and Sons, Sussex, England, pp. 1–34.

Zeng G., Zhou J., Dong H., Bai Y., An Q., Li L. and Zhou Y. (1990) Studies on the ecological effects of fluoride in atmosphere on plants. *China Env. Sci.* **10**, 263–268 (in Chinese).

Zheng Y. (1991) The effects of acid deposition on Nanshan *Pinus massoniana* in Chongqing. *Chongqing Env. Sci.* **13**, 24–32 (in Chinese).
Zheng Y. and Chen S. (1991). Assessment of the economic losses due to acid precipitation in Chongqing terrestrial ecosystems. *Atmos. Env.* (*China*) **6**, 45–51 (in Chinese).
Zheng Y., Last F.T., Xu Y. and Meng M. (1996) The effects of air pollution climate in Chongqing on four species of vegetable. *Chongqing Env. Sci.* **18**, 29–32 (in Chinese).
Zheng Y., Stevenson K.J., Barrowcliffe R., Chen S., Wang H. and Barnes J.D. (1998) Ozone levels in Chongqing: a potential threat to crop plants commonly grown in the region? *Env. Pollut.* **99**, 299–308.
Zhou J., Dai Z., Wu Z., Yu F. and Xu Y. (1993) The effects of SO_2 + NO_2 on superoxide dismutase activity and leaf injury of tomato plant. *Env. Sci.* (*China*) **13**, 429–432 (in Chinese).

CHAPTER 7

AIR POLLUTION IMPACTS ON VEGETATION IN TAIWAN

B.H. Sheu and C.P. Liu

1. Introduction

Taiwan is an island situated in the subtropical latitudes of the Pacific Ocean at the periphery of the Asian continent. As such, Taiwan experiences climatic extremes associated with the dominant wintertime continental airmass and the summertime Pacific high (Liu *et al.*, 1993). In addition, Taiwan lies between two branches of the warm Japanese current. It has a distinctly subtropical oceanic climate because of the ocean winds, frequent rains (annual average precipitation is about 2500 mm) and typhoons. The warm climate, plentiful moisture, great altitudinal relief, and complex topography of the land makes the flora and vegetation of Taiwan exceedingly luxuriant and diversified (Peng *et al.*, 1994).

Taiwan also has one of the highest levels of economic activity in the world, and as a result, is experiencing growing anthropogenic emissions of sulphur dioxide (SO_2), nitrogen oxides (NO_x), fluorides (HF) and other air pollutants. This has resulted in air pollution becoming an important environmental issue in recent decades, in part due to the impacts of pollutants on vegetation. Air pollution can affect a plant directly by damaging canopy foliage (e.g. via gaseous pollutant uptake), indirectly through its effects on the soil (e.g. via contamination caused by trace metal deposition), or by a combination of both direct and indirect mechanisms. An example of a pollutant that can act as described later is SO_2 which can cause damage via stomatal uptake as well as by causing soil acidification via wet deposition of atmospherically oxidised sulphur compounds (Wellburn, 1996). This section will describe air pollution impacts on vegetation in Taiwan through documentation of field evidence and experimental research. The steps that have been taken to improve air quality in Taiwan are also examined.

1.1. *Emissions*

There are now 71 air quality monitoring stations in Taiwan, with 57 stations for general use, three for industrial monitoring, two for park monitoring, four to provide background information, and five for traffic monitoring. Data collected over the past four years (1994 to 1997) (Tables 1 to 8) (EPA, 1998) clearly indicate that PM_{10}

Table 1. Summary of annual average concentrations for major air pollutants in north Taiwan (source: EPA, 1998).

	1994	1995	1996	1997
PM_{10} ($\mu g/m^3$)	59	56	50	53
O_3 (ppb)	19	20	21	21
SO_2 (ppb)	8	8	6	6
NO_2 (ppm)	0.03	0.03	0.02	0.03
CO (ppm)	1.1	1.0	0.9	0.9

Table 2. Summary of annual average concentrations for major air pollutants in central Taiwan (source: EPA, 1998).

	1994	1995	1996	1997
PM_{10} ($\mu g/m^3$)	76	71	71	69
O_3 (ppb)	20	20	23	23
SO_2 (ppb)	7	6	5	5
NO_2 (ppm)	0.03	0.02	0.02	0.02
CO (ppm)	0.9	0.8	0.8	0.7

Table 3. Summary of annual average concentrations for major air pollutants in specific cities that are especially polluting located in south Taiwan (source: EPA, 1998).

	1994	1995	1996	1997
PM_{10} (μg/m^3)	95	94	88	84
O_3 (ppb)	23	23	25	26
SO_2 (ppb)	13	13	11	10
NO_2 (ppm)	0.03	0.03	0.03	0.02
CO (ppm)	0.8	0.8	0.8	0.7

Table 4. Summary of annual average concentrations for major air pollutants at general use stations in Taiwan (source: EPA, 1998).

	1994	1995	1996	1997
PM_{10} (μg/m^3)	74.2	69.2	65.2	64.1
O_3 (ppb)	21.2	21.1	22.8	23.3
SO_2 (ppb)	8.1	7.9	6.3	5.9
NO_2 (ppb)	24.4	24.3	22.7	23.2
CO (ppm)	0.86	0.79	0.74	0.76

Table 5. Summary of annual average concentrations for major air pollutants in industrial monitoring stations in Taiwan (source: EPA, 1998).

	1994	1995	1996	1997
PM_{10} (μg/m^3)	69.2	66.7	61.9	60.9
O_3 (ppb)	–	–	–	–
SO_2 (ppb)	19.6	18.8	13.8	22.3
NO_2 (ppb)	25.1	23.9	22.0	24.6
CO (ppm)	–	–	–	–

Table 6. Summary of annual average concentrations for major air pollutants at park monitoring stations in Taiwan (source: EPA, 1998).

	1994	1995	1996	1997
PM_{10} (μg/m^3)	22.7	19	19.3	20.7
O_3 (ppb)	36.4	36.6	35.9	37
SO_2 (ppb)	1.6	1.6	1.4	1.6
NO_2 (ppb)	2.2	2.2	2.2	2.2
CO (ppm)	0.32	0.24	0.23	0.26

Table 7. Summary of annual average concentrations for major air pollutants at background information stations in Taiwan (source: EPA, 1998).

	1994	1995	1996	1997
PM_{10} (μg/m^3)	63.1	60.6	58	55.6
O_3 (ppb)	28.4	26.7	27.6	27.6
SO_2 (ppb)	5.8	5.6	4.8	5.4
NO_2 (ppb)	14.4	15.1	15.2	15.7
CO (ppm)	0.49	0.46	0.43	0.46

Table 8. Summary of annual average concentrations for major air pollutants in traffic monitoring stations in Taiwan (source: EPA, 1998).

	1994	1995	1996	1997
PM_{10} (μg/m^3)	120.1	77.7	61.5	72.1
O_3 (ppb)	–	–	–	–
SO_2 (ppb)	17.8	14.9	12.5	9.3
NO_2 (ppb)	54.8	52.5	49.1	38.3
CO (ppm)	5.82	1.44	4.47	2.31

(suspended particulate matter diameter < 10 mm) and ozone (O_3) are largely responsible for the instances of poor air quality that occur in Taiwan. This is evident through consideration of values recorded using the Pollutant Standard Index (PSI) (described fully in Sec. 3) which is the index used to describe air quality in Taiwan. For example, during 1996, 62% and 37% of the days exceeding this index were due to elevated PM_{10} and O_3 levels, respectively. Although air quality in Taiwan has improved dramatically in recent years, occasions when PSI values exceed 100 still occur for approximately 6% of the year (1996). This is considered a relatively high exceedance rate especially when compared to the target exceedance levels established for Taiwan. Different target exceedances have been established as follows, a short-term target for the year 2001 of below 3%; a mid-term target for 2006 of below 2% and a long-term target for 2011 of below 1.5%.

The 1996 emission data for different pollutants estimated the amount of PM_{10} to be 620 ktons with 59% from construction sites and roads; SOx emissions were estimated at 440 ktons with 79% from industrial combustion; NOx emissions were estimated at 600 ktons with 65% from mobile sources; non-methane hydrocarbon (NMHC) emissions were estimated at 830 ktons with 47% from mobile sources and finally carbon monoxide (CO) and Pb were 2130 and 4400 ktons, respectively, with approximately 85% of both originating from mobile sources (CTCI, 1996 and 1997). During 1997, the major pollutants responsible for exceedances of the PSI value of 100 were O_3 (52.4 days) and PM_{10} (46.9 days) whilst only a few days were in exceedance due to high SO_2 and CO concentrations. Compared with the 1996 data, the air quality in 1997 showed great improvement in concentrations of PM_{10} and SO_2; however, O_3 concentrations were found to have increased in urban areas. Based on the information collected from the air quality stations, the major air pollutants in Taiwan were PM_{10} and O_3 (EPA, 1998).

1.2. *Vegetation*

Taiwan is remarkable for its extraordinarily high floristic diversity. Nearly 4000 vascular plants have been found on this small island which has an area of only 36,000 km^2. Most native and endemic plants of

Table 9. Altitudinal vegetation zones and their temperature ranges in central Taiwan.

Altitudinal zone	Vegetation zone	Elevation (m a.s.l.)	Mean annual temp. (°C)
Alpine	*Juniperus-Rhododendron*	3600–4000	< 5
Subalpine	*Abies*	3100–3600	5–8
Upper montane	*Tsuga-Picea*	2500–3100	8–11
Montane	*Quercus*	1500–2500	11–17
Submontane	*Machilus-Castanopsis*	500–1500	17–23
Foothill	*Ficus-Machilus*	0–500	> 23

Taiwan are found in the forested areas, which cover more than half of the land area. As a result of the existence of substantial mountainous areas throughout this small island (where altitudes reach 4000 m a.s.l.), there is a three-dimensional variation in climate in Taiwan (Su, 1984a). The effect of temperature differences due to the range in elevation across the country accounts for much of the species diversity that exists, and is responsible for the differentiation of forest vegetation zones. These zones are characterised by dominant plant groups that can be associated with specific temperature ranges as described in Table 9 (Su, 1984b).

For the same reasons that determine the diversity of vegetation in the mountainous regions, there are also many kinds of agricultural crop types used in farming. These are described in Table 10.

2. Impacts of Air Pollution

Over the past 30 years, rapid industrialisation has resulted in Taiwan experiencing problems with air pollution that have affected not only human health but have also decreased the health and productivity of Taiwan's vegetation. The first case of air pollution injury to vegetation was recorded in the late 1960s when weather fleck symptoms were found on tobacco leaves in the field and were found to be caused by elevated O_3 concentrations (Street *et al.*, 1971). In addition, during the mid-1970s, observations were made of the widespread occurrence

Table 10. A summary of the percentage productivity of agricultural crop types in Taiwan.

Products of agriculture	Productivity (%)
Fruit	35.04
Betel palm	24.64
Mango	7.53
Pineapple	7.43
Pear	6.56
Banana	4.50
Citrus fruit	4.41
Prune	3.74
Grape	3.48
Others	37.71
Vegetables	23.61
Watermelon	10.56
Bamboo	9.32
Mushroom	6.15
Garlic	4.59
Radish	3.25
Others	63.10
Rice	21.71
Special crop	10.55
Sugarcane	26.22
Tea	23.55
Peanut	17.39
Sugarcane (food)	12.17
Tobacco	11.03
Others	9.65
Universal crop	4.00
Corn (food)	29.21
Corn (feed)	27.52
Sweet potato	25.59
Others	17.68
Others	5.54

Source: Department of Agriculture and Forestry in Taiwan, 1999.

of marginal scorch on banana leaves that were subsequently associated with fluoride pollution (Su *et al.*, 1978). As a consequence of these initial observations, a number of studies have been conducted to assess the impact of pollutants on vegetation. Several complete investigations on the effects of primary pollutants on vegetation have culminated in the publication of two pictorial atlases in 1984 and 1985 (Sun 1984; Lee and Lee, 1985). The publication of these atlases was regarded as an important milestone in the study of air pollution impacts on vegetation in Taiwan and has resulted in continuing activity in this field of research. The following sections describe some of the key observational and experimental studies that have recently been performed to assess pollutant impacts on vegetation in Taiwan.

2.1. *Field Evidence*

Ozone (O_3)

As described above, the first case of ozone injury to vegetation in Taiwan was recorded in the late 1960s and was observed by two American scientists using the tobacco variety Bel-W3 as an ozone bioindicator (Street *et al.*, 1971). In collaboration with the Taiwan Tobacco Research Institute, these scientists found that by passing polluted ambient air through a charcoal filter, and thereby removing O_3 from the air supply, the occurrence of weather flecks on the leaves of tobacco plants can be prevented (Sung *et al.*, 1973). This was seen as evidence that O_3 was at least partly responsible for the occurrence of these symptoms of visible injury. This first instance of visible injury occurred approximately 15 years prior to the establishment of the Environmental Protection Administration of the Taiwan Government and before any other pollutants had been investigated by the Taiwanese authorities or environmental scientists. However, other studies conducted at a similar time showed that tobacco vein-banding mosaic virus and phosphorus deficiency might produce similar visible injury symptoms to those attributed to O_3 pollution (Wu and Sung, 1973). The uncertainty caused by this observation postponed further studies investigating plant injury caused by O_3 for about 20 years (Sun, 1993a).

Recent evidence has shown that O_3 is one of the major air pollutants on the island. Data from monitoring networks located close to urban areas have frequently recorded elevated O_3 levels with values sometimes in exceedance of 120 ppb. The occurrence of these O_3 episodes renewed scientists' interest in studying the effects of this pollutant. For example, Sun (1993b) reported that leaf-bleaching injury on leafy sweet potato was occasionally noticed in the field in the basin area of north Taiwan, Taipei. The necrotic lesions observed usually only appeared on the upper surface of the sweet potato leaves. These findings initiated a two-day survey, the results of which indicated that O_3 might be affecting sweet potato and spinach over a large area of the Taipei Basin. Lin and Yang (1996) reported a very important effect of photochemical pollutants on horticulture in south Taiwan. A survey of tropical fruits, vegetables and ornamental crops found that many crops showed symptoms specifically associated with O_3 injury whilst crops such as cucumber, muskmelon, flowers, vegetables, guava, and Indian jujube all showed signs of injury that could be attributable more generally to poor air quality.

Sulphur dioxide (SO_2)

During the period of 1991 to 1993, five monitoring stations were set up in the field near Linkow Power Plant, Kwang-Yin Industrial Park and Tung-Hua Textile Plant to assess the effect of air pollution on the agricultural areas of northwestern Taiwan. Potted gladiolus and guava were used as indicator plants and subjected to the polluted ambient air. The degree of injury suffered by these plants was determined by measuring the area of leaf damage; this was also related to the sulphur content of the leaves. The results showed that the leaf injury observed in gladiolus was not only caused by SO_2 but was also affected by the northeast monsoon. At Kwangyin Industrial Park, the leaf injury of gladiolus was inversely related to the distance from the industrial park, but not related to the sulphur content of air. It was concluded that other pollutants emitted from the industrial park and carried by the northeast monsoon may have resulted in the leaf injury observed. At Tung-Hwa Textile Plant, long bleached lesions were observed on plants, and related to the waste gases emitted from the plant, however, no direct relationship between leaf injury and leaf sulphur content was recognised (Huang *et al.*, 1994).

Nitrogen oxides (NO_x)

To date, there have been no recorded instances in Taiwan of NO_x pollution causing plant injury in the field. However, peroxyacetyl nitrate (PAN), which acts as a sink for NO_x, was found to quite severely affect plant health and as a consequence has been the subject of further investigation. During the late 1980s, PAN-type symptoms were first observed on sensitive plants. An investigation by Sun (1993a) conducted between 1990 and 1993, revealed that PAN was affecting vegetation covering areas of approximately 500 km^2 in the Taipei area, 400 km^2 in the Taichung area, and 1200 km^2 in the Kaohsiung area. These three major metropolitan areas cover about one-seventh of the total plain area of the island. This clearly indicated the severity of the problem of photochemical smog in Taiwan. Bronzing and silvering on lower leaf surfaces of sword-leaf lettuce and round-leaf lettuce were first observed in Taipei, Taiwan in 1989. Later, PAN-type symptoms were also found in Taichung and Kaohsiung metropolitan areas on more lettuce cultivars including pan-leaf lettuce. Usually, typical symptoms developed within one day of a calm sunny day during which heavy smog had occurred. Symptoms occurred on the tip of young leaves, the mid-section of new fully developed leaves, and the base portion of mature leaves.

Fluorides (*HF*)

Gaseous and particulate forms of HF tend to be emitted from factories associated with the production of bricks, ceramics and glass. In Taiwan, because these factories are often close to farm areas, this pollutant often affects crops and suburban vegetation. For example, Fig. 1 shows distinct patterns of chlorosis and necrosis along the leaf tips and margins of field grown eggplant leaf (*Solanum melongena* L. var. *esculentum* Nees.) near a ceramic factory. Similar symptoms were observed in *Eucalyptus robusta*, one of the most common vegetation types found in suburban areas.

Fig. 1. Fluoride-induced leaf injury symptoms of field grown eggplant (*Solanum melongena* L. var. *esculentum* Nees.) located near a ceramic factory in Taiwan.

Field experiments performed by Hseih (1993) near ceramic factories found that the responses of plants to HF differed greatly among different plant species. Long-term exposures to approximately 1 ~ 2 ppb HF resulted in some plants exhibiting symptoms of chlorosis or burn at the leaf tip and around the leaf margins. However, some tolerant species did not show any symptoms even though high concentrations of fluorine were measured in the plants after relatively long-term exposures to the pollutant. As HF accumulates in plants, higher concentrations of fluorine are usually found in the lower canopy plant leaves. As such, the analysis of these lower leaves may provide more reliable data for the detection of HF pollution. In addition, a two-year investigation by Lee *et al.* (1996) found that the HF content of plants around a brick factory was consistently high compared to HF contents of plants located at a distance from the polluted area. The HF content of sampled plants was found to be closely related to the distance from the pollution source. Banana (*Musa sapientum*) and betel nut (*Areca cathecu*) were found to be the most sensitive species to HF pollution.

In contrast, annuals such as *Bidens bipinnata, Erigeron sumatrensis, Erechitites valerianaefloia, Ageratum hostonianum, Amaranthus viridis* and *Digitaria adscendens* accumulated more than 200 μg/g of HF without exhibiting any abnormal symptoms. These annual weed species are commonly found in Taiwan and have the potential to be used as biomonitors for atmospheric HF pollution.

Suspended particulate matter (SPM)

According to a report of Yu and Chang (1997), suspended particulate matter (SPM) has also been shown to influence plant performance in urban areas. Investigations showed that dust severely reduced the photosynthetic rate of roadside plants. In addition, significant depositions of dust and smoke particles have been observed on some fruit trees in the vicinity of urban roads.

2.2. *Experimental Research*

Ozone (O_3)

Liu and Young (1996) confirmed that widespread high O_3 levels existed over most of western Taiwan. The station with the most frequent occurrence of hourly mean O_3 concentrations exceeding the national standard of 120 ppb was at Ping Tung, in south Taiwan, with exceedances recorded 134 times between July 1993 and December 1994 and 85 times throughout the whole of 1995. Yu and Chang (1997) recorded similar elevated O_3 concentration values in a study that also involved fumigating nine native groundcover vegetation types with 100 ppb O_3 for four hours per day. After two days fumigation, the *Begonia* species had exhibited necrosis attributable to water-logging of the unfolded leaves; *Bidens* species exhibited white spotting on mature leaves, *Wedelia* species exhibited purple flecks on some older leaves, and the *Ageratum* species displayed white spotting on some leaves. However, no visible injury symptoms were observed in the species *Dicliptera, Alternathera, Emilia* and *Calyptocarpus.*

Sun (1993b) performed O_3 fumigation experiments in six continuously stirred tank reactors (CSTRs), results showed that leafy sweet potato cv. Changhua was sensitive to O_3 when exposed to concentrations higher than 200 ppb for two hours. Spinach was more sensitive to O_3 than sweet potato as symptoms were exhibited at concentrations as low as 100 ppb for two hours. Typical leaf injury symptoms, similar to those found in the field, were observed on leafy sweet potato and spinach plants after exposing plants to 200 ppb O_3 for two hours in CSTRs. These results confirmed that O_3 was the primary factor causing plant injury in northern Taiwan.

Sulphur dioxide (SO_2)

Hsieh (1993) found that the responses of plants to SO_2 differed among plant species. Under fumigations using 300 ~ 600 ppb of SO_2, peanut, sweet potato, water spinach, cabbage and banana did not show any signs of visible injury. In contrast, different sizes and forms of brown spots appeared on the leaves of lowland rice, sesame, morning glory (heart-leaf type), gladiolus and guava after several days of this fumigation regime. After every eight-hourly period of SO_2 fumigation performed over 30 consecutive days, a substantial concentration of soluble SO_4 could be detected in the mature leaves of the upper canopy of these sensitive species. This suggests that analysis of soluble SO_4 concentration in these upper canopy leaves can be a useful method for the detection of long-term exposure to high SO_2 concentrations. Sheu (1994) studied the response of seedlings of *Schima superba* (one of the species recommended for agricultural planting in Taiwan) to well-defined SO_2 concentrations to determine whether SO_2 affects the physiology and growth of this species. After four weeks of exposure to 325 ppb of SO_2 for seven hours per day from 9:00 to 16:00, the photosynthetic rate of seedlings was significantly decreased.

Nitrogen Oxides (NO_x)

The visible injury symptoms observed on lettuce leaves as described in Sec. 2.1 and assumed to be caused by exposure to PAN were simulated after exposure of lettuce plants to man-made PAN at phytotoxic concentrations in CSTRs in the greenhouse. The visible symptoms were entirely different to those resulting from controlled O_3 exposures performed using the same CSTR fumigation apparatus. Results also showed that the degree of sensitivity of five varieties of lettuce exposed to PAN were highly variable (Sun, 1993a).

Suspended particulate matter (SPM)

In order to understand the effect of SPM on crop growth, SPM samples from cement, building works or coal plants were collected using dustfall jars. Estimations of the total SPM from cement and coal plants indicated that these industries were responsible for producing very high levels of SPM pollution. The pH of cement, soil and coal dust SPM solution was 12.55, 6.89 and 5.54 respectively and the conductivity was 1380, 133 and 414 $\mu S/cm^3$ respectively. The germination rate of spinach seeds decreased when treated with all SPM solutions. Germination rates of rice, soybean and lettuce seeds were also decreased on exposure to high concentrations (40 and 50%) of SPM solutions. The growth and photosynthetic rate of seedlings of these species were all affected by high SPM concentrations, with spinach found to be particularly sensitive (Fan, 1993a and 1993b).

3. Applicability of European and North American Air Quality Standards

The Taiwan Air Quality Monitoring Network (TAQMN) has 71 air quality monitoring stations as described previously in Sec. 1.1. Local air quality information is communicated to the public using a PSI designed to standardise pollutant concentration information for ease of understanding. This PSI is based on exposure indices used to define North American Air Quality Standards (AQS). The monitoring data collated into the PSI format include information describing concentration levels for criteria pollutants namely, particulate matter (PM_{10}), sulphur dioxide (SO_2), nitrogen dioxide (NO_2), carbon monoxide (CO) and ozone (O_3). For each pollutant, a sub-index is calculated from a segmented linear function that transforms ambient concentrations to a scale from 0 to 500. The breakpoints for the PSI are listed in Table 11.

Table 11. Air quality in Taiwan described using PSI values.

PSI value	24-hr. SPM (mg/m3)	24-hr. SO_2 (ppm)	8-hr. CO (ppm)	1-hr. O_3 (ppm)	1-hr. NO_2 (ppm)
50 (good)	50	0.03	4.5	0.06	-
100 (moderate)	150	0.14	9	0.12	-
200 (unhealthy)	350	0.30	15	0.2	0.6
300 (very unhealthy)	420	0.60	30	0.4	1.2
400 (hazardous)	500	0.80	40	0.5	1.6
500 (hazardous)	600	1.00	50	0.6	2.0

- No index values are reported for these concentration levels because there are no short-term standards defined.

Air quality forecasting is only made for individual pollutants, appropriate PSI values for pollutants occurring in combination (i.e. during photochemical smog episodes where O_3 and PAN co-occur at high concentrations) have to date not been considered. In addition, there are still no data available that would indicate the degree of damage that may occur to vegetation for each of the different PSI classes.

4. Future Research

In Taiwan, air pollutant emission standards have not been established for agricultural regions, this is in part explained by the fact that only very limited research has been targeted towards systematically defining "no-effect thresholds" for different vegetation types. Robust exposure-response relationships are an essential pre-requisite to defining acceptable levels of air quality and consequently targeted emission reduction strategies.

However, it is clear from the observational and experimental studies described for Taiwan in this section, that ambient levels of air pollution are significantly affecting the local vegetation. It is also evident that the expression of injury symptoms induced by different types of pollution are extremely varied and dependent upon a number of factors including species type and variety, concentrations and duration of pollutant exposures and prevailing climatic conditions. Therefore, it is not advisable to use only visible symptoms as a tool for diagnosis of injury caused by air pollution. Ideally, long-term field data should be collected that considers the impacts of air pollutants both individually and in combination to establish a complete database of air pollution impacts on vegetation.

5. Conclusions

The Air Pollution Control Act (APCA) was first established in Taiwan in 1975. The law covers three objectives which address: (1) emission standards for both stationary and mobile sources, these standards are becoming more stringent each time values are reviewed; (2) clean fuel criteria, including low sulphur content in fuel and gradual elimination of leaded gasoline; and (3) monitoring and inspection systems. In 1992, the Act was amended to include economic incentives. The Environmental Protection Administration (EPA) is responsible for the enforcement of the APCA and has also been given responsibility for enforcing the Country Implementation Plan (CIP), additional legislation which has similarly been introduced in order to control pollutant emissions. The CIP includes four objectives designed specifically to control SO_2 emissions: (1) to supply low-sulphur fuel and de-sulphurised fuel; (2) to decrease sulphur content in diesel; (3) to install fuel gas de-sulphurisation devices; and (4) to collect SO_2 emission fees. Besides these SO_2-related objectives, there were also four objectives related to controlling NO_x emissions: (1) to impose more stringent emission standards on new vehicles and new stationary sources; (2) to enforce on-road and off-road inspections; (3) to improve control technologies; (4) to collect NO_x emission fees. According to data collected by the CIP, emissions of SO_2 and NO_x in 2000 were 0.34 million tons and 0.75 million tons, respectively. When compared to emission figures recorded during the baseline year 1991, it can be seen that SO_2 emissions have decreased by 0.25 million tons (−42%) whilst NO_x emissions have increased by 0.15 million tons (+25%) (Liu *et al.*, 1996).

Results from the air pollution research that have been conducted in Taiwan over the last 20 years have been applied in a number of different programmes. These include: (1) developing methods to use indicator plants for biomonitoring within air pollution studies; (2) screening for tolerant tree species that may be effective in absorbing pollutants in high-risk areas; and (3) resolving disputes related to the extent of plant damage caused by ambient levels of air pollution. It looks highly probable that air pollution will continue to affect the vegetation of Taiwan for at least the next few decades, since per capita consumption and the associated increases in emissions have been steadily growing in recent years. As such, there is an urgent need for the effects of various air pollutants on the diverse and abundant vegetation of Taiwan to be comprehensively studied in the near future (Sun, 1995).

References

Chung Ting Kung Ch'eng (CTCI) (1996) *Technical Support for Execution of Air Quality Improvement/Maintenance Plan for Each County and Air Quality Caring Capacity Assessment Planning.* Taiwan (Chinese).

Chung Ting Kung Ch'eng (CTCI) (1997) *Technical Support for Execution of Air Quality Improvement/Maintenance Plan for Each County and Stationary Source Control Activity Planning.* Taiwan (Chinese).

Department of Agriculture and Forestry (1999) *Taiwan Agriculture Statistics.* Taiwan.

Environmental Protection Administration (EPA) (1998) *The Annual Assessment Report of the Air Pollution Control in Taiwan Area For 1997.* Taiwan (Chinese).

Fan C.N. (1993a) Effect of particulate matter in air on plant growth I. The morphology and the effect of particulate matter in air on the germination and seedling growth of crops. In *Workshop on Effects of Air Pollution and Agrometeorology on Crop Production in Taiwan.* Taiwan Agricultural Research Institute, Taiwan, pp. 201–216 (Chinese).

Fan C.N. (1993b) Effect of particulate matter in air on plant growth II. The effect of particulate matter in air on the germination and photosynthesis rate of crops. In *Proceedings of the Workshop on the Relationships Between Atmospheric Quality and Agricultural Management.* Taiwan, pp. 93–109 (Chinese).

Hsieh C.F. (1993) Responses of plants to the pollutants of fluorides, sulfur oxides and chlorides. In *Workshop on Effects of Air Pollution and Agrometeorology on Crop Production in Taiwan.* Taiwan Agricultural Research Institute, Taiwan, pp. 106–124 (Chinese).

Huang Y.H., Lin Y.Y. and Liao C.H. (1994) *Monitoring of Air Pollution in Agricultural Areas of Northwestern Taiwan.* Bulletin of the Taoyuan District Agricultural Improvement Station Number 19 (Chinese).

Lee Y.H. and Lee K.C. (1985) *Symptom Diagnosis of Air Pollution to Vegetation.* Taiwan Vegetation Protection Center, Taiwan (Chinese).

Lee Y.H., Shyu T.H. and Chiang M.Y. (1996) The fluoride accumulation of plants around a fluoride pollution source. In *Proceedings of the Symposium on Effects of Agroclimate, Air Pollution and Acid Rain on Agricultural Production and the Strategy in Response,* ed. Yang C.M. Taiwan Agricultural Research Institute, Taiwan, pp. 88–96.

Lin C.C. and Yang S.H. (1996) The effect of atmospheric qualities on horticultural crops in southern Taiwan. *Chinese J. Agromet.* **3**(4), 183–196 (Chinese).

Liu C.M., Buhr M., Hsu K.J. and Merrill J.T. (1993) Long-rang transport of ozone to southern Taiwan. In *Proceedings of International Conference on Regional Environment and Climate Changes in East Asia.* Taiwan, pp. 17–21.

Liu C.M. and Young C.Y. (1996) The state of ozone pollution and the control strategy. In *Proceedings of the Symposium on Effects of Agroclimate, Air Pollution and Acid Rain on Agricultural Production and the Strategy in Response,* ed. Yang C.M. Taiwan Agricultural Research Institute, Taiwan, pp. 9–32.

Liu C.M., Young C.Y. and Su W.C. (1996) Scenarios of SO_2 and NOx emission in Taiwan. In *Proceedings of International Conference on Acid Deposition in East Asia.* Environmental Protection Administration, Taiwan, pp. 409–426.

Peng C.I., Kuo C.M. and Yang Y.P. (1994) Botanical diversity and inventory of Taiwan. In *Biodiversity and Terrestrial Ecosystems. Institute of Botany,* eds. Peng C.I. and Chou C.H. Academia Sinica Monograph Series No. 14, Taiwan, pp. 75–86.

Street O.E., Sung H.Y., Wu H.Y. and Menser H.A. (1971) Studies on weather fleck of tobacco in Taiwan. *Tobacco Sci.* **15**, 128–131 (Chinese).

Su H.C., Ke W.S., Juang T.Y., Hwang M.D. and Hwang S.C. (1978) Study on the reason of marginal necrosis of banana — especially the effect of fluorides. *Taiwan Banana Res. Special Issue* **21**, 1–21 (Chinese).

Su H.J. (1984a) Studies on the climate and vegetation types of the natural forests in Taiwan. (I) Analysis of the variation in climate factors. *Quart. J. Chinese Forest.* **17**(3), 1–14 (Chinese).

Su H.J. (1984b) Studies on the climate and vegetation types of the natural forests in Taiwan. (II) Altitudinal vegetation zones in relation to temperature gradient. *Quart. J. Chinese Forest.* **17**(4), 57–73 (Chinese),

Sun E.J. (1984) *Environmental Pollution and Identification of Its Adverse Effect and A Pictorial Atlas.* Taiwan Environmental Protection Administration (Chinese).

Sun E.J. (1993a) Effects of peroxyacetyl nitrate on lettuce plants in Taiwan. *Plant Pathol. Bull.* **2**, 33–42 (Chinese).

Sun E.J. (1993b) Ozone injury to leafy sweet potato and spinach in northern Taiwan. In *Proceedings of the Workshop on the Relationships between Atmospheric Quality and Agricultural Management* Taiwan, pp. 111–126.

Sun E.J. (1995) Effects of air pollution on vegetation in Taiwan. *Plant Protect. Bull.* **37**, 142–156 (Chinese).
Sung C.H., Chen H.H. and Wu J.K. (1973) Air pollution related to tobacco weather fleck. In *Annual Report Tobacco Research Institute Taiwan Tobacco and Wine Monopoly Bureau.* Taichung, Taiwan, pp. 38–41 (Chinese).
Wellburn A. (1996) *Air Pollution and Climate Change: The Biological Impact,* 2nd Ed. Longman, Singapore.
Wu J.K. and Sung C.H. (1973) Study on environmental factors in relation to the occurrence of weather fleck in tobacco. In *Annual Report of Tobacco Research Institute Taiwan Tobacco and Wine Monopoly Bureau.* Taichung, Taiwan, pp. 42–50 (Chinese).
Yu W.C. and Chang Y.S. (1997) A study on the tolerance of air pollution in several Taiwanese wildflowers. *Garden Bull.* **4**(1), 67–87.

CHAPTER 8

AIR POLLUTION IMPACTS ON VEGETATION IN INDIA

M. Agrawal

1. Introduction

Air pollution has become an extremely serious problem in India due to the rapid increases in industrialisation and urbanisation that have occurred over the last two decades. In the initial years of industrialisation, air pollution was identified as a problem restricted to areas around point sources. More recently, the problem has spread to remote rural areas supporting large productive agricultural land (Singh *et al.*, 1990). The rapidly increasing population also means more land is needed for agriculture, industry and housing. India currently has one of the fastest developing economies and on gross output basis ranks amongst the top ten industrialised nations. India's population is expected to reach the billion mark by the end of this century.

The use of fossil fuel in power generation and the transport sector is emerging as the biggest environmental problem. Fossil fuel consumption has increased from 75 million tonnes per year in 1964 to 245 million tonnes per year in 1990 (Shrestha *et al.*, 1996). The number of vehicles has increased from 1.86 million in 1971 to 32 million in 1996 and is expected to increase further to 53 million by the year 2000 (Varshney *et al.*, 1997). In this decade, the problem of urban air pollution has attracted special attention due to large increases in the urban population and the fact that industries are also concentrated in the cities. Emissions from heavily loaded and badly maintained automobiles, open disposal and burning of municipal solid wastes and domestic combustion of coal are additional sources of urban air pollution. The recent emphasis of economic liberalisation in the country is likely to aggravate the air pollution problem in the next century.

1.1. *Emissions*

The major air pollutants of concern are identified as sulphur dioxide (SO_2), nitrogen dioxide (NO_2), ozone (O_3), hydrogen fluoride (HF) and particulate matter in form of cement dust, coal dust, fly ash, heavy metal particles, etc. The National Environmental Engineering Research Institute (NEERI), Nagpur and the Central Pollution Control Board (CPCB) are preparing a database on national ambient air quality providing data for a wide cross-section of different industrial, geographic and climatic conditions. The annual average SO_2 concentration ranged from 10 to 40 $\mu g\ m^{-3}$ in most of the regions of the country (Agarwal *et al.*, 1999). Industrial belts and metropolitan cities showed annual average SO_2 concentrations ranging from 60 to 85 $\mu g\ m^{-3}$. In a case study around a 1500 MW thermal power plant situated in Obra (U.P.), SO_2 concentrations of 120 and 66 $\mu g\ m^{-3}$ were recorded at 0.5 and 6 km, respectively in the direction of the prevailing wind (Singh *et al.*, 1990). Seasonal patterns were apparent with higher SO_2 averages during the winter and lower averages during the summer and rainy seasons. Annual increases in SO_2 concentrations were greatest in the northern region.

The annual average NO_2 concentrations range from 10 to 90 $\mu g\ m^{-3}$ in different parts of the country. NO_2 concentrations are especially high at metropolitan cities. Traffic density has been found to contribute between 52 and 72% of the total estimated NO_x in Mumbai, Delhi and Chennai (Agarwal *et al.*, 1999). Southern regions of the country were found to experience the minimum NO_x concentrations. In eastern and western regions, NO_x emissions resulted in annual mean concentrations above the standard (60 $\mu g\ m^{-3}$) at many monitoring stations. High NO_x concentrations were even recorded in small towns of the northeast. Annual average NO_2 concentrations in different areas of

Varanasi city situated in eastern gangatic plain of Uttar Pradesh ranged from 19 to 59 $\mu g\ m^{-3}$ (Pandey *et al.*, 1992). Seasonal variations of NO_x concentrations show sharp peaks during winter months.

High concentrations of the secondary air pollutant O_3 have also been reported for some parts of the country. Ground level one-hour mean O_3 concentrations in Delhi were recorded between 20 and 273 $\mu g\ m^{-3}$ (Varshney and Aggarwal, 1992). Singh *et al.* (1997) have reported that eight-hour mean O_3 concentrations in Delhi exceeded the World Health Organisation's (WHO) mean standard of 100 to 200 $\mu g\ m^{-3}$ by 10 to 40%. In Varanasi city, annual average O_3 concentrations varied between 16 to 48 $\mu g\ m^{-3}$ (Pandey and Agrawal, 1992). The range of two-hour mean O_3 concentrations were 21 to 160 $\mu g\ m^{-3}$. Annual average O_3 concentrations of 27 ppb at Pune and ten-day averages of 15 ppb at Nilgiri Biosphere forests located in South India were reported by Khemani *et al.* (1995). Seasonally, O_3 concentrations were found to be highest during the summer; diurnally, maxima were found to occur during the early afternoon in winter and during the late afternoon in summer (Pandey *et al.*, 1992).

Significant levels of HF have been reported in the vicinity of aluminium factories. For example, the annual average HF concentrations at 0.5, 1, 2, 3, 5, 11 km in the northeast direction and 32 km north of HINDALCO Aluminium Factory, Renukoot were recorded as 3.35, 3.49, 2.41, 0.83, 0.44 and 0.06 $\mu g\ m^{-3}$, respectively during 1989–1990 (Narayan *et al.*, 1994).

At present, particulate pollution is the most important problem in the country. Total suspended particulate matter (SPM) emissions in Delhi were estimated at around 115,700 tonnes per annum in 1990, this value is projected to increase to 122,600 tonnes per annum by 2000 (NEERI, 1991a). Annual average SPM levels of 400 $\mu g\ m^{-3}$ were recorded in Delhi with maximum levels having remained in excess of 1000 $\mu g\ m^{-3}$ since 1987. The southern and northeast regions of India tend to have lower levels of SPM than other regions (Anon, 1997). In Calcutta, annual average SPM concentration have been found to be above 350 $\mu g\ m^{-3}$ since 1987 (Anon, 1995). A major proportion of industrial SPM emissions is due to coal burning power plants (NEERI, 1991b). The transport sector has doubled its share in the total SPM emissions between 1970 to 1990.

1.2. *Vegetation*

Forests and grasslands are two natural vegetation types in the country. India's forest cover was estimated at 63.34 million ha in 1997 accounting for 23% of the geographical area. Indian forests are classified as tropical (up to 1000 m altitude), montane subtropical (between 1000 and 1600 m altitude), temperate (above 1600 m altitude on mountains of Himalayas and Nilgiris) and alpine (between 3000 and 4000 m altitude). Tropical forests are classified into tropical wet evergreen forests (western coast, northeast and Andamans islands); tropical moist semi-evergreen forests (Northern Assam, Bengal, parts of Orissa); tropical moist deciduous forests (Kerela, Karnataka, Southern M.P., eastern U.P., Bihar, Bengal and Orissa); and tropical dry deciduous forests in drier parts of U.P., Bihar, Maharashtra, northern M.P., Punjab, Tamilnadu and Andhra Pradesh. Most of the industrial activities and urban areas lie within tropical and subtropical forests. The dominant species of these forests are *Dipteriocarpus, Artocarpus, Mangifera, Terminalia, Azadirachta, Shorea, Tectona* etc.

The total food grain production in India has increased from 50.8 million tonnes during 1950–1951 to 199 million tonnes in 1996–1997. The cultivated land area is about 140 to 142 million ha with further increases not envisaged (Paroda, 1999). Wheat (68.68 million tonnes), rice (81.20 million tonnes) and oil seeds (25.16 million tonnes) are the major crops of the country. Pulses form an integral part of the vegetarian diet in the country as they are a rich source of protein. The total area under pulses has varied from 22 to 24 million ha over the last few years, giving an almost stable production of 12 to 14 million tonnes. Chickpea, pigeon pea, mungbean, urdbean and lentil are major pulses in order of decreasing area sown. As a result of intensification of agriculture to ensure food security for the rapidly increasing population, soil resources are becoming limited. Macro- and micronutrient deficiencies are now surfacing more prominently. Water will be a scarce resource in the future with the availability of good quality water decreasing in the future. The resurgence of pests and diseases and increased environmental pollution are other factors leading to declining crop productivity.

2. Impact of Air Pollution

Adverse impacts of air pollution on plants around industrial sources and metropolitan cities have been reported from various parts of the country.

2.1. *Field Evidence*

Visible injury

Spatial and temporal variations in visible injury symptoms on trees, shrubs and herbs around large point sources have been reported (Dubey, 1990; Singh *et al.*, 1990; Pandey, 1978). Based on the extent of foliar injury (irregular chlorosis and necrosis on leaf surfaces) and distance of occurrence, i.e. near, far and farthest from the thermal power plant, dry tropical forest vegetation has been grouped into various categories of susceptibility (Pandey, 1978; Table 1).

Defoliation of branches and mortality of apical buds leading to asymmetrical tree canopies have been reported around cement factories (Singh *et al.*, 1990). Leaf injury in the form of bifacial chlorotic and necrotic lesions on *Cassia fistula* and *Carissa carandas* has been reported from heavily polluted zones of Varanasi city (Pandey and Agrawal, 1994a). Tip burning of grain crops and bifacial chlorosis and necrosis of leguminous crops have been reported around aluminium factories and thermal power plants (Singh *et al.*, 1990; Rao *et al.*, 1990). The areas where visible injury symptoms have been recorded have annual SO_2 and NO_2 concentrations above 60 to 80 $\mu g\ m^{-3}$. Bambawale (1986) has also recorded evidence of O_3 injury to potato crops in the Panjab.

Yield response

A number of field studies have been carried out around industrial sources of pollution to evaluate yield responses of plants. Crop yield declined from 10 to 50% in response to levels of SO_2 (75 to 139 $\mu g\ m^{-3}$), NO_2 (4.8 to 110 $\mu g\ m^{-3}$) and particulates (306 to 762 $\mu g\ m^{-3}$) in the area (Table 2). Maximum reduction in yield was

Table 1. Relative sensitivity of some plant species occurring around Obra thermal power plant (from Pandey, 1978).

Tolerant	Moderately tolerant	Sensitive
Acacia catechu Willd.	*Anogeissus latifolia* Wall.	*Aegle marmelos* Correa.
Cassia tora L.	*Boerhaavia diffusa* L.	*Bauhinia tomentosa* L.
Cyperus rotundus Miq.	*Butea frondosa* Wall.	*Boswellia serrata* Roxb.
Dicanthium annulatum (Forskal) Stapf	*Cassia fistula* L.	*Buchanania lanzan* Spreng.
Diospyros melanoxylon Blume	*Desmodium triflorum* DC.	*Gardenia turgida* Roxb.
Eragrostis tenella Nees	*Eclipta alba* Hassk.	*Grewia tiliaefolia* Vahl.
Euphorbia hirta L.	*Hardwickia binata* Roxb.	*Madhuca indica* J.F. Gmel.
Lagerstroemia parviflora Roxb.	*Malvestrum tricuspidatum* A. Gray	*Miliusa tomentosa* (Roxb.) J. Sinclair
Saccharum munja Roxb.	*Phyllanthus simplex* Retz.	*Nyctanthes arbor-tristis* L.
Tephrosia purpurea Pers.	*Scoparia dulcis* L.	*Phaesolus* sp.
Zizyphus jujuba Lam.	*Sida acuta* Burm.	*Phyllanthus emblica* Gaertn.
Zizyphus nummularia DC.	*Vernonia cinerea* Less.	*Setaria glauca* Beauv.

observed for *Pisum sativum* (40 to 50%) and minimum in *Brassica campestris* and *Sesamum indicum* (10 to 15%). In the case of wheat, yield (g m^{-2}) was reduced by 47, 43, 33, 41 and 29%, respectively, at distances of 1.5 km SE, 3.0 km SE, 5.0 km SE, 2.0 km E and 3.0 km NE from a 1500 MW thermal power plant in Obra, Uttar Pradesh (Table 3). Multiple regression attempted between concentrations of SO_2, NO_2, TSP and dust fall versus wheat yield loss showed significant positive coefficient between SO_2 and yield loss whereas others have negative coefficients indicating that SO_2 is the most critical factor affecting yield loss of wheat around thermal power plants (Table 4).

Table 2. Yield reductions (%) in different crop plants growing in the Obra, Dala and Renukoot areas.

Crop	Range of reduction		Sensitivity range
	Minimum	Maximum	
Triticum aestivum (Wheat)	30	50	S
Hordeum vulgare (Barley)	25	40	S
Cicer arietinum (Gram)	10	15	R
Brassica campestris (Mustard)	10	15	R
Pisum sativum (Pea)	40	50	S
Zea mays (Maize)	20	30	I
Phaseolus mungo (Urd)	30	50	S
Oryza sativa (Rice)	30	40	S
Cajanus cajan (Arhar)	20	30	I
Sorghum vulgare (Jowar)	15	20	R
Sesamum indicum (Til)	10	15	R

S: Sensitive (> 30); I: Intermediate (15–30); R: Resistant (< 15).

Table 3. The effect of the Obra thermal power plant on *Triticum aestivum* L. Values in parenthesis are % reduction from the control site.

Parameter	Distance (km) and direction from source					Control
	1.5 SE	3.0 SE	5.0 SE	2.0 E	3.0 NE	22.0 N
SO_2 (μgm^{-3})	139.9	100.6	76.4	74.6	38.4	20.08
NO_2 (μgm^{-3})	110.4	76.2	69.1	48.1	23.1	10.3
TSP (μgm^{-3})	764.2	385	275	306	106	42.1
Dust fall rate ($g\ m^{-2}\ day^{-1}$)	3.88	2.75	1.89	1.98	1.04	0.43
Biomass 75 days ($g\ plant^{-1}$)	1.45 (72%)	2.3 (55%)	2.8 (46%)	2.6 (50%)	2.95 (43%)	5.2
Yields ($g\ m^{-2}$)	205 (47%)	220 (43%)	259 (33%)	221 (41%)	274 (29%)	389

Table 4. Coefficient between independent variables versus yield using multiple regression (Singh *et al.*, 1990).

Independent variable	Coefficient
Constant	11.77
SO_2	1.14
NO_2	−0.48
TSP	−0.033
Dust fall	−11.49

Field transect studies

In India, pollution damage to dry tropical forest has been reported from areas around thermal power plants, aluminium factories, open cast mining and other industries (Dubey, 1990; Singh *et al.*, 1990; Rao *et al.*, 1990), where high concentrations of SO_2, NO_2 and SPM were recorded. Reductions in stem perimeter and leaf weight of tree species were reported around Nagda Industrial Complex (Pawar, 1982) and in the Betul forest area around Satpura thermal power plant (Dubey *et al.*, 1982). Differential responses of plants to pollutants was attributed to the differential adaptive potentials of plants under field conditions (Rao and Dubey, 1990). *Dalbergia sissoo* was found to be tolerant to pollution due to its ability to enhance antioxidant levels in leaves which resulted in faster sulphite oxidation reducing toxicity.

A two-year study conducted in the urban environment of Varanasi using transplants of one shrub (*Carissa carandas*) and two tree species (*Delonix regia* and *Cassia fistula*) to evaluate the response pattern to existing ambient air quality showed significant negative correlations between pollutant concentrations and parameters such as plant height, basal diameter, canopy area, biomass and chlorophyll, ascorbic acid and nitrogen contents in leaves (Pandey and Agrawal, 1994a). Similarly, transect studies around thermal power plants, cement factories and other industries have also shown adverse effects on plants. Leaf area injury was found to be directly correlated with dust deposition and foliar sulphur content of *Mangifera indica* and *Tectona grandis* plants growing in the prevailing wind direction from a petroleum refinery at Barauni, Bihar (Prasad and Rao, 1985).

Foliar sulphur (S) analysis indicated that tree species present at sites experiencing higher annual SO_2 accumulated higher sulphate (SO_4^{2-}-S) than those at sites with low SO_2 levels (Agrawal and Singh, 1999). Evergreen plants showed a gradual increase in SO_4^{2-}-S throughout the year, whereas in deciduous plants higher magnitudes of increase occurred with the onset of new leaves during summer. Foliar SO_4^{2-}-S/organic S ratio increased more at SO_2 stressed locations than the control sites. This ratio was found to be a useful indicator of the state of forest trees under sulphur pollution stress.

Field studies investigating community structural pattern have also shown that the number of individuals and species richness were lower at polluted compared to unpolluted sites (Singh *et al.*, 1994; Narayan *et al.*, 1994). Species diversity and evenness were inversely related, whereas concentration of dominance was directly related to the pollution load. This clearly indicates an increase in the proportion of resistant plants showing a tendency towards a definite selection strategy of plants in response to air pollution. Tree density and canopy cover gradually increased with decreasing pollution load around an aluminium factory in Renukoot (Narayan *et al.*, 1994). Communities were heterogenous at distantly situated sites because ambient conditions favoured the survival, growth and regeneration of existing vegetation and also the establishment of new arrivals. Significant reductions in peak standing biomass of grassland vegetation have been reported in response to ambient pollutant levels around the Obra thermal power plant (Pandey, 1993).

The average leaf area of plants in the vicinity of power plants and other polluted sites was smaller than those growing at less polluted sites (Agrawal *et al.*, 1993). Total chlorophyll content and specific leaf area were reduced considerably at sites receiving higher pollution loads (Agrawal and Agrawal, 1989). The physiological status of the plants as reflected by metabolite contents such as starch, protein and ascorbic acid was significantly lower in plants growing closer to emissionsources compared to those growing at distant sites (Rao *et al.*, 1990).Chlorophyll content and dry weight of leaves also declined in response to fluoride accumulation in leaves around aluminium factories (Lal and Ambasht, 1981; Singh *et al.*, 1990).

To evaluate the effect of air pollution on crops under field conditions, plants can either be grown in pots and then transferred to different sites, or grown in field plots. Field studies have clearly demonstrated that areas receiving higher pollution load show significant decline in shoot length, leaf area, chlorophyll content, biomass accumulation and yield of wheat compared to the control site (Ayer and Bedi, 1991). Field studies with *Oryza sativa* showed significant reductions in panicle length, dry weight of filled grain and grain yield of three varieties located close to a fertiliser plant experiencing ambient conditions at Baroda, Gujrat when compared to those grown at a relatively clean site (Anbazhagan *et al.*, 1989). The average peak levels of SO_2 and NO_2 measured near the source were 144 and 210 $\mu g\ m^{-3}$, respectively.

In a study to evaluate the impact of cement dust on paddy, spatial variation in foliar dust deposition and yield were correlated (Purushothamanan *et al.*, 1996). The average yield loss in paddy due to a cement dust polluted environment was estimated at 19.56%. A detailed study on *Oryza sativa* growing at different sites around Dala cement factory in U.P. showed that vegetative and reproductive parts accumulated significantly lower biomass at sites receiving higher dust load (Singh *et al.*, 1990). Vegetative biomass was reduced by 44% and reproductive biomass by 60% at 1 km as compared to the biomass observed at 10 km distance from the factory. Grain yield was negatively correlated with dust load in the area. Energy content was 4.6 k cal g^{-1} dw and 3 k cal g^{-1} dw, respectively at 10 and 1 km from the cement factory. Field studies with *Cicer arietinum, Glycine max* and *Cajanus cajan* also showed reductions in plant height, total chlorophyll and fresh and dry weight of plants growing at polluted compared to control sites (Varshney *et al.*, 1997).

In view of the potential impact of urban air pollution on peri-urban agriculture, pot-grown plants of *Vigna radiata* var Malviya Jyoti and *Spinacia oleracea* var All Green were exposed in ambient air to the time of harvest. Both plants showed variable response to ambient air depending upon the pollutant levels (Tables 5 and 6). Sites receiving the highest SO_2, NO_2 and O_3 concentrations showed minimum values

Table 5. Mean pollutant concentrations and different plant characteristics of *Vigna radiata* var. Malviya Jyoti grown at different sites in peri-urban area of Varanasi city during summer.

Site	Mean pollutant concentration (ppb)			*Vigna radiata* var. Malviya Jyoti			
	SO_2	NO_2	O_3	Photosynthesis (μmol CO_2 m^{-2} s^{-1})	Biomass (g $plant^{-1}$)	No. of seeds $plant^{-1}$	Yield (g $plant^{-1}$)
1	13.3	31.1	55.7	8.34 ± 0.11	11.58 ± 0.66	130.38 ± 10.77	4.2 ± 0.19
3	8.05	11.7	9.7	10.03 ± 0.08	17.28 ± 1.33	181.33 ± 13.86	6.4 ± 0.59
4	17.18	31.9	25.1	8.13 ± 0.11	10.18 ± 0.41	90.17 ± 6.49	3.21 ± 0.27
9	32.2	80.1	58.5	5.26 ± 0.25	5.33 ± 0.41	45.28 ± 3.30	1.64 ± 0.13

Table 6. Mean pollutant concentrations and different plant characteristics of *Spinacia oleracea* var. All Green grown at different sites in peri-urban area of Varanasi city during summer.

Site	Mean pollutant concentration (ppb)			*Spinacia oleracea* var. All Green		
	SO_2	NO_2	O_3	Photosynthesis (μmol CO_2 m^{-2} s^{-1})	No. of Leaves	Biomass (g $plant^{-1}$)
1	13.3	31.1	55.7	5.61 ± 0.10	6.38 ± 0.13	0.45 ± 0.03
3	8.05	11.7	9.7	6.18 ± 0.01	7.75 ± 0.17	0.66 ± 0.04
4	17.18	31.9	25.1	5.79 ± 0.19	6.25 ± 0.19	0.46 ± 0.02
9	32.2	80.1	58.5	3.55 ± 0.33	4.69 ± 0.18	0.18 ± 0.02

of net photosynthesis, biomass and yield. Reductions in biomass and yield of plants growing at the minimum pollution load (site 3) and the highest pollution load site (site 9) were compared and showed that yield of *V. radiata* was reduced by 79% and biomass of *S. oleracea* was reduced by 72.8% at the more polluted location.

2.2. *Experimental Research*

Plants response to air pollution can also be studied by simulating pollutant concentrations either in closed top or open top chambers under field conditions. Most of the investigations in closed top chambers have been carried out using 1 to 1.5 m^3 chambers with SO_2, NO_2, HF and O_3. SO_2, NO_2 and HF are generally produced chemically while O_3 can be generated using a standard ozonator. The desired concentrations of the pollutants in the chamber are obtained by diluting with incoming air from a blower. Plants used for fumigation were either pot or plot grown. Pollutant monitoring was achieved either by using chemical methods or in some cases by using sophisticated instruments. Wheat has been the most extensively studied crop plant for air pollution impact assessments followed by rice, maize broad bean, soybean, mustard, etc. Of the different pollutants, SO_2 is the most studied for response evaluation of plants.

Closed top chamber

Exposure of *Triticum aestivum* var. J-24 to 261 $\mu g\ m^{-2}$ SO_2 for 2-hrs daily for 80 days resulted in reductions of 8, 55 and 44% in dry weight (g $plant^{-1}$), weight of 1000 grains (g) and weight of grains (kg m^{-2}) (Ayer and Bedi, 1986). Singh and Prakash (1995) reported reductions of 16, 22 and 27% in weight of 1000 grains, 19, 25 and 30% in yield (g $plant^{-1}$) and 25, 31 and 37% in dry weight (g $plant^{-1}$) of *T. aestivum* var. Sharbati Sonora exposed to 320, 667 and 1334 $\mu g\ m^{-3}$ SO_2 respectively for four hours daily for 100 days. Similarly *Oryza sativa* var. Ratna showed respective reductions of 29.4 and 46.9% in biomass, 31.2 and 43.7% in number of tillers and 58.8 and 64.7% in number of ears when exposed to 653 and 1306 $\mu g\ m^{-3}$ SO_2 for 1.5 hours daily for

30 days (Nandi *et al.*, 1985). *Hordeum vulgare* var. K-192 showed lower reductions than wheat, the reductions being 7.3 and 20% in biomass and 19 and 33% in weight of 1000 grains when exposed to 313 and 653 μg m^{-3} SO_2 for four hours daily for 60 days (Chand *et al.*, 1989). *Zea mays* L. var. American Sweet corn showed reductions of 12 and 20% at 261.2 μg m^{-3} and 25 and 73% at 1306 μg m^{-3}, respectively in weight of 1000 grains (g) and yield (kg 100 m^{-2}) when exposed to SO_2 for one hour daily for 73 days (Ayer and Bedi, 1990). *Cicer arietinum* was found to be relatively resistant to SO_2. On comparing the overall sensitivity of crop plants studied with respect to SO_2 in closed top chambers, it appears that the order of sensitivity is *Vicia faba* > *Vigna sinensis* > *Oryza sativa* > *Vigna radiata* > *Panicum miliaceum* > *Cicer arietinum* > *Brassica campestris* (Agrawal *et al.*, 1991). In the case of SO_2 phytotoxicity, physiological and biochemical changes precede the development of visible injury symptoms.

Lycopersicon esculentum var. Pusa Ruby when exposed to 261 μg m^{-3} SO_2 for four hours daily for 50 days showed a 28.3% increase in height, whereas at 522 μg m^{-3} SO_2, height declined by 16.8%. NO_2 also increased plant height by 26 and 70%, respectively at 1306 and 522 μg m^{-3} for four hours daily for 40 days. However, the combination of NO_2 and SO_2 only increased the height by 7.5% (Pandey and Agrawal, 1994a and b).

Simulation experiments conducted to understand the dose-response pattern of *Oryza sativa* to two levels of HF showed no linear relationship. Reductions in biomass accumulation between plant age day 20 and 60 varied from 1 to 33% and 7 to 34% at HF concentrations of 1.0 and 2.0 μg m^{-3} for two hours daily for 50 days (Narayan, 1992).

Exposure of different crop plants to O_3 at 156 μg m^{-3} for 1.5 hours daily for 30 days resulted in percent foliar injury of 29, 28 and 39% in *O. sativa*, *P. miliaceum* and *V. faba*, respectively. The reductions in biomass were 18, 42, 34 and 49%, respectively in *C. arietinum*, *O. sativa*, *P. miliaceum* and *V. faba* for the same O_3 dose (Agrawal, 1982).

Open top chamber studies

Wheat and soybean cultivars have been those most extensively studied using open top chamber facilities. Physiological and biochemical characteristics have been evaluated at realistic concentrations of SO_2 and O_3 (Tables 7 and 8). SO_2 concentrations of 157 μg m^{-3} for eight hours daily for 70 days resulted in reductions of yield by 15, 17 and 25% and in biomass by 8, 6 and 4% in *Triticum aestivum* var. M234, HP1209 and M213, respectively (Deepak, 1999; Table 7). Similarly *Glycine max* showed yield reductions of 16 and 19%, respectively in cultivars PK 472 and Bragg at similar SO_2 concentrations. Other response patterns are shown in Table 7. At higher concentrations of SO_2 i.e. 390 μg m^{-3} for four hours daily for 35 days, yield reductions were found to be 41, 61, 42 and 57%, respectively for *T. aestivum* var. M 213, M206, M37 and M234 (Rajput, 1993). *G. max* showed greater reductions in biomass and yield compared to *T. aestivum* at higher SO_2 dose.

To evaluate the influence of O_3 on morphological and physiological characteristics, biomass accumulation and yield attributes, wheat and soybean plants were exposed to 137 and 196 μg m^{-3} O_3 for four hours daily from germination to physiological maturity (Singh, 1998). The data clearly showed a linear dose-response relationship with respect to yield, biomass and photosynthesis (Table 8). The relationship clearly indicates that O_3 had a more extreme impact on *G. max* as

Table 7. Percent changes (+ increase; – decrease) of specific plant characteristics compared to control plants at age 70 days after exposure to 157 μg m^{-3} SO_2 for eight hours daily for 70 days.

Characteristics	Wheat			Soybean	
	M234	HP 1209	M213	PK472	Bragg
Photosynthesis	–.016	+1.76	–10.29	–19.0	–13
WUE	–16	–16	–28	–28	–2
Starch	–6	–11	–12	–7	–9
Protein	–12	–14	–11	–17	–13
SO42-S	+67	+47	+45	+72	+89
Leaf area	–7	–6	–9	–16	–2
Biomass	–8	–6	–4	–16.7	–19.17
Harvest index*	+3.8	+7.2	+10.6	+12.94	+5.26
Yield*	–15	–17	25	21	19

*At the time of harvest.

Table 8. Correlation coefficient and linear regression between selected parameters of control and ozone treated plants of *Glycine max* and *Triticum aestivum.*

Correlation	Plant and Variety	Correlation coefficient	Regression equation $y = a + bx$
Dose vs. Yield	*G. max* PK 472	$-0.795^{\dagger}$	$y = 121.31 - 1.55x$
	G. max Bragg	$-0.916^{\dagger}$	$y = 194.36 - 1.84x$
	T. aestivum M234	-0.685^{*}	$y = 216.70 - 1.25x$
	T. aestivum HP 1209	-0.652^{*}	$y = 187.31 - 1.46x$
Dose vs. Biomass	*G. max* PK 472	$-0.916^{\dagger}$	$y = 11.56 - 0.13x$
	G. max Bragg	$-0.994^{\dagger}$	$y = 14.49 - 0.12x$
	T. aestivum M234	$-0.937^{\dagger}$	$y = 9.62 - 0.08x$
	T. aestivum HP 1209	$-0.865^{\dagger}$	$y = 4.49 - 0.04x$
Photosynthesis vs. Biomass	*G. max* PK 472	$0.826^{\dagger}$	$y = 4.11 + 0.57x$
	G. max Bragg	$0.868^{\dagger}$	$y = 5.99 + 0.70x$
	T. aestivum M234	$0.911^{\dagger}$	$y = 5.64 + 0.19x$
	T. aestivum HP 1209	$0.793^{\dagger}$	$y = 2.40 + 0.13x$

Level of significance: $^{*} = p < 0.05$; $^{\dagger} = p < 0.01$.

compared to *T. aestivum.* The data depicted in Table 9 clearly demonstrate the intraspecific variation in response pattern of wheat and soybean cultivars to O_3 treatment.

Only a very few pollutant fumigation studies have been carried out for tree species in India. Six-month-old tree saplings of *Mangifera indica, Psidium guajava* and *S. cuminii* exposed to 0.08 ppm (208 $\mu g\ m^{-3}$) and 0.15 ppm (390 $\mu g\ m^{-3}$) SO_2 for four hours daily for three months showed reductions in starch, protein, ascorbic acid, total chlorophyll and carotenoid contents and an increase in proline and phenol contents and peroxidase activity (Table 10) (Pandey, 1993). Biomass accumulation was reduced more in above ground compared to below ground parts. Maximum adverse effects in different parameters were observed in *M. indica* which has also showed greater sensitivity to thermal power plant emissions under field conditions.

In an attempt to understand the interactive effects of HF and SO_2, *Cynodon dactylon,* an important forage species, was exposed to 1.0 $\mu g\ m^{-3}$ HF, 200 $\mu g\ m^{-3}$ SO_2, singly and in combination daily for two hours

Table 9. Percent reductions of selected plant characteristics compared to control plant at age 70 days after exposure to 137 (T1) and 196 (T2) μg m^{-3} O_3 for four hours daily for 70 days.

Parameters	Soybean				Wheat			
	PK472		Bragg		HP1209		M234	
	T1	T2	T1	T2	T1	T2	T1	T2
Photosynthesis	19.8	40.4	25.6	32.4	29.0	41.0	36.0	43.0
Total chlorophyll content	26.6	34.1	11.5	36.9	19.8	42.1	15.0	20.9
Ascorbic acid content	15.0	15.0	1.1	4.0	5.8	19.6	24.9	34.6
Plant length	20.0	23.0	6.5	17	11.6	17.1	1.87	4.1
Leaf area	20.0	23.0	21.0	24.0	20.1	34.1	15.3	27.8
Total biomass	19.6	26.8	15.0	20.8	6.5	24.7	13.2	20.0
Wt. of 1000 seeds*	3.0	21.0	3.0	18.0	14.0	16.0	3.3	14.4
Yield*	13.9	33.5	10.0	25.0	8.0	17.0	4.7	15.5

*At the time of harvest.

from 50 to 170 days age in open top chambers. Negative influence of these exposures was found with respect to photosynthetic pigments, metabolites and enzyme activity, growth characteristics and biomass accumulation (Narayan, 1992). However, HF and SO_2 in combination had antagonistic effects. When compared with crops such as *Oryza sativa* and *Vigna radiata*, *C. dactylon* showed very low levels of sensitivity to these pollutants.

3. Applicability of European and North American Air Quality Standards

Ambient air quality standards in India were introduced under Section 16(2) (h) of the Air Act (1981). In April 1994, national ambient air

Table 10. Percent increase (+) or decrease (–) in various parameters of tree species exposed to different concentrations of SO_2 for 90 days.

Parameters	*M. indica*		*P. guajava*		*S. cuminis*	
	T_1	T_2	T_1	T_2	T_1	T_2
Chlorophyll	−44	−52	−35	−50	−20	−31
Ascorbic acid	−55	−57	−52	−58	−57	−60
Phenol	+50	+82	+46	+70	+30	+46
Proline	+74	+140	+43	+70	+30	+52
Peroxidase	+120	+170	+95	+135	+46	+85
Ave. sulphur	+14	+20	+8	+14	+12	+17
Shoot length	−13	−16	−10	−15	−3	−5
No. of leaves	−7	−14	−9	−11	−2	−12
Biomass	−15	−23	−13.5	−21	−7	−13

T_1 = 0.08 ppm for 2 hours daily; T_2 = 0.15 ppm for 2 hours daily.

Table 11. National ambient air quality standards (CPCB, 1997).

Pollutant	Time weighted average	Concentration in ambient air (μg m^{-3})		
		Industrial	Rural and residential	Sensitive
Sulphur dioxide (SO_2)	Annual average*	80	60	15
	24 h†	120	80	30
Oxides of nitrogen (NO_2)	Annual average*	80	60	15
	24 h†	120	80	30
Suspended particulate matter (SPM)	Annual average*	360	140	70
	24 h†	500	200	100

*Annual arithmetic mean of minimum 104 measurements in a year, taken for a week, 24-hourly at uniform interval.

†24-hourly/8-hourly values should meet 98% of the time in a year.

quality standards were revised and new standards were established for three designated regions, namely industrial, residential and rural and other sensitive areas (Table 11). These standards are primarily health-based and therefore are not ideal for assessing pollutant impacts on vegetation. Comparisons of Indian standards with similar air quality standards in Europe and North America show that, overall, Indian standards appear to be less stringent as higher values are set. However, direct comparison of these values is complicated by a number of factors that are known to be important in modifying pollutant impacts to vegetation in India compared to Europe and North America. For example, the climate of the country plays an important role in influencing seasonal patterns of plant sensitivity to air pollution, as well as the formation of secondary air pollutants such as O_3. Furthermore, in India there exist a large number of small and medium-sized industries using limited control technologies to reduce emissions. The combination of both these factors results in Indian vegetation types being exposed to different pollutant mixtures and concentration levels as compared to exposure patterns common in Europe and North America. In addition, large differences exist between Indian species and cultivar-type sensitivity to different pollutants; agronomic practices also vary across different regions of India. As such, it is generally considered inadvisable to apply North American and European standards to Indian conditions due to the complexities of differing pollution climates, receptor types and growing conditions.

However, it is imperative that appropriate air quality standards are established to target emission reductions, especially in high-risk areas. For example, due to the rapid growth of urban centres, air pollution is now having a major influence on the productivity and nutritional status of crops grown in peri-urban and rural agricultural areas. As such, comprehensive national air quality standards need to be introduced to protect different species and varieties, and they should consider factors likely to modify plant response to pollutants, such as seasonality, soil type and local/regional air pollutant exposure patterns.

4. Future Research

In India, risk assessments of air pollution impacts to both forests and agricultural crops have been limited to field studies performed under *in situ* conditions. Long-term field campaigns using pollutant exclusion techniques need to be performed in a coordinated manner in different parts of the country to fully assess the damage to agricultural crops and forest trees attributable to air pollution. Fumigation experiments to develop dose-response functions of pollutants separately and in combination are also required. Evaluation of the air pollution climate in different parts of the country is currently only performed around industrial areas and urban centres. There is an urgent need for additional monitoring centres to be located across rural and semi-urban areas of the country. Most Indian forest vegetation types are still unexplored in relation to their response to air pollution. The areas currently considered to be at a relatively low risk from air pollution impacts also need to be identified and investigated, due to the chances of long-range transport of primary pollutants and likely formation of secondary pollutants like O_3 and acid rain. SPM is an important air pollutant in India, and proper evaluation of vegetation responses to this pollutant should be considered as important as assessing impacts associated with what tend to be considered the key gaseous pollutants.

India is a vast country with varying climatic zones and vegetation types, hence a national programme to fully assess the impact of air pollution on the different vegetation types that occur across India should be initiated. The development of pollution abatement policies that use air quality standards established both for agriculture and forestry should be a priority for future scientific research.

5. Conclusions

National air quality data clearly show that SO_2 and SPM are currently the most important pollutants in India. The western and northern parts of the country are more polluted and also have a higher risk of adverse air pollution impacts occurring in the future. Coal-based electric generation and transportation are the most important contributors to air pollution. Meteorological conditions in most parts of the country are favourable to O_3 formation which is emerging as a major threat to crop loss due to the long-range transport of precursor emissions.

A substantial body of observational and experimental evidence has been collected within India to show that current levels of air pollution are impacting significantly on the local vegetation. Both annual crops and perennial plants show variable degrees of susceptibility to present levels of air pollution. Response patterns vary with climatic conditions, agronomic practices, cultivars and pollutant combinations. The effects of air pollution may also have significant economic and social impacts. More detailed national studies are necessary to explore high and low risk zones of air pollution in different regions of the country so that control policies can be developed to reduce vegetation damage.

References

Agrawal M. (1982) *A Study of Phytotoxicity of Ozone and Sulphur Dioxide Pollutants.* Ph.D. Thesis, Banaras Hindu University, Varanasi, India.

Agrawal M. and Agrawal S.B. (1989) Phytomonitoring of air pollution around a thermal power plant. *Atmos. Env.* **23**(4), 763–769.

Agrawal M., Singh S.K., Singh J and Rao D.N. (1991) Biomonitoring of air pollution around urban and industrial sites. *J. Env. Biol.*, 211–222 (special issue).

Agrawal M., Singh J., Jha A.K and Singh J.S. (1993) Coal and coal-based problems in a dry tropical environment, India. In *Advances in Trace Elements in Coal and Coal Combustion residues,* eds. Keefer R.F. and Sajwan K.S. Lewis Publishers, Inc., Geogia, USA, pp. 27–57.

Agarwal A., Narain S. and Srabani S. (1999) *State of India's Environment.* The Citizens Fifth Report, Part I National Overview, Centre for Science and Environment, New Delhi.

Agrawal M. and Singh J. (2000) Impact of coal power plant emission on the foliar elemental concentrations in plants in a low rainfall tropical region. *Env. Monitoring Assess.* **60**(3), 261–282.

Anbazhagan M., Krishnamurthy R. and Bhagawat K.A. (1989) The performance of three cultivars of rice plants grown near to and distant from a fertilizer plant. *Env. Pollut.* 58(2/3), 125–137.

Anon (1995) *National Ambient Air Quality Statistics of India, 1992.* Central Pollution Control Board, New Delhi.

Anon (1997) *Ambient Air Quality — Status and Statistics, 1992 to 1995,* Central Pollution Control Board, New Delhi.

Ayer S.K. and Bedi S.J. (1986) Effect of artificial fumigation of SO_2 on *Triticum sativum* L var. J-24 (wheat). *Indian J. Air Pollut. Control* **7**(2), 75–87.

Ayer S.K. and Bedi S.J. (1990) Effect of artificial fumigation of sulphur dioxide on growth and yield of *Zea mays* L. var. American sweet corn. *Pollut. Res.* **9,** 33–37.

Ayer S.K. and Bedi S.J. (1991) Effect of industrial air pollution on *Triticum aestivum* L. var. J-24 (wheat). *Procs. Nat. Acad. Sci. India* **61(B)II**, 223–229.

Bambawale O. (1986) Evidence of ozone injury to a crop plant in India. *Atmos. Env.* **20**(7), 1501–1503.

Chand S., Yadav N. and Singh V. (1989) Long term effect of SO_2 on growth and yield of *Hordeum vulgare* L. cv. K-192. *Procs. Nat. Acad. Sci. India* **59**(B), 119–125.

CPCB (1997) *National Ambient Air Quality Standards.* Central Pollution Control Board, New Delhi.

Deepak S.S. (1999) *Interactive Effects of Elevated Levels of Carbon Dioxide and Sulphur Dioxide on Selected Plants.* Ph.D. Thesis, Banaras Hindu University, Varanasi, India.

Dubey P.S., Trivedi L., Shringi S.K. and Wagela D.K. (1982) *Pollution Studies on Betul Forest Area Due to Satpura Thermal Power Station Aerial Discharges.* Final Report submitted to Department of Environment and Forest, Government of India (19/27/78).

Dubey P.S. (1990) *Study and Assessment of Plant Response Against Air Pollution in Industrial Environments.* Final technical report submitted to Ministry of Environment and Forests, New Delhi (14/260/85 MAB/RE).

Khemani L.T., Momin G.A., Rao P.S.P., Vijay Kumar R. and Safai P.D. (1995) Study of surface ozone behaviour at urban and forested sites in India. *Atmos. Env.* **29**(16), 2021–2024.

Lal B. and Ambasht R.S. (1981) Impairment of chlorophyll content in the leaves of *Diospyros melanoxylon* in relation to fluoride pollution. *Water Air Soil Pollut.* **16**(3), 361–365.

Nandi P.K., Agrawal M. and Rao D.N. (1985) Photosynthetic potential and growth of SO_2 exposed *Oryza sativa. J. Biol. Res.* **5**(1), 13–17.

Narayan D. (1992) *Impact of Aluminium Factory Emission on Vegetation and Soil.* Ph.D Thesis, Banaras Hindu University, Varanasi, India.

Narayan D., Agrawal M, Pandey J. and Singh Jyoti (1994) Changes in vegetation characteristics downwind of an aluminium factory in India. *Ann. Bot.* **73**, 557–565.

NEERI (1991a) *Air Pollution Aspects of Three Indian Megacities I: Delhi.* National Environmental Engineering Research Institute, Nagpur, India.

NEERI (1991b) *Air Pollution Aspects of Three Megacities II Bombay.* National Environmental Engineering Research Institute, Nagpur, India.

Pandey S.N. (1978) *Effects of Coal Smoke and Sulphur Dioxide Pollution on Plants.* Ph.D. Thesis, Banaras Hindu University, Varanasi, India.

Pandey J. and Agrawal M. (1992) Ozone concentration variabilities in a seasonally dry tropical climate. *Env. Int.* **18**, 515–520.

Pandey J., Agrawal M., Khanam N. Narayan Deo and Rao D.N. (1992) Air pollutant concentrations in Varanasi, India. *Atmos. Env.* **26**(B), 91–98.

Pandey S. (1993) *Effect of Air Pollution on Plants in the Vicinity of Thermal Power Plant.* Ph.D thesis, Banaras Hindu University, Varanasi, India.

Pandey J. and Agrawal M. (1994a) Evaluation of air pollution phytotoxicity through transplants in a seasonally dry tropical environment. *New Phytol.* **126**, 53–61.

Pandey J. and Agrawal M. (1994b) Growth responses of tomato plants to low concentration of sulphur dioxide and nitrogen dioxide. *Sci. Horti.* **58**, 67–76.

Paroda R.S. (1999) *For a Food Secure Future.* The Hindu Survey of Indian Agriculture, Chennai.

Pawar K. (1982) *Pollution Studies in Nagda Area Due to Birla Industrial Complex Discharges.* Ph.D. Thesis, Vikram University, Ujjain, India.

Prasad B.J. and Rao D.N. (1985) Phytomonitoring of air pollution in the vicinity of a petroleum refinery. *Env. Conservation* **12**(4), 351–354.

Purushothamanan S., Mukundan K. and Viswanath S. (1996) The impact of cement kiln dust on rural economy. A case study. *Ind. J. Agric. Econ.* **51**(3), 407–411.

Rajput M. (1993) *Plant Responses to Sulphur Dioxide at Varying Soil Fertility.* Ph.D. Thesis Banaras Hindu University, Varanasi, India.

Rao D.N., Agrawal M. and Singh J. (1990) *Study of Pollution Sink Efficiency, Growth Response and Productivity Patterns of Plants with Respect to Flyash and* SO_2. Final Technical Report submitted to Ministry of Environment and Forest, Government of India (MOE/141/266/85).

Rao M.V. and Dubey P.S. (1990) Biochemical aspects (antioxidants) for the development of tolerance in plants under low levels of ambient air pollutants. *Env. Pollut.* **64**, 55–66.

Shrestha R.M., Bhattacharya S.C. and Malla S. (1996) Energy use and sulphur dioxide emissions in Asia. *J. Env. Manag.* **46**, 359–372.

Singh J.S., Singh K.P and Agrawal M. (1990) *Environmental Degradation of the Obra-Renukoot-Singrauli Area, India and Its Impact on Natural and Derived Ecosystems.* Final Technical Report submitted to Ministry of Environment and Forest, Government of India (14/167/84 MAB EN-21 RE).

Singh J., Agrawal M. and Narayan D. (1994) Effect of power plant emissions on plant community structure. *Ecotoxicology* **3**, 110–122.

Singh S.P. and Prakash G. (1995) Ecophysiological impact of SO_2 pollution on crop plants. *Adv. Plant Sci.* **8**(1), 1–19.

Singh A., Sarin S.M., Shanmugam D., Sharma N., Attri A.K. and Jain W.K. (1997) Ozone distribution in the urban environment of Delhi during winter months. *Atmos. Env.* **31**, 3421–3427.

Singh E.R.A. (1998) *Effect of Ozone Pollution on Selected Crop Plants.* Ph.D. Thesis, Banaras Hindu University, Varanasi.

Varshney C.K. and Aggarwal M. (1992) Ozone pollution in the urban atmosphere of Delhi. *Atmos. Env.* **26**B(3), 291–294.

Varshney C.K., Agrawal M., Ahmad K.J., Dubey P.S. and Raza S.H. (1997) *Effect of Air Pollution on Indian Crop Plants.* Final Report, ODA Project Imperial College of Science, Technology and Medicines, UK.

CHAPTER 9

AIR POLLUTION IMPACTS ON VEGETATION IN PAKISTAN

A. Wahid

1. Introduction

Pakistan is an important agrarian country of South Asia comprising four provinces: Punjab, Sindh, Baluchistan and North-West Frontier Province (NWFP), with a total geographical area of 803,943 km^2 populated by 140 million inhabitants (Fig. 1). Recently, urbanisation, industrialisation and transportation (especially motor traffic) have increased at an unprecedented scale in Pakistan owing to the rapid population growth and the movement of people from agrarian societies to cities (Wahid *et al.*, 1995a and 1997b; Shamsi *et al.*, 2000). Pollution control is limited due to both technical and socio-economic reasons. As such, emission levels are rising rapidly in the region with the subsequent occurrence of deteriorating air quality and serious health-related problems. For example, respiratory and cardiovascular ailments, inflammatory and permeability responses, neurobehavioural effects, irritation to eyes, nose, throat and headache, are all health-related problems which have been associated with elevated exposure to nitrogen oxides (NO_x), sulphur dioxide (SO_2), ozone (O_3), carbon monoxide (CO), and lead in the major cities of Pakistan (WHO/UNEP, 1992; Wahid, 1999; Wahid and Marshall, 2000).

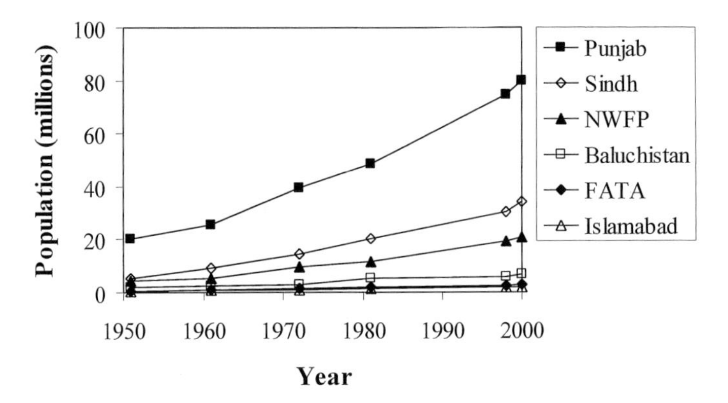

Fig. 1. Population increases in Pakistan by region between 1951 and 2000 (Anonymous, 2000).

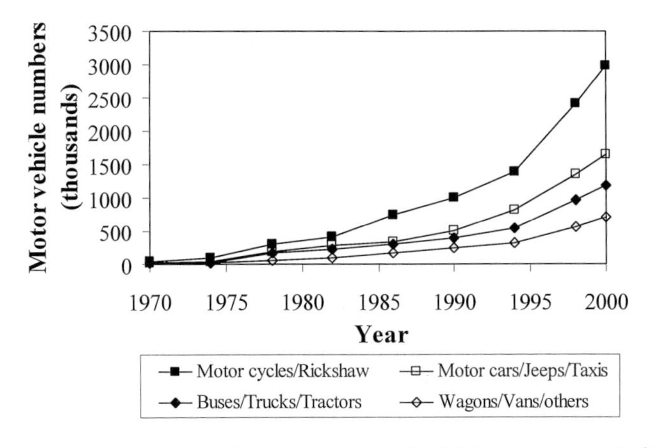

Fig. 2. Increases in the number of motor vehicles in Pakistan between 1970 and 2000 (Anonymous, 2000).

1.1. *Emissions*

The deterioration in air quality in certain Pakistan regions has predominantly been due to a tremendous increase in the number of motor vehicles (Fig. 2) and similar increases of the various types of industry in the major cities of Pakistan. Motor vehicles and industrial emissions primarily include NO_x, SO_2, volatile organic compounds (VOCs), CO, carbon dioxide (CO_2), ammonia (NH_3), lead, smoke and dust (Wahid and Marshall, 2000). NO_x and VOCs are the primary pollutant precursors of secondary air pollutants such as O_3 and PAN. The formation of such pollutants is favoured under the climatic conditions of Pakistan with frequent occurrences of a high number of sunshine hours, high temperatures, low humidity and relatively still air (Wahid *et al.*, 1995a and b). Due to the processes involved in secondary pollutant formation, elevated concentrations of pollutants such as ozone tend to be associated with rural rather then urban areas, with concentration levels determined according to the direction of the prevailing wind, meteorological and other topo-graphical conditions (Kafiat *et al.*, 1994).

Lahore is an important historical city of Pakistan and is also the capital of the Punjab province with a population of 7.65 million people. In Lahore, there are more than 750,000 automobiles registered as emitting visible smoke and this number is rapidly increasing. Lahore city is receiving 33,460 tonnes of NO_2, 2518 tonnes of SO_2, 4812 tonnes of SPM, 57,570 tonnes of hydrocarbons, 486 tonnes of aldehydes and 246 tonnes of CO per annum. Of these total emissions in and around Lahore, emissions from automobile exhausts make up 18.7% of SPM emissions, 92.8% of CO emissions, 89% of VOC emissions, 75.2% of NO_x emissions, 50% of SO_2 emissions, and 100% of aldehyde emissions. Similarly, the other cities of Punjab are also badly affected by noxious vehicular smoke as well as unchecked toxic industrial emissions, dust (from various origins), prolonged construction and developmental activities and a reduction in vegetation cover (EPA, 1998; Wahid, 1999). Figures 3(a) to (c) show the seasonal concentration profiles of NO_2, O_3 and SO_2 between 1992 and 2000 on alternate years in Lahore. Figure 3(d) describes SPM concentrations in different urban/industrial zones in Lahore for every other year between 1990 and 2000.

In the city of Lahore, elevated levels of NO_2 are strongly associated with the nature of the anthropogenic emission source as well as being determined by geographical factors. A detailed survey of NO_2 concentrations was carried out in the major cities of Punjab (Kafiat *et al.*, 1994), and showed that NO_2 concentrations were higher at all sites in Lahore compared to the other Punjab cities. This may be due to the high NO_x emissions ($NO + NO_2$) resulting from the large numbers of motor vehicles on Lahore's roads (Table 1).

Table 1. Annual mean NO_2 concentrations (ppb) at various sites in major cities of Punjab, Pakistan.

Site	City						
	Lahore	*Rawalpindi*	*Faisalabad*	*Multan*	*Gujranwala*	*Sargodha*	*Islamabad*
Road crossings on main roads and highways	51	38	26	23	21	18	15
Roadsides sites on busy main roads	39	33	23	21	18	15	–
Roadside sites in commercial areas	45	24	21	19	19	16	11
Public transport depots	58	32	30	26	20	14	–
Railway stations	59	36	–	23	18	13	–
Industrial areas	33	16	15	15	14	14	6
Residential areas (modern)	14	13	11	11	10	7	5
Residential areas (old)	14	11	10	9	9	6	–
Public parks and gardens	7	6	6	5	5	4	3
Rural sites	5	4	4	4	4	3	–

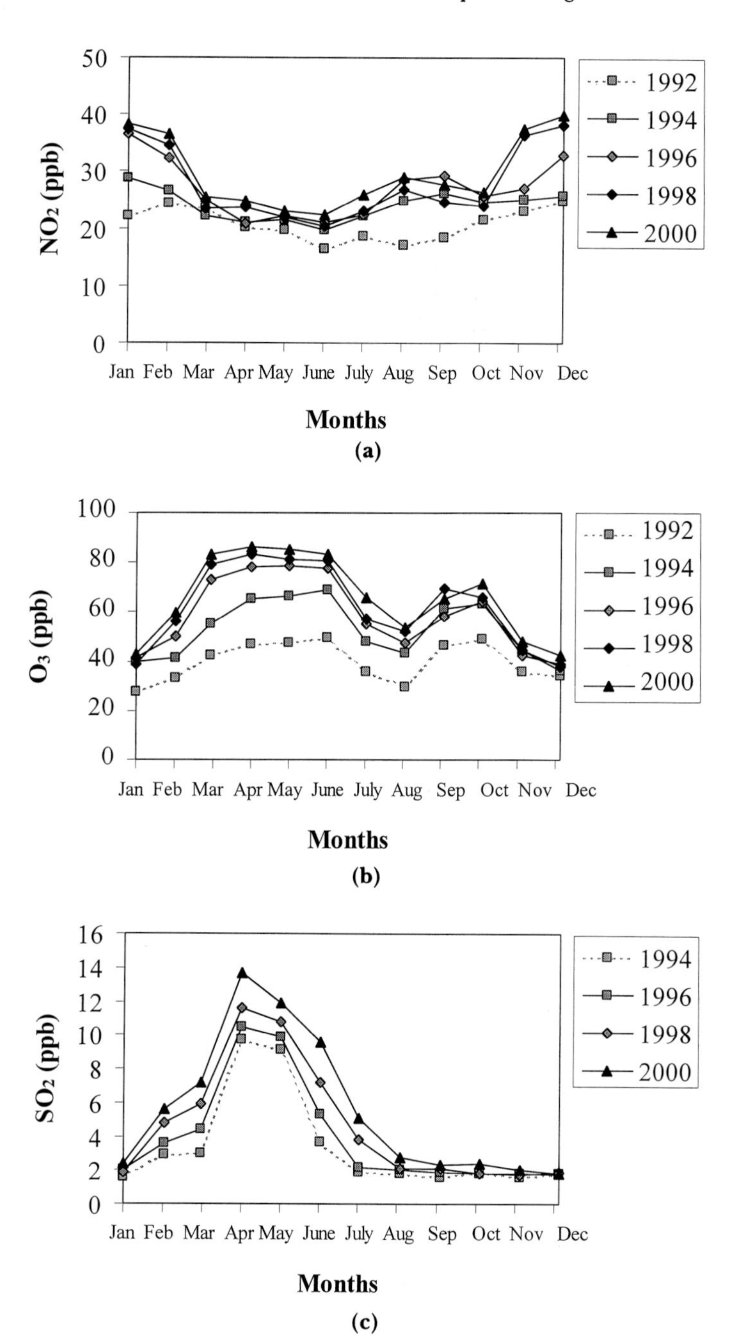

Fig. 3. Seasonal concentration profiles for **(a)** NO_2, **(b)** O_3, **(c)** SO_2, and **(d)** SPM concentrations for a number of different years in Lahore, Pakistan (EPA, 1998; Wahid, *pers. comm.*).

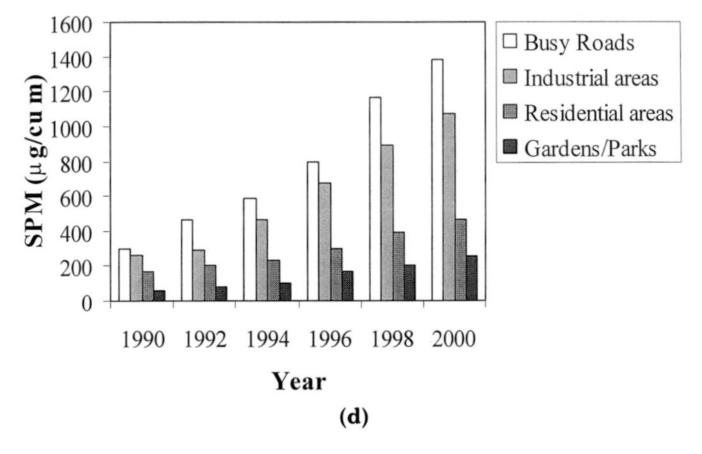

Fig. 3. (*Continued*)

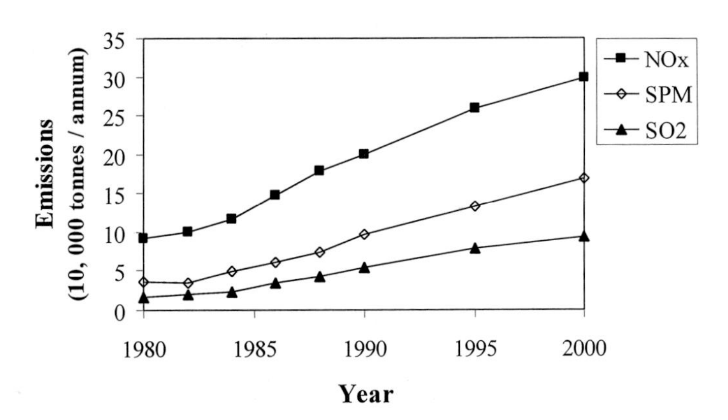

Fig. 4. Motor vehicle emissions trends between 1980 and 2000 in Karachi, Pakistan (WHO/UNEP, 1992; Wahid, *pers. comm.*).

Karachi is the largest city in Pakistan and is also experiencing rapid growth in its population, numbers of motor vehicles, industrial activity (especially oil refineries, metallurgical and power generation plants) and concomitant urban spread. Karachi's motor vehicle numbers are growing 2.5% faster than the human population with associated increases in emissions (Fig. 4); as a result of all these factors, the air quality of Karachi is rapidly deteriorating. In Karachi, 85% of SO_2 emissions are emitted by oil-fired power stations, NO_x is emitted in

Table 2. Emission values of NO_x, SO_2 and SPM by sector for 1989 in Karachi, Pakistan.

Sources	Emissions (tonnes/annum)		
	NO_x	*SO_2*	*SMP*
Power generation	19,000	65,100	1200
Industrial/commercial fuel consumption	600	50	360
Domestic fuel consumption	2500	3150	20,230
Industrial processes	10	4310	54,000
Traffic/transport	19,350	4200	7450
Waste burning	9000	200	23,000
Total	50,460	77,010	106,240

almost equal proportions from traffic/transport and power generation (approx. 38% each) with open waste burning being the third largest source of NO_x, contributing approx. 18% to the total. The main sources of anthropogenic emissions of SPM are industrial processes (approx. 51%), mainly from iron/steel/coke production and other construction/developmental activities (WHO/UNEP, 1992).

The annual emissions of various pollutants are summarised in Table 2. Only limited information is available regarding O_3 concentration trends in Karachi. Ghauri *et al.* (1991) reported diurnal maximum O_3 concentrations of 38–42 ppb at an inland site compared to 24 ppb at a coastal site in Karachi.

1.2. *Vegetation*

Pakistan is predominantly an agrarian country whose diverse climatic regions are capable of supporting many different types of vegetation and a wide variety of different ecosystems. Terrestrial ecosystems in Pakistan are classified into 18 distinct ecological zones with nine major vegetative zones. These range from the permanent snowfields and cold deserts of the mountainous north to the arid sub-tropical zones of the Sindh and Balachistan provinces; from the dry temperate coniferous forests of the inner Himalayas to the tropical deciduous forests of the Himalayan foothills, the steppe forests of the Suleiman Range and the thorn forests of the Indus plains; and from the swamps and riverine communities of the Indus and its tributaries to the mangrove forests of the Indus delta and Arabian Sea coast. The coast of Pakistan forms the northern boundary of the Arabian Sea. Coastal ecosystems include numerous deltas and estuaries with extensive inter-tidal mudflats and their associated wetlands, sandy beaches, rocky shores, mangroves and sea grasses. The seas of Pakistan are the richest in phytoplankton and zooplankton in the Arabian Sea Region. Over 5600 species of vascular plants have been recorded, including both native and introduced species. The families with the largest number of species are the Compositeae (649 spp.), Poaceae (597 spp.), Papilionaceae (437 spp.), Brassicaceae (250 spp.) and Cyperaceae (202 spp.). Among the lower plants, there are at least 189 pteridophytes (ferns and their allies).

Among the cereals, wheat and rice are the most important crops in terms of their area of cultivation, production and consumption. Punjab is the dominant agricultural province; wheat and rice are grown on thousands of hectares of cultivated land in this province, and comprise 72% and 55%, respectively of the total cultivated land area of the country. Punjab is also famous for the production of fine aromatic rice and exports more than one million tonnes of rice per annum, this represents 10% of the total world rice trade. Wheat grains are used as a main staple food in the country (Wahid *et al.*, 1995a–c; Wahid, 1999). Tobacco, cotton and potatoes are also important cash crops in Pakistan and are grown all over Punjab and Sindh. Soybean is regarded as a "miracle crop" due to its good quality seed oil and high quality protein. Although this crop is native to NWFP province it is also grown in Punjab (Wahid *et al.*, 1998b and 2001).

In recent years, due to the ever-increasing number of motor vehicles and industries in the mega-cities of Pakistan, air pollution has become a very serious problem. The harmful effects of air pollution on a number of different crop species have been extensively studied and reduction in productivity is dramatic (Maggs *et al.*, 1993 and 1995; Wahid *et al.*, 1994, 1995a–c, 1997a and b, 1998b, 2001 and 2001a–d; Nasim *et al.*, 1995; Bajwa *et al.*, 1997; Wahid, 1999; Shamsi *et al.*, 2000). The nutritional quality of bread is also altered due to atmospheric pollutants (Wahid *et al.*, 1998a). These studies have identified the Punjab province of Pakistan as an area within which vegetation is significantly at risk from air pollution, and one that is likely to face an increase in the potential threat of air pollution on agricultural production in the future. Vegetation types not classified as agricultural species, but also found on farmland, such as forests, scrub and planted trees, cover 4.2 million hectares of land or 4.8% of the country. However, no research has yet been carried out on the effects of air pollution on these important semi-natural species.

2. Impacts of Ambient Air Pollution on Plants

Ambient air pollution is an all pervasive phytotoxin. Chronic exposure of plants to pollutants under appropriate conditions can result in quite considerable invisible injury symptoms such as reductions in crop growth and yield. Acute exposure can result in visible injury such as chlorosis or necrosis of leaves of sensitive plant species (Kafiat *et al.*, 1994; Shamsi *et al.*, 2000).

2.1. *Visible Injury to Plants*

The effects of air pollution on plant growth and symptoms of visible injury have only been investigated in Pakistan in recent years. The large surface area of terrestrial plants acts as a natural sink for gaseous pollutants, since they are constantly exposed in the natural environment. The degree of injury to susceptible plants can be directly related to the pollutant concentration and duration, plant variety and species, climatic conditions and edaphic factors.

In Lahore, the distributions of O_3 and NO_x both in and around the city were monitored during 1993 and 1994. O_3 distributions were monitored using the Bel-W3 tobacco variety and measured using the potassium iodide method. NO_x concentrations were measured using NO_2 diffusion tubes (Kafiat *et al.*, 1994). Bel-W3 is a biomonitor that has been used across the world to indicate O_3 levels. The plant showed visible symptoms of O_3 damage under the climatic conditions of Pakistan. In this study, a consistent relationship between O_3 and NO_2 monitored at each site was apparent, with an inverse relationship between the two gases (Table 3). This can in part be explained by the

Table 3. Summary of measured pollutant concentrations and associated leaf injury indices obtained using Bel-W tobacco plants in Lahore, Pakistan.

Site type	Site description	Distance from city centre (km)	NO_2 (ppb)*	O_3 (ppb)†	Leaf injury index (%)‡
Urban	Rang Mahal (City centre)	0	58	56	5
	Engineering University	1	39	57	7
	Balal Ganj	1	36	60	7
Semi-urban	Cantonment	6	28	60	19
	Punjab University	7.5	27	70	19
Rural	Rakh Dera Chahl	30	8	75	27
	Bath village	31	8	72	20

*Weekly means; †6 h mean (1000–1600 h) once/week; ‡Leaf injury to leaves of cv. Bell-W3.

scavenging reaction between O_3 and nitric oxide (NO) where $NO + O_3 \Leftrightarrow NO_2 + O_2$.

2.2. *Experimental Research on Crops*

The ambient gaseous pollutants of greatest concern for plant growth and yield include O_3, SO_2 and NO_x. Open-top chambers have been used extensively in Europe and North America to study the effects of air pollutants on vegetation and they are particularly suited to developing countries due to their moderate cost, portability and ease of maintenance. In addition, studies carried out in Pakistan have found that the environmental conditions inside the chambers do not alter greatly compared to the ambient environment (Wahid *et al.*, 1995a–c; Wahid, 1999).

Extensive experimental studies have been carried out in Lahore using open-top chambers on local cultivars of wheat, rice, chickpea, mungbean and soybean crops since 1991. Crop growth was determined throughout the growth period during each experiment by weekly measurements of several vegetative growth parameters (viz. plant height, number of leaves, number of tillers/shoots and senescent leaves, etc.). In general, plants exposed to ambient air pollution have fewer

Fig. 5. The effect of air filtration on vegetative growth of Pakistan rice cv. Irri-6 grown in Lahore, Pakistan.

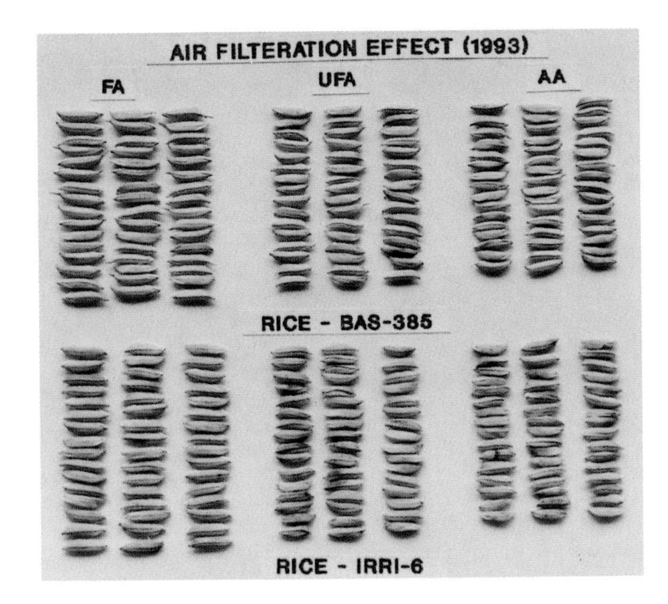

Fig. 6. The effect of air filtration on grain size of Pakistan rice varieties Bas-385 and Irri-6 grown in Lahore, Pakistan.

Fig. 7. The effect of filtration on Pakistan wheat cv. Chak-86 during the 1992–1993 growing season in Lahore, Pakistan.

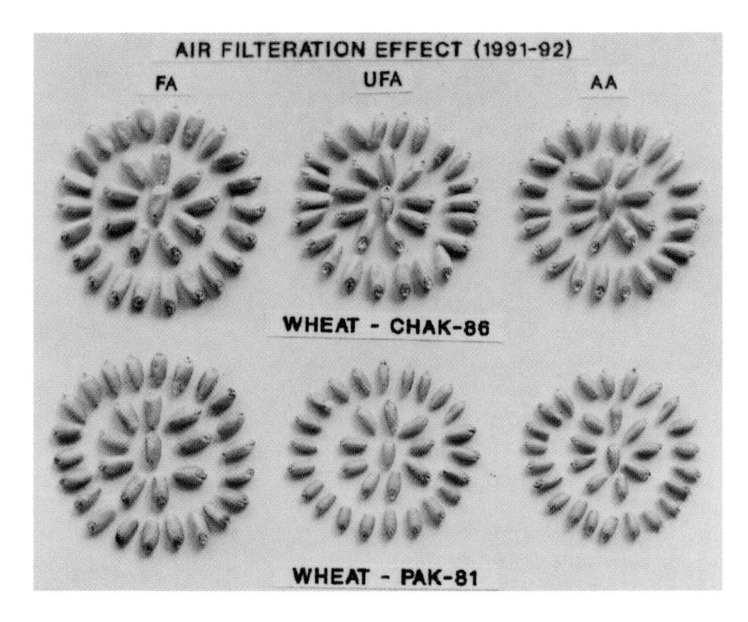

Fig. 8. The effect of air filtration on grain size of Pakistan wheat cv. chak-86 and Pak-81 grown in Lahore, Pakistan.

numbers of tillers/shoots, a lesser number of total leaves and experience an accelerated rate of leaf senescence compared to control plants grown in pollution-free air. A reduction in various reproductive parameters (viz. ears/spikelets/pods, seeds per ear/spikelets/pods, seed weight per plant and 1000 seed weight) in exposed plants could be attributed to the combined decrease in the above-mentioned growth parameters (see Figs. 5 to 8). Recent experimental studies that have recorded yield losses as a result of exposure to air pollution for several different crop species are summarised in Table 4.

More recently, investigations using the chemical "EDU" (N-[2-{2-oxo-1-imidazolidinyl}ethyl]-n2 phenylurea) have been carried out in Pakistan. EDU is known to have ozone protectant properties and has been widely used in North America and Western Europe to determine whether ozone is impacting on crops, both in terms of visible injury

Table 4. A summary of investigations of ambient air pollution effects for several agricultural crops of Pakistan investigated using open-top chambers.

Crops	Pollutants (ppb)		Yield losses (%)	References
	O_3*	NO_2†		
Wheat (6 varieties)	33–85	23–30	29–47	Maggs *et al.* (1993 and 1995); Wahid *et al.* (1994, 1995a and c, 1997a, 1998a and 2001a); Wahid (1999); Shamsi *et al.*
			(2000)	
Rice (5 varieties)	35–60	13–25	28–42	Wahid *et al.* (1994, 1995b, 1997a and 2001b); Maggs *et al.* (1995); Wahid (1999); Shamsi *et al.*
			(2000)	
Soybean (2 varieties)	64	29	37–46	Wahid *et al.* (1997b, 1998b and 2001)
Chickpea (3 varieties)	59	38	23–27	Wahid *et al.* (2001c)
Mungbean (2 varieties)	66	31	26–34	Wahid *et al.* (2001d)

*O_3 seasonal mean (6 h/day seasonal mean); †NO_2 seasonal mean (on weekly basis).

and growth reductions. Plants of soybean (*Glycine max* L.) were grown with and without the EDU treatment given as a soil drench at a suburban site, a rural site and a rural road site in the vicinity of the city of Lahore. The development and yield of the plants were determined in two experiments during 1996 and 1997. The first experiment was performed immediately after the monsoon and the other during the following spring. Concentrations of NO_2 and oxidant (O_3) were measured at each site. At the same time, an open-top chamber study, also investigating soybean, was being conducted at the suburban site. At the suburban site, the effect of EDU on soybean yield as compared to control plants (i.e. plants without EDU application) was similar to the effects of air-filtration on yield recorded within the open-top chamber study. At both the rural sites, the control plants (i.e. those without EDU) showed large yield reductions when compared with the EDU- treated plants and also when compared with the control plants at the suburban site (see Figs. 9 and 10). This effect was seen in both experiments (i.e. both post-monsoon and spring investigations), however the results showed that for all locations, the greatest yield reductions occurred during the pre-monsoon rather than the

Fig. 9. Synergistic visible injury symptoms to soybean cv. NARC-I resulting from road-side exposure to $O_3 + NO_2 + SO_2$ at a rural site in Lahore, Pakistan.

Fig. 10. Pakistan Soybean cv. NARC-I showing protective effect of EDU at a roadside rural site in Lahore, Pakistan.

Table 5. Effects of ambient ozone on soybean yield grown in field conditions in and around the city of Lahore, Pakistan as assessed using the EDU method.

Site	Distance from Punjab University (km)	Year	Pollutants (ppb)		Yield Losses (%)
			O_3*	NO_2†	
Punjab University, Lahore (Semi urban area)	0	1996‡	40	14	32
		1997¶	63	26	53
Kala Shah Kaku (Roadside Rural area)	28	1996	48	27	62
		1997	70	34	74
Rakh Dera Chahl Village (Remote rural area)	30	1996	49	2	49
		1997	75	1	64

‡Post-monsoon season; ¶Pre-monsoon season.
(Source: Wahid *et al.*, 1997b and 2001).

post-monsoon season. Oxidant concentrations were also greater at both of the rural sites compared with the suburban site. The results imply that ozone is a widespread pollutant and may be causing significant crop losses in rural areas around Lahore (Table 5).

Only very few investigations into the effects of air pollution on vegetation have been carried out in Pakistan. In addition to the studies described above, Ahmad *et al.* (1986) investigated pollutants from vehicle exhausts and industrial activities and found that certain plant species exhibited reductions in chlorophyll and protein contents when exposed to these pollutants as compared to plants grown under similar conditions at an unpolluted site. Another study (Iqbal, 1987), reported that growth of *Trifolium repens* L. at low and medium levels of sulphate and nitrate was increased slightly on exposure to the pollutants individually, with pollutant concentrations of SO_2 and NO_2 at 0.068 ppm. However, when the plants were exposed to these pollutant concentrations in combination, plant growth was significantly decreased. Qadir and Iqbal (1991) working at Karachi found significant reductions in various growth parameters of *Pongamia pinnata* (49.16%) and *Albezia lebbeck* (23.27%) grown from polluted seeds as compared to control. In a similar study (Shafiq and Iqbal, 1999), fruit length, breadth and weight were significantly reduced in some roadside trees (*Pongamia pinnata, Cassia siamea* and *Peltophorum pterocarpum*) growing along busy roads of Karachi. A recent study (Iqbal and Shafiq, 2001) carried out on some trees (*Carissa carandas, Azadirachta indica* and *Delonix regia*) have revealed significant reduction in plant cover, height and number of leaves due to heavy cement dust pollution in Karachi city.

3. Applicability of North American and European Air Quality Standards

A number of studies have been conducted on local cultivars of Pakistan crops. These studies are described in the following text and show the large variability in sensitivity to air pollution that exists both between different species but also between cultivars of the same species. These data emphasise one of the main difficulties in applying North America or European air quality guidelines or standards to Pakistan or other developing country regions. Namely, European or North American dose-response relationships may not protect the most sensitive species or cultivars growing in other regions around the world.

Table 6. Summary of experimental pollutant concentrations and exposure duration used for wheat and rice cultivar screening.

Pollutant	*Duration*	*Concentration*	*Assessment method*
(a) For wheat:			
Ozone	8 h/day for 7 days	82 nl l^{-1}	Visible damage
Sulphur dioxide	24 h/day for 3 days	330 nl l^{-1}	Visible damage
Nitrogen dioxide	24 h/day for 12 days	397 nl l^{-1}	Photosynthetic efficiency
Fluoride	24 h	25, 50 and 100 μl l^{-1}	Visible damage
(b) For rice:			
Ozone	8 h/day for 5 days	50, 100 and 200 μl l^{-1}	Visible damage
Sulphur dioxide	24 h/day for 5 days	100 and 200 μl l^{-1}	Visible damage

The sensitivities of locally grown Pakistan cultivars of wheat, rice, soybean, chickpea, *Trifolium* spp. and tobacco to O_3, SO_2, NO_2 and F pollutants were determined in closed chambers at Imperial College, London. This was achieved by fumigating seedlings with controlled concentrations of pollutants for lengths of time sufficient to produce visible symptoms (Table 6).

It is clear from Table 7 that the sensitivities of the different Pakistani wheat cultivars to each type of pollutant vary significantly. However, no clear association between the relative sensitivities to different pollutants were observed, making the identification of cultivar sensitivity (or tolerance) to a range of pollutants very difficult.

For rice, leaves were assessed in terms of the percentage area showing visible damage and the number of leaves falling into one of three damage classes: low (0–5%), medium (6–25%) and high (> 25%). IRRI-6 was chosen as a relatively sensitive cultivar to O_3 and Basmati-385 was chosen as a relatively resistant one (Table 8).

Table 7. Summary of Pakistani wheat cultivar sensitivities to major pollutants.

Wheat cultivar	Pollutants			
	Visible damage			*Fv/Fm*
	% Leaf length damaged			*% of FA*
	O_3	SO_2	F^-	NO_2
Kohi-Noor-83	28	76	19	100.5
Pari-73	41	74	33	99.7
Lyallpur-73	51	77	20	100.4
Chakwal-86	12	84	34	101.3
Rawal-87	19	87	8	100.9
Faisalabad-83	64	84	19	100.1
Punjab-81	89	92	32	99.7
Pak-81	60	84	18	100.9
Barani-83	92	90	28	99.9
Faisalabad-85	23	92	24	101.9
B.W.P.-79	*	96	15	101.1
Sandal	*	100	19	101.5
L.U.-26S	*	89	33	100.9
Blue Silver	*	100	25	101.4
Punjab-85	*	92	20	98.8

*Cultivars not screened for ozone sensitivities.

Table 8. Pakistani rice cultivar sensitivities to ozone and sulphur dioxide.

Rice cultivar	Pollutants	
	O_3	SO_2
	% leaves damage categories	
	High	*Medium + high*
Kashmir Basmati	0.28	2.17
Basmati-370	1.16	0.5
IRRI-6	6.28	3.52
Basmati-385	0	13.4
Basmati Pak	2.12	4.36
Basmati-198	0.03	11.23

Notes: Percentage leaves within each category are the means over all pollutant concentrations.

Commercial Pakistan tobacco cultivars, obtained from the Pakistan Tobacco Board, were screened for their sensitivities to ozone and comparisons were made with the American tobacco cultivar Bel-W3 (sensitive) and Bel-B (resistant). The results of screening (at 80 nl l^{-1} O_3 for eight hours per day for seven days) are shown in Table 9.

Other species that seem particularly sensitive to air pollution are those belonging to the leguminoseae family. Seeds of seven species belonging to this family were obtained from the Ayub Agricultural Research Institute, Faisalabad, and plants were fumigated with 80 nl l^{-1} O_3 for eight hours per day for five days. After fumigation, the number of leaves falling into several distinct classes were counted and expressed as a percentage of the total number of leaves on each plant. The results are summarised in Table 10. Results indicate a marked difference in the sensitivities of the species. *Medicago* and *Trifolium* spp. showed significantly higher levels of damage than *Cicer, Glycine* and *Lathyrus* spp.

Table 9. Ozone sensitivities of native Pakistani tobacco cultivars in comparison with Bel-W3 and Bel-B.

Tobacco cultivars	Damage class		
	Low (0–5%)	*Moderate (0–25%)*	*High (> 25%)*
Kango Baster	46	37	8
K399	78	14	6
MA2	81	19	0
Coker-206	91	9	0
Kakuba	84	16	0
ITB-2-12	52	28	15
KHG-11	34	37	27
Galpao	87	13	0
PBD6	43	19	28
NC13	49	45	19
Coker-48	97	3	0
Paraguay-49	85	15	0
TI-1112	33	34	30
Speight-G28	42	39	10
Coker-176	51	33	8
Bel-W3	20	32	45
Bel-B	85	13	1

Table 10. Ozone sensitivities of native Pakistani legume species.

Legume species	Cultivar	% Leaves falling within each class						
		I *(0%)*	*II* *(1–3%)*	*III* *(4–10%)*	*IV* *(11–25%)*	*V* *(26–50%)*	*VI* *(51–55%)*	*VII* *(> 75%)*
Medicago scutella	Sava	53	21	11	8	4	3	0
Medicago rugosa	Paraponta	40	20	10	9	8	9	5
Medicago trankatula	Cyprus	40	18	12	10	9	9	2
Trifolium alexandrianum	Comm.	41	10	12	16	13	2	6
Cicer arietinum	Noor-91	92	2	2	2	1	0	0
	Paidar-91	93	2	1	2	2	0	0
	Punjab-91	96	2	1	0	1	0	0
Glycine maximus	Bragg	100	0	0	0	0	0	0
	Crawford	100	0	0	0	0	0	0
	Williams	100	0	0	0	0	0	0
	Swat	100	0	0	0	0	0	0
Lathyrus odoratus	Comm.	100	0	0	0	0	0	0

After initial determination of their sensitivities to pollutant gases, the different cultivars from each experiment were grown under field conditions in Pakistan to determine the effects of ambient air pollution on their growth and yield. The different cultivars showed similar sensitivities under Pakistani field conditions as were recorded earlier under UK conditions

4. Future Research Priorities

In view of the work conducted to date investigating the impacts of air pollution on vegetation in Pakistan, future research priorities should involve the following:

1. Open-top chamber experiments and EDU experiments need to be conducted at more rural locations in Punjab to ascertain the real effects of air pollution on a wide range of crop types.
2. A fumigation facility in Pakistan would be useful both in evaluating the sensitivities of a wide range of plants and also to provide a training facility for future researchers in this field.
3. Additional monitoring of the distribution of O_3, NO_2, SO_2 and F_l pollutant concentrations both in urban and rural areas proximal to the major cities of Pakistan and in agricultural areas located in more remote rural areas is urgently required. Such monitoring could be achieved using physico-chemical techniques and by usage of indicator plants to help predict crop yield losses.
4. Continuous air pollution monitors would aid understanding of pollutant concentration build-up in urban as well as rural locations and would be helpful in establishing air quality standards.

5. Conclusions

The results of the open-top chamber and EDU experiments, along with the monitoring of pollutants in and around the city of Lahore, have clearly shown that ambient levels of ozone and nitrogen dioxide are increasing rapidly and impacting on crop growth and yield. On the basis of extensive studies on many important crops in Pakistan, it has been found that drastic yield reductions (30–50%) in staple crops can be attributed to the air pollution in peri-urban and rural areas. Immediate action is necessary to mediate these effects. Important crops should be screened against noxious pollutants through air-filtration or fumigation experiments, resistant species and varieties can then recommended to farmers in an attempt to enhance yields. Unfortunately, the understanding of the adverse effects of ambient air pollution on fauna and flora is still little known amongst agricultural policy-makers of Pakistan due to lack of environmental education and awareness.

Acknowledgements

I wish to thank the generous financial support of the European Commission via the Directorate General XII Scientific Cooperation Programme, Brussels, through two research contracts (CI1-CT90-0789 and CI1-CT90-0865), and UK Department for International Development Research Project No. R6289, for making successful investigations on the impacts of air pollution on Pakistani crops; the results of which became a part of this report. It is also pleasure to pay my deep sense of gratitude to my worthy teachers: Professor S.R.A. Shamsi (Punjab University, Pakistan), Professor J.N.B. Bell (Imperial College, London) and Professor M.R. Ashmore (University of Bradford) for their inspiring guidance and continuous encouragement throughout the course of these studies. Finally, I am deeply indebted to my worthy parents and Mrs. Anne Elliott (Biology Department, Imperial College, London) for all their time, care and love.

References

Ahmad S., Ismail F. and Majid J. (1986) Effect of atmospheric pollution on chlorophyll and protein contents of some plants growing in Karachi region, *Pakistan J Sci. Indus. Res.* **29**(6), 464–467.

Anonymous (2000) Federal Bureau of Statistics. Economic and Urban Advisory Wing, Government of Pakistan, Islamabad.

Bajwa R., Ahmad S., Uzma M., Nasim G. and Wahid A. (1997) Impact of air pollution on mungbean, *Vigna radiata* (L.) grown in open top chamber system in Pakistan. I. Effects on vegetative growth and yield, *Sci. Khyber* **10**(2), 37–50.

EPA (1998) State of air pollution in Lahore. In *Environmental Pollution and EPA*. Environmental Protection Agency, Government of the Punjab, Lahore, Pakistan, pp. 1–9.

Ghauri B.M.K., Salam M. and Mirza M.I. (1991) Surface ozone in Karachi. In *Ozone Depletion: Implications for the Tropics*, ed. Illas M. Uni. Sci. Malaysia and UNEP, pp. 169–177.

Iqbal M.Z. (1987) Growth of white clover (*Trifolium repens*, L.) exposed to low concentrations of SO_2 and NO_2 at different levels of sulphate and nitrate nutrients, *Pakistan J. Sci. Indus. Res.* **30**(3), 221–223.

Iqbal M.Z. and Shafiq M. (2001) Periodical effect of cement dust pollution on the growth of some plant species, *Turk. J. Bot.* **25**, 19–24.

Kafiat U., Shamsi S.R.A., Maggs R., Ashmore M.R. and van der Eerden L. (1994) *Air Pollution Monitoring in the Punjab Using Plants as Bionindicators and Static Physico-Chemical Techniques.* Final Report, European Commission Research Project No. CI1-CT90-0865, Brussels. University of the Punjab, Lahore and Imperial College, London.

Maggs R., Wahid A., Ashmore M.R. and Shamsi S.R.A. (1993) The impact of ambient air pollution on wheat in the Punjab, Pakistan. In *Effects of Air Pollution on Agricultural Crops in Europe*, eds. Jager *et al.* Air Pollution Research Report 46, European Commission, Brussels, pp. 571–574.

Maggs R., Wahid A., Shamsi S.R.A. and Ashmore M.R. (1995) Effects of ambient air pollution on wheat and rice yield in Pakistan, *Water Air Soil Pollut.* **85**, 1311–1316.

Nasim G., Saeed S, Wahid A. and Bajwa R. (1995) Impact of air pollution on growth, yield and vesicular arbuscular mycorrhizal status of wheat, *Triticum aestivum* var. Pak-81. *Biota* **1**(1/2), 91–111.

Qadir N. and Iqbal M.Z. (1991) Growth of some plants raised from polluted and unpolluted seeds, *Int. J. Env. Studies* **39**, 95–99.

Shafiq M. and Iqbal M.Z. (1999) Effects of autovehicular exhaust on pods and seeds of some roadside trees, *ECOPRINT* **6**(1), 35–40.

Shamsi S.R.A., Bell J.N.B., Ashmore M.R., Maggs R., Kafiat U. and Wahid A. (2000) The impacts of air pollution on crops in developing countries — a case study in Pakistan. In *Pollution Stress: Indication, Mitigation and Conservation*, eds. Yunus M. *et al.* Kluwer Academic Press, The Netherlands.

Wahid A. *Personal Communication.* Environmental Pollution Research Laboratory, Department of Botany, Government College, Lahore, Pakistan.

Wahid A., Maggs R., Sahmsi S.R.A., Bell J.N.B. and Ashmore M.R. (1994) *The Impact of Ambient Air Pollution on Agriculture Crops in Punjab.* Final Report, European Commission Research. Project No. CI1-CT90-0789, Brussels. Punjab University, Lahore and Imperial College, London.

Wahid A., Maggs R., Shamsi S.R.A., Bell J.N.B. and Ashmore M.R. (1995a) Air pollution and its impacts on wheat yield in the Pakistan, Punjab, *Env. Pollut.* **88**, 147–154.

Wahid A., Maggs R., Shamsi S.R.A., Bell J.N.B. and Ashmore M.R. (1995b) Effects of air pollution on rice yield in the Pakistan Punjab, *Env. Pollut.* **90**(3), 323–329.

Wahid A., Maggs R., Shamsi S.R.A., Bell J.N.B. and Ashmore M.R. (1995c) Effects of air pollution on the growth and yield of some wheat varieties grown in open-top chambers in the Punjab, Pakistan. In *Proceedings of the 4th International Symposium/Workshop on the Applications of Molecular Biological Research in Agriculture, Health and Environment*, ed. Riazuddin S. National Centre of Excellence in Molecular Biology, University of the Punjab, Lahore, Pakistan, pp. 161–168.

Wahid A., Shamsi S.R.A., Bell J.N.B. and Ashmore M.R. (1997a) Effects of ambient air pollution on the yield of some wheat and rice varieties grown in open-top chambers in Lahore, Pakistan. *Acta Scientia* **7**(2), 141–152.

Wahid A., Milne E., Marshall F.M., Shamsi S.R.A., Ashmore M.R. and Bell J.N.B. (1997b) *The Impacts and Costs of Air Pollution on Agricultural Crops in the Developing Countries.* Final Report, UK Department for International Development Research Project No. R6289. Punjab University, Lahore and Imperial College, London.

Wahid A., Kafiat U. and Shamsi S.R.A. (1998a) Effects of atmospheric pollution on the nutritional quality of wheat grains in Lahore, *Acta Scientia* **8**(1/2), 115–130.

Wahid A., Nasir M.G.A., Anwar M.H. and Shamsi S.R.A. (1998b) *Impact of Ambient Air Pollution on Growth and Yield of Soybean in Lahore.* Final Report. National Research Project, Punjab University, Lahore, Pakistan.

Wahid A. (1999) *Studies on the Impacts of Ambient Air Pollution on the Growth and Yield of Wheat and Rice.* Ph.D. thesis, University of the Punjab, Pakistan and Imperial College, London.

Wahid A. and Marshall F.M. (2000) A preliminary assessment of street seller's exposure to carbon monoxide in Lahore, *Acta Scientia* **10**(1), 1–15.

Wahid A., Milne E., Shamsi S.R.A., Marshall F.M. and Ashmore M.R. (2001) Effects of oxidants on soybean growth and yield in the Pakistan Punjab, *Env. Pollut.* **113**(3), 271–280.

Wahid A., Bell J.N.B. and Marshall F.M. (2001a) The growth and yield responses of four wheat cultivars of Pakistan to ambient air pollution, *Env. Pollut.* (submitted).

Wahid A., Bell J.N.B. and Marshall F.M. (2001b) Growth and yield losses in three cultivars of rice due to O_3 and NO_2 pollution in Pakistan, *Env. Pollut.* (submitted).

Wahid A., Bell J.N.B., Ahmed S. and Marshall F.M. (2001c) Evidence of yield reduction in chickpea (*Cicer arietinum*) due to gaseous pollutants in Pakistan (in preparation).

Wahid A., Ahmed S., Bell J.N.B. and Marshall F.M. (2001d) Air pollution-induced losses in productivity in some mungbean (*Vigna radiata*) varieties of Pakistan (in preparation).

WHO/UNEP (1992) *Urban Air Pollution in Megacities of the World.* Blackwell Publishers, Oxford, pp. 114–123.

CHAPTER 10

AIR POLLUTION AND VEGETATION IN EGYPT: A REVIEW

N.M. Abdel-Latif

1. Introduction

The rapid increase in air pollutant emissions poses an increasing threat to agriculture in Egypt. In the following sections, data are presented showing that pollutant concentrations in many rural areas around greater Cairo and within the Nile Delta region, are in exceedance of the phytotoxic levels set to protect vegetation in developed countries. Field evidence indicates that air pollutants can occur in peri-urban and rural areas in Egypt at concentrations sufficient to damage sensitive crops. However, the impact of air pollution on agriculture has not yet been fully assessed due to the limitation of data sets and experimental facilities. Furthermore, air pollution monitoring has tended to focus on urban areas; as such there is an urgent need for monitoring to be performed on a regular basis in rural areas. The acquisition of such air pollution concentration data and the establishment of air quality guidelines, developed in consideration of local cropping patterns and climatic conditions, would enable assessments of current and future impacts of air pollution on Egyptian agriculture.

The following sections do not make any reference to the impacts of air pollution on Egyptian tree species. This is primarily due to the fact that hardly any forested areas actually exist in Egypt; the largest tree communities are the date-palm trees located in oases of the Western desert and along the north coast of Sinai. These areas are far away from pollution sources and therefore not to be considered at risk from damage by air pollutants. However, numerous tree species are scattered throughout agricultural lands and along the Nile as well as in towns and cities across Egypt, including many rare species and old trees in the public gardens of Cairo and Alexandria. To date, no known observations or experimental investigations have been recorded or performed to indicate whether ambient pollution levels are impacting on these trees.

1.1. *Emissions*

The rapid increase in urbanisation and industrial development, as well as the increase in motorised vehicular traffic, has resulted in a significant decrease in air quality in Egypt. The high rates of consumption of fossil fuel necessitated by these activities, coupled with only limited efforts to control emissions, have resulted in increased atmospheric loads of air pollutants, particularly in urban areas. These have been found to cause deleterious effects on Egyptian vegetation and ecosystems around large cities, such as Cairo and Alexandria, and along the main highways across the Delta region. Egypt is a populous country and has a rapid annual growth rate. According to the 1996 census, the population in Egypt has increased dramatically to 61.4 million with an annual growth rate of 2.1% (SIS, 1998). Moreover, the urban population, as a percentage of the total population, was estimated to increase from 45% in 1995 to 54% in 2015, with an annual growth rate of 2.5% (UNDP, 1998).

The main Egyptian industries are cement, iron and steel, coke, textiles, motor vehicle manufacturing, chemical-fertiliser production, refrigerator manufacturing, oil and food processing (in the greater Cairo area); petrochemical, chloralkali, paper and urea fertiliser production (in the Alexandria area), and textiles, chemical, fertiliser and pesticide production (in the Delta region). Most of the industrial activities in greater Cairo are scattered within or near to agricultural lands, which are the main source of the vegetable produce supplied to Cairo. Cairo, being the capital of Egypt, is the busiest urban, commercial and industrial centre in Africa and has been the main location

Table 1. Annual fuel use in greater Cairo (10^3 tonnes).

	90/91	91/92	92/93	93/94	94/95	95/96
Petrol	1016	989	944	945	996	1021
Diesel	1115	1139	1154	1222	1323	1446
Natural gas	6127	6921	6945	8735	9228	9896

Source: Shakour *et al.* (1998).

for the vast urban-industrial-vehicular expansion that has occurred since the 1960s. Cairo is responsible for over half of Egypt's energy consumption (WHO/UNEP, 1992), with Table 1 showing how the consumption of a number of different fuels has increased over recent years.

In greater Cairo the number of vehicles increased by 10% per annum between 1980 and 1990 (Güsten *et al.*, 1994), with almost one million vehicles registered in this area in 1990/1991 (WHO/UNEP, 1992; Güsten *et al.*, 1994).

Over the past 40 years, the Air Pollution Department of the National Research Centre (NRC) has been responsible for a number of projects and research studies that have been conducted to monitor atmospheric pollutants and their effects on the environment in Egypt (e.g. Abdel Salam, 1963; Abdel Salam and Sowelim, 1967a and b; Hindy, 1973; Shakour, 1982 and 1992; Ali and Nasralla, 1986; Hindy *et al.*, 1988a and b; Ibrahim, 1992). In addition, since the 1980s, long-term air pollution monitoring programmes have been performed in Egypt by the Ministry of Health (MoH), in co-operation with the World Health Organization (WHO), and the Egyptian Environmental Affairs Authority (EEAA), in co-operation with Egyptian universities and research centres and many international agencies. Unfortunately, no published data describing the state of air quality over recent years are available from the MoH or the EEAA programmes. However, it has been possible to collate air quality data from studies conducted in the Air Pollution Department of the NRC as well as from other Egyptian sources. Tables 2 and 3 show monitored concentrations of what are considered to be the major pollutants in

Table 2. Annual and seasonal concentrations of gaseous pollutants in Egypt.

Year/Month	Site description	ppb	Source
	Sulphur Dioxide (SO_2)		
Greater Cairo			
Dec'82–Nov'84	Industrial	30	Hindy *et al.* (1988a)
	Rural	20	
Sep'83–Aug'84	Urban	90	Nasralla *et al.* (1986)
Nov'87–Jan'88	Industrial (cultivated lands)	60	Ali (1993)
	Industrial (cultivated lands)	34	
May'89–Apr'90	Residential	6	Shakour (1992)
Jul'91–Dec'91	Residential (new centre)	7	Shakour and El-Taieb (1992)
	Residential (new centre)	9	
	Residential (populated centre)	16	
Jul'93–Dec'95	Residential	14	Shakour and Zakey (1998)
Sep'94–Aug'95	Industrial	44	Shakour *et al.* (1998)
Alexandria			
Oct'94–Nov'97	Urban (city centre)	105	Hassan (1999)
	Sub-urban	35	
	Residential	16	
	Nitrogen Oxides (NO_x)		
Greater Cairo			
Nov'87–Jan'88	Industrial (cultivated lands)	72	Ali (1993)
	Industrial (cultivated lands)	42	
May'89–Feb'90	Residential (new centre)	42	Shakour (1993)
	Residential (new centre)	21 (NO_2)	
Jul'91–Dec'91	Residential (new centre)	22 (NO_2)	Shakour and El-Taieb (1992)
	Residential (new centre)	25 (NO_2)	
	Residential (populated centre)	41 (NO_2)	

Table 2. (*Continued*)

Year/Month	Site description	ppb	Source
Sep'94–Aug'95	Industrial	84 (NO_2)	Shakour *et al.* (1998)
Alexandria			
Oct'94–Nov'97	Urban (city centre)	194	Hassan (1999)
	Sub-urban	45	
	Residential	9	
	Ozone (O_3)		
Greater Cairo			
Mar'91	Industrial (upwind)	41 (8–18 h)	Güsten *et al.* (1994)
	Industrial (downwind)	50 (8–18 h)	
	Residential	60 (8–18 h)	
	Residential	67 (8–18 h)	
	Residential	62 (8–18 h)	
Dec'91–Sep'95	Industrial (summer)	78 (½ h)	Farag *et al.* (1993)
	Urban (summer)	65 (½ h)	
	Residential (summer)	56 (½ h)	
Alexandria			
Oct'94–Nov'97	Urban (city centre)	155 (9–17 h)	Hassan (1999)
	Sub-urban	77 (9–17 h)	
	Residential	100 (9–17 h)	

Egypt, i.e. sulphur dioxide (SO_2), nitrogen oxides (NO_x), ozone (O_3), and smoke and suspended particulate matter (SPM). Industrial and urban activities in the greater Cairo and Alexandria areas are the main sources of these pollutants.

Table 2 shows that high concentrations of all the major pollutants have been recorded over the last two decades. The concentrations of SO_2 in the atmosphere of industrial and urban areas, both in Greater Cairo and Alexandria, are higher than those recorded in residential areas. This would indicate that the combustion of fuel used in different methods of transportation is a major source of SO_2 in Egypt. Concentrations of NO_x recorded in many residential areas are as high as those recorded in industrial areas within Greater Cairo.

Table 3. Annual and seasonal concentrations of smoke and SPM in Egypt.

Year/Month	Site description	Smoke ($\mu g\ m^{-3}$)	SPM ($\mu g\ m^{-3}$)	Source
Jun'77–May'78	Industrial	127	–	Abdel-Salam *et al.* (1981)
	Residential	62	–	
Dec'82–Nov'84	Industrial (1983)	100	567	Hindy *et al.* (1990)
	Residential (1983)	74	394	
	Industrial (1984)	89	475	
	Residential (1984)	76	316	
Sep'83–Aug'84	Urban (city centre)	240	–	Nasralla *et al.* (1986)
	Residential (moderate)	104	–	
Nov'87–Jan'88	Industrial (cultivated lands)	–	680	Ali (1993)
		–	528	Ibrahim (1992)
1989	Industrial/Residential	89	1036	
May'89–Apr'90	Residential	–	331	Shakour (1992)
Dec'89–Feb'90	Residential (new centre)	–	195	Shakour and Hindy (1992)
1989	Residential (moderate)	86	602	Abdel-Latif (1993)
	Residential (upwind)	99	548	
	Urban (commercial)	100	633	
	Urban (city centre)	180	700	
1990	Residential (moderate)	–	579	Abdel-Latif (1993)
	Residential (upwind)	–	374	
	Urban (commercial)	–	515	
	Urban (city centre)	–	651	
1994	Residential	–	366	Shakour and El-Taieb (1995)
1992–1997	Residential	–	625	Hassanien *et al.* (pers. comm.)

The reported concentrations of O_3 over the last decade clearly show that O_3 levels in many residential areas exceed concentrations in industrial areas, confirming the regional nature of this pollutant under Egyptian conditions. Table 2 also shows that concentrations of

SO_2, NO_x and O_3 in the urban centre of Alexandria are significantly higher than those found in greater Cairo. Table 3 shows that high concentrations of smoke and SPM have been recorded over the last few decades in Egypt, and that concentrations of these pollutants have been steadily increasing since the early 1980s.

1.2. *Vegetation*

There is no doubt that food supply is a growing problem in Third World countries resulting predominantly from the rapid increase in population size. Since the 1980s, policy makers in Egypt have given special attention to the agricultural sector in order to meet the food requirements of the coming century, with policies focussing on optimising the utilisation of all available agricultural resources. The government has already begun to implement many large national agricultural projects, which are intended to increase real annual development in this sector by 4.1%, mainly through land reclamation and cultivation (SIS, 1998).

Egyptian agriculture is characterised by the limited cultivable area located predominantly along the River Nile, which owing to very low rainfall, represents almost the only source of irrigation water in Egypt. The cultivable area accounts for 3.7% of the total area of Egypt (1 million km^2) at 0.06 ha per capita (FAO, 1997). Agriculture in Egypt can be considered one of the most intensive agricultural systems in the world in terms of production per unit area, with land being cultivated two or three times a year (Abdel Hakim and Aboumandour, 1994). For example, in 1990 the area under crops was 2,607,000 ha while the area harvested was 4,861,000 ha, with an intensification rate of 1.86 (Abdel Hakim and Aboumandour, 1994). The main growing seasons in Egypt are during the winter (November to May), summer (March/April to September) and Nili (May to October). The crops listed in Table 4 represent the main agricultural crops grown in Egypt for staple foods and for industry. According to the area occupied by each group of crops, cereals and fodder crops occupied the largest part of the area harvested in the period 1970 to 1991, followed by fibre plants (mainly cotton), and vegetables (Abdel Hakim and Aboumandour, 1994). The area occupied by cereals, given as a

Table 4. Main agricultural crops in Egypt.

Cereals	Fodder	Legumes	Oil plants	Industrial	Vegetables	Fruit
Wheat (W) Maize (S; N) Rice (S; N)	Berseem/ Clover (W)	Beans (W) Lentil (W)	Soya (S) Sesame (S)	Cotton (S) Sugar cane (S)	Tomato (All) Potato (All) Onion (All)	Orange Grapes Mango

S = Summer; W = Winter; N = Nili; All = All seasons.

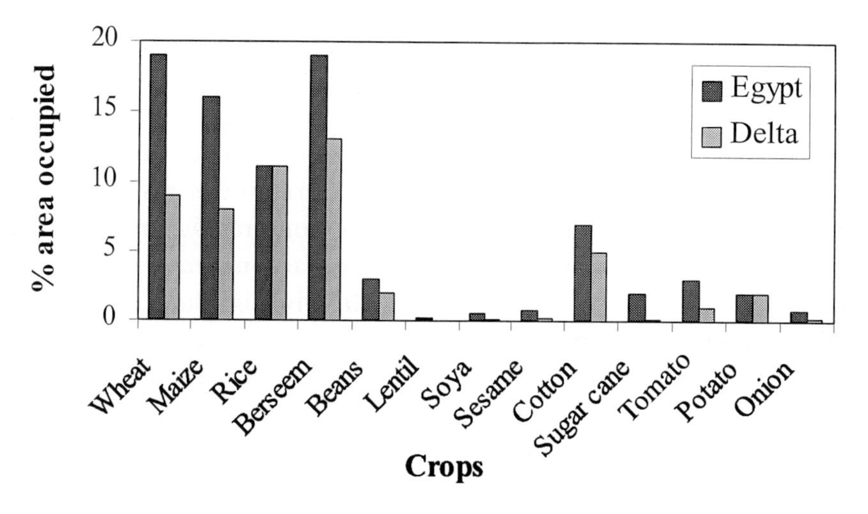

Fig. 1. The percentage land area occupied by important crops in the Delta region as compared to Egypt as a whole for the year 1996 (source: MALR, 1996).

percentage of the total harvested area, has increased from 41.8% in 1970/74 to 46.2% in 1990/91. According to statistics for 1996 provided by the Ministry of Agriculture, the crops listed in Table 4, with the exception of fruits, occupied more than 80% of the total cultivated area over different agricultural seasons (MALR, 1996).

The major area of cultivated land in Egypt is located within the Nile Delta which consists of 13 Governorates. Figure 1 shows the percentage land area occupied by specific crops in this Delta region as compared to Egypt as a whole. Many agricultural rural areas in this Delta region are located close to roads and/or near industrial activities and consequently experience high levels of air pollution

(Ali, 1993; Nasralla and Ali, 1984). The rural agricultural areas of greater Cairo (found within the Cairo, Giza and Qalubia Governorates) suffer particularly low air quality as a result of widespread industrial activities that occur within these regions. These rural areas of greater Cairo and the nearby governorates are especially important since they are the main sources of vegetables for residents of the cities of Cairo and Giza. In addition, the prevailing local climatic conditions of these regions can lead to the transport of pollution up to several kilometres from the industrial and urban centres. This results in the occurrence of air pollution episodes in rural areas located significant distances from the greater Cairo region (Hindy, 1991).

To encourage geographical re-distribution of the population and economic development of different activities, several new economic-based cities were established during the 1980s and 1990s. The first residential/industrial cities were built in the desert area around greater Cairo. The growth of industrial activities in these new communities progressed fairly rapidly with electronic, textile and ceramic industries being amongst the typical activities. Further development in these cities is expected in the future. It is understood that these cities and their activities were environmentally planned according to policy developed to protect the Egyptian environment against hazards; however, no information is available about air quality in and around these new communities.

2. Impacts of Air Pollution

2.1. *Field Evidence*

Most of the studies concerned with assessing the effects of air quality on vegetation in Egypt have been performed over the last 20 years. These have tended to investigate the impacts of the ambient air pollution associated with the greater Cairo area and the main roads within the Nile Delta region. Metal accumulation and subsequent effects on plant growth provided the main focus of many of these studies, in response to the high concentrations of heavy metals (in SPM) recorded in the atmosphere during this period (see Table 3). Investigations that have been performed under field conditions have shown that for many areas within greater Cairo, the atmospheric pollution load is causing damage to Egyptian vegetation. The main effects that have been observed are the appearance of visible injury symptoms and biomass reductions.

Among these studies, Ali (1993) examined the effects of industrial air pollution on the growth of clover/berseem (*Trifolium pratense*) and Egyptian mallow (*Malva parviflora*) between November 1987 and January 1988. Transplanted clover and Egyptian mallow plants were distributed at two heavily polluted sites (sites I and II), both in agricultural areas close to industrial sites in the Shoubra El-Khaima area (where industrial activities include ferrous, non-ferrous, ceramic and textile industries and a thermal power station). A third control site was located at a rural area 16 to 17 km upwind of this industrial area. During the study, high concentrations of different pollutants were recorded in the atmosphere of the Shoubra El-Khaima industrial area (Table 5). Furthermore, concentrations of more than 100 ppb O_3 (hourly means) were recorded in this area during the study period in a separate investigation reported by Nasralla (1991).

Table 5. Concentrations of air pollutants at Shoubra El-Khaima recorded between Nov. '87 and Jan. '88.

	SO_2 (ppb)	NO_x (ppb)	TSP ($\mu g/m^3$)	Pb ($\mu g/m^3$)	Cd ($\mu g/m^3$)
Site I	60	72	680	1.8	0.08
Site II	34	42	528	0.8	0.04

The main visible injuries that were observed on the leaves of plants grown at the two polluted sites were necrosis, red spots and chlorosis; 60% and 19% of clover leaves and 54% and 7% of Egyptian mallow leaves were injured at site I and site II, respectively. Comparisons of specific plant characteristics were made between plants grown at the upwind rural site and those at the two polluted sites. Results clearly indicated that total chlorophyll content, leaf numbers and dry weight (biomass) were reduced in plants growing at the polluted sites as a consequence of the poor air quality of the industrial area (Table 6).

Table 6. Percentage reductions in the growth of plants exposed at Shoubra El-Khaima compared to control plants grown at an unpolluted site.

	Total chlorophyll		Leaf number		Shoot dry weight	
	Site I	Site II	Site I	Site II	Site I	Site II
Clover	65	12	50	9	68	56
Mallow	29	13	0	0	54	31

Table 7. Concentrations of TSP and total sulphation (T.S.) in Cairo from Jan '90 to Apr '90.

Sites	TSP ($\mu g\ m^{-3}$)	T.S. (mg SO_2 100 cm^{-1} 30 $days^{-1}$)
Site 1 (suburban)	Not measured	15
Site 2 (residential; downwind)	564	17
Site 3 (residential; commercial)	549	20
Site 4 (city centre)	644	22

In addition to these effects, clover and Egyptian mallow plants grown at the polluted site accumulated high concentrations of Pb and Cd in their vegetative parts. The author attributed the loss in the productivity of the investigated species to the environmental pollution that was prevalent over the agricultural land of the Shoubra El-Khaima industrial area.

To examine the impact of the different air pollutants in Cairo on plant growth, lettuce (*Lactuca sativa*) and barley (*Hordeum vulgare*) plants were transplanted and exposed to ambient air at four sites of differing air quality in Cairo city (Abdel-Latif, 1993). The pollutants monitored at these sites during the exposure period (January to April 1990) were total suspended particulates (TSP) and SO_2 (measured as total sulphation), the results are shown in Table 7. Total sulphation is a technique used to determine SO_2 concentrations in ambient air over a prolonged period of time, and depends on the reaction of SO_2 with lead dioxide (PbO_2) to form lead sulphate (ASTM, 1987).

Chlorosis and necrotic lesions were the main visible symptoms observed on damaged leaves at all sites. Moreover, there were reductions in dry weight, total chlorophyll content, total carbohydrates, and total nitrogen contents in barley and lettuce plants exposed at sites 2, 3 and 4 compared to sites 1, the less polluted site. In addition, the exposed plants at sites 3 and 4 accumulated high concentrations of sulphur, Pb, Cd and other heavy metals in their tissues.

Hassan *et al.* (1995) assessed the impact of ozone on the growth and yield of local varieties of radish (*Raphanus sativus* L. Cv. Balady) and turnip (*Brassica rapa* L. Cv. Sultani) at suburban and rural sites in the Alexandria area using the anti-ozonant *N-[2-(2-oxo-1-imidazolidinyl) ethyl]-N'phenylurea* (EDU). The mean six-hour concentration of oxidants over the experimental period, estimated using the neutral buffered KI method, was 54.8 ppb at the suburban site and 66.9 ppb at the rural site. The main pollutant effects were the appearance of visible injury symptoms (chlorotic spots) on the upper leaf surface of exposed plants and reductions in plant biomass. These effects were recorded for radish at both sites; and for turnip at the rural site only. Application of EDU increased the growth of radish more than turnip at the rural site, while the growth of radish only was increased by EDU at the other site. This study proved that levels of ambient O_3 in Egypt are high enough to have significant impacts on the growth and yield of local varieties of vegetable crops, even at a time of year when O_3 levels are relatively low.

2.2. *Experimental Research*

Due to the lack of experimental equipment and facilities required for fumigation investigations, to date only a few studies have examined the growth and physiological responses of different Egyptian plant species to pollutants under controlled conditions. For example, in the case of ozone, it has only been in recent years that the sensitivities of different Egyptian species and varieties have been investigated. Studies conducted by Hassan *et al.* (1994 and 1995) assessed the impact of ozone, at realistic concentrations for Egypt, on cultivars of two vegetable crops using exposure facilities at Imperial College in the United Kingdom.

Exposure to 80 ppb O_3 for eight days, in closed fumigation chambers, significantly reduced the photosynthetic rate of Egyptian varieties of radish (*Raphanus sativus* L. Cv. Balady) and turnip (*Brassica rapa* L. Cv. Sultani) plants by 28% and 15%, respectively (Hassan *et al.*, 1994). Under ozone stress, 40% and 23% reductions in the yield of radish and turnip were found, respectively. The authors also reported that stomatal conductance of radish increased while that of turnip decreased due to fumigation with O_3. Exposure of these varieties to O_3 in another experiment using open-top chambers (OTCs) (Hassan *et al.*, 1995), verified the results of the closed chamber fumigation experiment and proved that levels of ambient O_3 in Egypt are high enough to affect the growth and yield of local vegetation.

More recently, Hassan (1998) exposed Egyptian varieties of wheat (*Triticum aestivum* L.), tomato (*Lycopersicon esculentum* Mill) and broad bean (*Vicia faba*) to 68 ppb O_3 (8-h d^{-1}) in semi-open top chambers from the early growth seedling stage. Severe foliar injury symptoms developed over leaves of all crop types in the form of chlorotic stippling and necrotic spots, with a significant increase in the number of injured leaves with increasing cumulative exposure. The rates of photosynthesis were reduced following O_3 exposure in bean and tomato plants by 26% each and by 16% in wheat, while the effect of O_3 on stomatal conductance was very variable in the three species. Furthermore, overall growth and yield of the exposed crops were reduced.

Egyptian grown varieties of tomato (*Lycopersicon esculentum* Mill) were exposed to O_3 concentrations, within the range recorded in Egypt, in open-top chambers (OTCs) at the Federal Agricultural Research Centre (FAL) in Braunschweig, Germany (Hassan *et al.*, 1998 and 1999). The O_3 impacts observed included reductions in photosynthetic rate, stomatal conductance and chlorophyll content. In addition, the number and weight of fruits of the exposed plants were found to decrease by 23% and 44%, respectively (Hassan *et al.*, 1998 and 1999).

A widely grown summer vegetable, Jew's mallow (*Corchorus olitorius*), was exposed to 80 ppb O_3 in a recent study at Imperial College in the United Kingdom. Fumigation with ozone caused a large reduction in leaf stem and root biomass, in addition to foliar necrosis and accelerated leaf senescence during the first growth cycle. The re-growth of cut plants was inhibited during the re-growth cycle, even in the absence of O_3. Moreover, the effect of ozone on Jew's mallow was greater when plants were exposed from germination, rather than from seedling establishment (Abdel-Latif, 2001).

Deposited particulate matter over plant leaves is known to affect plant growth, not only by reducing gaseous exchange and light reaching the leaves, but also because of its metallic constituents, such as Pb and Cd, which have harmful effects on many physiological processes. In an experimental study to examine the effect of Cd on an Egyptian variety of broad bean (*Vicia faba*), Ibrahim (1992) treated experimental plants with different levels of Cd (10, 25 and 100 mg l^{-1}) either through irrigation or spraying over leaves. The recorded results showed that total chlorophyll, total carbohydrate, nitrogen content and dry weight of treated pants were reduced compared to control plants and this reduction in growth was positively correlated with the increase in Cd level. The author concluded that Cd is phytotoxic to broad bean at levels as low as 10 mg Cd l^{-1} in irrigating or sprayed solution.

Evidence from the limited number of fumigation and field experiments described above would suggest that ambient levels of air pollution (as described in Tables 2 and 3), and in particular O_3, are causing damage to sensitive crop species over a significant proportion of Egypt's agricultural land. The actual spatial extent and magnitude of these impacts is difficult to quantify for two main reasons. Firstly, since exposure-response relationships for local Egyptian species and varieties have not been systematically defined and secondly, due to the limited monitoring network, especially in suburban and rural areas, resulting in a lack of local and regional pollutant concentration data. The use of controlled fumigation techniques can greatly facilitate interpretation of field observations to understand the response mechanisms to pollutants, both individually or in combination. However, it should be recognised that exposure regimes, even under field situations, may alter the physiological sensitivity of vegetation.

3. Applicability of European and North America Air Quality Standards

In recent years, environmental issues have increasingly been a focus of concern for the government in Egypt. Substantial political support has been given to the enforcement of environmental protection law (Law No. 4 of 1994) and other applicable legislative rules. Such legislation has been developed by the Ministry of Environmental Affairs (MEA) and the Egyptian Environmental Affairs Agency (EEAA) to protect the Egyptian environment against hazards to the community and public hygiene. Many projects have also been implemented in the greater Cairo area with international agencies such as USAID and the Danish International Development Agency (DANIDA) with the objective of improving Cairo's air.

Egypt has fixed ambient air quality standards for many gaseous and particulate compounds. The Egyptian standards for sulphur dioxide, nitrogen dioxide, ozone, smoke and suspended particulate matter set by Law No. 4 of 1994 are given in Table 8. These ambient air quality standards are based purely on human health impacts, and in general, values are significantly higher than WHO guidelines and damage thresholds developed in Europe to protect vegetation and crop plants from injury by air pollution. Only the Egyptian Air Quality Standards (EAQS) for SO_2 and NO_2 are consistent with North American Air Quality Standards (NAAQS). For example, the daily and annual limit values of SO_2, set by Law No. 4 of 1994, are similar to NAAQS established to provide adequate protection against adverse effects on humans and vegetation. The hourly and daily limit values for NO_2 standards are lower than the acceptable maximum limits applied in North America

Table 8. Egyptian Standards for Air Quality — Law No. 4 of 1994.

Pollutant	SO_2	NO_2	O_3	Smoke	SPM
Standards ($\mu g\ m^{-3}$)	350 (hour) 150 (24 h) 60 (year)	400 (hour) 150 (24 h)	200 (hour) 120 (8 h)	150 (24 h) 60 (year)	230 (24h) 90 (year)

Source: Law No. 4 of 1994, 4th ed. (1998).

(i.e. 434 μg m^{-3} and 199 μg m^{-3}, as 1-hour and 24-hour standards, respectively). It may be advisable to revise the current Egyptian standards of O_3 and SPM so that these are also similar to developed country standards and as such provide protection to vegetation as well as human health.

The national pollutant concentration data given in Tables 2 and 3 shows that the concentrations of O_3, NO_2 and SPM recorded during the last decade in Egypt exceed phytotoxic levels in many agricultural areas around greater Cairo and within the Nile Delta region. SO_2, which is unlikely to be present in significant concentrations in rural areas, was recorded at high concentrations in the atmosphere above cultivated land adjacent to industrial areas in greater Cairo. These levels of air pollution could potentially result in damage to sensitive crops in this region, and as such, the rural areas of these governorates near greater Cairo should be considered high-risk areas in terms of air pollution and vegetation damage. However, as air pollution monitoring or emission data are very limited and inconsistent outside greater Cairo and Alexandria, it is currently not possible to perform quantified risk assessments based on national data.

The current environmental policy with regards to air pollution is to concentrate on controlling emissions from different sources and much progress has been made in this respect. For example, unleaded gasoline is now available throughout greater Cairo and natural gas is already used extensively in various industries and about 80% of thermal power stations. In addition, filters have been installed to control discharge from cement factories and studies are being performed to establish new emission standards for automobiles. These enormous efforts may be beneficial in improving air quality by reducing particulate matter and smoke emissions to levels close to current EAQS. However, gaseous pollutants may be expected to rise with increasing industrial development, vehicular traffic and electricity demands. Furthermore, Egypt's prevailing climatic conditions predominantly include warm anticyclone conditions and high humidity for most of the year, these are ideal conditions for the formation and accumulation of different pollutants, especially O_3. This means that high concentrations of ozone may result from high rates of emission precursor pollutants such as nitrogen oxides and hydrocarbons.

4. Future Research

Over recent decades, most of the air pollution studies performed in Egypt concentrated on investigating ambient air quality, and focussed mainly on the greater Cairo region. These studies were conducted on a short-term basis due to problems associated with monitoring techniques and research facilities. For similar reasons, no filtration studies have been carried out to assess the impact of any of the major pollutants on plants. The first assessments of the impact of individual pollutants on Egyptian crops were performed at Imperial College in the United Kingdom in the early 1990s, when Hassan *et al.* (1994) investigated the response of Egyptian cultivars to O_3 under controlled conditions. Similar studies were later conducted under field conditions in Egypt (Hassan *et al.*, 1995). Since that time, limited fumigation studies have been carried out to examine the sensitivities of different Egyptian varieties to O_3. These have shown the potential for O_3 to cause damage to sensitive species and varieties at relatively low ozone exposures.

The establishment of dose-response relationships for Egyptian species and cultivars and under Egyptian conditions (i.e. local soil types, irrigation patterns, pollutant profiles and climatic conditions) are necessary to identify sensitive crop species and enable assessments of the impact of air pollution on Egyptian vegetation to be performed. Additionally, comprehensive monitoring of the major air pollutants should be performed in agricultural and rural locations to enable the identification of areas where vegetation may be at risk from damage by air pollution and to enable the assessment of crop yield losses.

Assessments of the impacts of SPM and heavy metals on crop losses should also be performed, since evidence suggests these pollutants are as important as gaseous pollutants in damaging vegetation. Heavy metal pollution is another threat to vegetation in Egypt and experimental results have indicated that heavy metals, such as lead, have a great impact on crops in many agricultural areas around greater Cairo and around major roads within the Delta Nile.

5. Conclusions

Air pollution problems in Egypt have increased dramatically since the 1960s as a consequence of the rapid increase in industrial and urban development. Both particulate and gaseous air pollutants, emitted mainly from industrial sources and motor vehicles, have reached high concentrations in greater Cairo, Alexandria and the Nile Delta region. Mean concentrations of all the major pollutants (SO_2, NO_x, O_3 and SPM) regularly exceed WHO guidelines and are likely to be at concentrations capable of adversely affecting the growth and yield of sensitive vegetation types. According to risk predictions for developing countries (Ashmore and Marshall, 1998), the danger posed to Egyptian vegetation is likely to continue in the future, in spite of recent efforts to improve air quality in Egypt.

The only experimental studies to investigate the impacts of air pollutants on vegetation have been performed to study the impacts of O_3 on specific Egyptian crop species and cultivars. Other pollutants have not been investigated due to the lack of facilities that are required for such studies. In Egypt, little is known about the ambient concentrations of different pollutants, especially in rural and agricultural areas. However, the occurrence of high emission rates coupled with field and experimental evidence describing the responses of important Egyptian crops such as tomato, radish, turnip and Jew's mallow, to relatively low concentrations of O_3, emphasise the increasing threat air pollution may pose to agriculture. The cultivated land around greater Cairo, where most industrial complexes are situated, is likely to be at high risk from air pollution. Areas across the whole of the Nile Delta region could potentially also be at risk from O_3 impacts, especially given the regional nature of this pollutant.

Introducing air quality guidelines to protect vegetation in Egypt could prove a useful tool in achieving current national priorities to improve air quality and thereby achieve self-sufficiency in food and limit damage to economic crops. However, the use of such tools would require the availability of air quality data from rural areas at a sufficient spatial and temporal resolution, as well as robust dose-response relationships for local Egyptian species and varieties. If this information was available it would be possible to move a step closer to making economic assessments of the damage caused to vegetation by air pollution. Such information is crucial to ensuring that appropriate air quality management strategies are adopted to efficiently improve local and regional air quality.

Acknowledgements

I gratefully acknowledge RAPIDC/SEI group for their great offer to produce this report, which has never been given for any research student in Egypt before. Also, I am indebted to Prof. Mike Ashmore, Bradford University, for his inspiration, valuable advice and untiring support. All my deepest thanks to all colleagues at National Research Centre and Alexandria University for data provided.

References

Abdel Hakim T. and Aboumandour M. (1994) *The Egyptian Agriculture Sector and Its Prospects for the Year 2000.* Final report, Study P280, September 1993. ECSC-EC-EAEC, Brussels and Luxembourg.

Abdel-Latif N.M. (1993) *An Investigation on Some Combustion Generated Pollutants Affected Plant Growth.* M.Sc. Thesis, Zagazig University, Zagazig, Egypt.

Abdel-Latif N.M. (2001) *Risk Evaluation for Air Pollution Effects on Vegetation in Egypt.* Ph.D. Thesis, London University, UK.

Abdel Salam M.S. (1963) Air pollution: some findings in Cairo studies. *Symposium on Occupational Health and Industrial Medicine.* Ministry of Health, Cairo, pp. 329–340.

Abdel Salam M.S. and Sowelim M.A. (1967a) Dust deposits in the city of Cairo. *Atmos. Env.* **1**, 211–220.

Abdel Salam M.S. and Sowelim M.A. (1967b) Dust-fall caused by the spring Khamasin storms in Cairo: a preliminary report. *Atmos. Env.* **1**, 221-226.

Abdel Salam M.S., Farag S.A. and Higazy N.A. (1981) Smoke concentration in the greater Cairo atmosphere. *Atmos. Env.* **15**, 157–161.

Ali E.A. and Nasralla M.M. (1986) Impact of motor vehicle exhausts on the cadmium and lead content of clover plants grown around Egyptian traffic roads. *Int. J. Env. Studies* **28**, 157–161.

Ali E.A. (1993) Damage to plants due to industrial pollution and their use as bioindicators in Egypt. *Env. Pollut.* **81**, 251–255.

Ashmore M.R. and Marshall F.M. (1998) Direct impacts of pollutant gases on crops and forests. In *Regional Air Pollution in Developing Countries,* eds. Kuylenstierna J. and Hicks W.K., Stockholm Environment Institute, York, UK, pp. 21–33.

ASTM (1987) Standard method for evaluating of total sulphation activity by the lead dioxide method. In *Annual Book of ASTM Standards,* Vol. 11.03, pp. 81–85.

FAO (1997) *The State of Food and Agriculture 1997.* FAO agriculture series No. 30. FAO, Rome.

Farag S.A., Rizk H.F.S., El-Bahnasawy R.M. and El-Meleigy M.I. (1993) The effect of pesticides on surface ozone concentration. *Int. J. Env. Educ. Inf.* **12**, 217–224.

Güsten H., Heinrich G., Weppner J., Abdel-Aal M.M., Abdel-Hay F.A., Ramadan A.B., Tawfik F.S., Ahmed D.M., Hassan G.K.Y., Cuitas T., Jeftic J. and Klasinc L. (1994) Ozone formation in the greater Cairo area. *Sci. Total Env.* **155**, 285–295.

Hassan I.A., Ashmore M.R. and Bell J.N.B. (1994) Effects of O_3 on the stomatal behaviour of Egyptian varieties of radish (*Raphanus sativus* L. Cv. Baladey) and turnip (*Brassica rapa* L. Cv. Sultani). *New Phytol.* **128**, 243–249.

Hassan I.A., Ashmore M.R. and Bell J.N.B. (1995) Effect of ozone on radish and turnip under Egyptian field conditions. *Env. Pollut.* **89**, 107–114.

Hassan I.A. (1998) Effect of O_3 on crop quality: A case study from Egypt. In *Responses of Plant Metabolism to Air Pollution and Global Change,* eds. De Kok L.J. and Stulen I. Backhuys Publishers, Leiden, pp. 323–327.

Hassan I.A., Bender J. and Weigel H-.J. (1998). Metabolic responses of Egyptian cultivars of tomatoes (*Lycopersicon esculentum* Mill) to low doses of O_3 and drought stress. In *6th Egyptian Botanical Conference.* Egyptian Society of Botany, Cairo University, Cairo, Egypt.

Hassan I.A. (1999) Air pollution in the Alexandria region, Egypt — I: An investigation of air quality. *Int. J. Env. Educ. Inf.* **18**, 67–78.

Hassan I.A., Bender J. and Weigel H-.J. (1999) Effect of O_3 and drought stress on growth, yield and physiology of tomato (*Lycopersicon esculentum* Mill. Cv. Balady). *Gratenbauwissenschaft,* **64**, 152–157.

Hassanien M.A., Rieuwerts J. and Shakour A.A. Seasonal and annual variations in air concentrations of Pb, Cd and PAHs in Cairo, Egypt. Pers. Comm.

Hindy K.T. (1973) *Effect of Types of Rocks Quarried in Aswan, Fayum and Around Cairo on Air Pollution in Nearby Inhabited Areas.* M.Sc. Thesis, Ain Shams University, Cairo, Egypt.

Hindy K.T., Farag S.A. and El-Taieb N.M. (1988a) Sulphur dioxide study in industrial and residential areas in Cairo. *Int. J. Env. Educ. Inf.* **7**, 29–47.

Hindy K.T., Farag S.A. and El-Taieb N.M. (1988b) Yearly cycle of water-soluble content of total suspended particulate matter in industrial and residential areas in Cairo. *Int. J. Env. Educ. Inf.* **7**, 121–135.

Hindy K.T., Farag S.A. and El-Taieb N.M. (1990) Monthly and seasonal trends of total suspended particulate matter and smoke concentration in industrial and residential areas in Cairo. *Atmos. Env.* **24B**, 343–353.

Hindy K.T. (1991) Study of alluvial soil contamination with heavy metals due to air pollution in Cairo. *Int. J. Env. Studies* **38**, 273–279.

Ibrahim Y.H. (1992) *Industrial Pollutants and Their Impact on Plant.* M.Sc. Thesis, Cairo University, Cairo, Egypt.

Law No. 4 of 1994 (1988) *Environmental Law of Arab Republic of Egypt*, 4th Ed. Cairo (in Arabic).

MALR (1996) *Agricultural Economic Report.* Economic Affairs Sector, Ministry of Agriculture and Land Reclamation, Cairo (in Arabic).

Nasralla M.M. and Ali E.A. (1984) Lead, cadmium and zinc pollution around Egyptian traffic roads. *Egyptian J. Occup. Med.* **8**, 197–210.

Nasralla M.M., Shakour A.A. and Ali E.A. (1986). Sulphur dioxide and smoke in subtropical Cairo atmosphere. *Trans. Egyptian Soc. Chem. Eng. (TESCE)* **12**, 1–14.

Nasralla M.M. (1991) *Impact of Shoubra El-Kheima Electricity Power Station on the Environment of Surrounding Residential Areas.* Technical report, National Research Centre, Cairo, Egypt.

Shakour A.A. (1982) *Study on Some Pollutants in the Cairo Atmosphere.* Ph.D. Thesis, Al-Azhar University, Cairo, Egypt.

Shakour A.A. (1992) Particulate matter and sulphur compounds in a residential area in Cairo. *Int. J. Env. Educ. Inf.* **11**, 191–200.

Shakour A.A. and El-Taieb N.M. (1992) Study of gaseous pollutants in new urban centre in Cairo. *The 5th International Conference on Environmental Contamination.* Morgs, Switzerland, pp. 207–209.

Shakour A.A. and Hindy K.T. (1992) Study of atmospheric contamination in recently developed residential area in Cairo. *The 5th International Conference on Environmental Contamination.* Morgs, Switzerland, pp. 200–203.

Shakour A.A. (1993) Nitrogen-containing compounds in an urban atmosphere. *Int. J. Env. Educ. Inf.* **12**, 137–148.

Shakour A.A. and El-Taieb N.M. (1995) Lead deposition on residential area in the vicinity of lead smelter in Cairo city. *International Conference of Heavy Metals in the Environment*, Hamburg, Vol. 2, pp. 213–216.

Shakour A.A. and Zakey A.S. (1998) The relationship between sulphur dioxide and sulphate in the Cairo atmosphere. *Central European J. Occup. Env. Med.* **4**, 82–90.

Shakour A.A., El-Taieb N.M., Mohamed S.K. and Zakey A.S. (1998) Investigation of air quality in the industrial area north Cairo. *Central European J. Occup. Env. Med.* **4**, 192–204.

SIS. (1998) *Annual Book.* State Information Services, Ministry of Information, Egypt (www.sis.gov.eg/public/yearbook98/html)

UNDP (1998) *Human Development Report 1998.* Oxford University Press, Oxford.

WHO/UNEP (1992) *Urban Air Pollution in Megacities of the World.* Blackwell, Oxford.

CHAPTER 11

AIR POLLUTION IMPACTS ON VEGETATION IN SOUTH AFRICA

A.M. van Tienhoven and M.C. Scholes

1. Introduction

The southern African region has considerable mineral resources, a rich biological diversity and wildlife heritage, and a warm sunny climate. As a result, mineral exploitation and tourism form the mainstay of most of the southern African economies. The sunny weather that helps to make the region an attractive tourist destination is caused by stable atmospheric conditions that occur for much of the year. However, the stable air layer also serves to trap air pollutants produced during mineral extraction and processing. The subsequent impacts of air pollution on natural ecosystems, agricultural production, and even human health in the region are still poorly understood.

The atmosphere over southern Africa is a crucial factor determining the extent and duration of air pollution in the region. Although South Africa is the subject of this section, the atmospheric circulation pattern straddles the entire sub-region and ignores political boundaries, so in effect, the entire region shares the same airmass. The semi-closed anticyclonic circulation system prevails over the sub-continent for about two-thirds of the year (Tyson, 1998). These climatic conditions have a considerable impact on the behaviour and fate of pollutants emitted into the atmosphere. Pollutants are trapped within a stable atmospheric layer in which vertical distribution is limited, especially in the winter months. The dispersion conditions for the highveld region of South Africa have been rated "as among the most unfavourable anywhere in the world" (Tyson *et al.*, 1988) and conditions are exacerbated by the high density of polluting industries in the region.

The high frequency of sunny days also promotes the photochemical transformation of trace gases to secondary pollutants such as ozone and peroxyacyl nitrate (PAN). Both primary and secondary products can thus be transported and recirculated over considerable distances within the sub-continent, before exiting over the oceans (Garstang *et al.*, 1996; Tyson, 1998).

In South Africa, air pollution was first identified as a potential threat to vegetation in 1988 (Tyson *et al.*, 1988). The commercial forests of the eastern escarpment were highlighted as a threatened resource because of their proximity to the heavily industrialised highveld. The term "highveld" is used to denote the interior plateau of South Africa and is generally at a mean altitude of 1700 metres above sea level. Recently, concerns have also been raised over the potential impacts on crop yields on the highveld (Marshall *et al.*, 1998). However, with increasing development and industrial expansion, the highveld is no longer the only area affected by air pollution in South Africa.

Air pollutants that could impact on vegetation include sulphur dioxide (SO_2), ozone (O_3), oxides of nitrogen (NO_x), and hydrogen fluoride (HF). The potential effects of these gaseous pollutants on vegetation in South Africa are the subject of this review. Although a substantial body of research has been performed to assess the potential impacts of acid deposition on South African ecosystems, acid deposition is not addressed here since this review concentrates only on the direct impacts of air pollution on vegetation.

1.1. *Emissions and Ambient Air Quality*

South Africa's considerable mineral wealth is exploited by numerous mining and industrial concerns, most of which are located in the Gauteng, Western Mpumalanga and Northwest provinces. This highly industrialised area is situated on the central highveld and

Table 1. Emissions (in metric tonnes per year) from major industries in South Africa and the highveld (Wells *et al.*, 1996).

	South Africa	Industrial highveld region	Highveld emissions as % of total
Particulates	331,339	285,405	86
SO_2	2,120,452	1,986,193	94
NO_x	1,004,716	913,486	91

is responsible for most (> 90%) of South Africa's scheduled emissions. Table 1 shows the emissions contributed by industry on the highveld in 1993.

At least 98% of South Africa's electricity is generated by Eskom, South Africa's electricity utility. This is more than half of all the electricity produced in Africa. Most of Eskom's electricity is generated by coal-fired power stations, which in 1994, burned 76.9 million tonnes of coal, and emitted 122,000 tonnes of particulates, 1,167,000 tonnes of SO_2, and 582,000 tonnes of nitrogen oxides (Rorich and Galpin, 1998). Other industries on the highveld include ferro-alloy works, steelworks and foundries as well as the petrochemical processing plants at Secunda and Sasolburg.

Although industrial emissions are generally a local concern, trace chemicals with industrial signatures have been shown to be transported over thousands of kilometres within the stable circulation system, even reaching as far as Kenya (Annegarn, 1997; Garstang *et al.*, 1996; Piketh *et al.*, 1998). The anthropogenic sulphur signal is most likely from industrial activities on the South African highveld. In particular, the emission stacks of the power stations are very tall (250 to 300 metres) to ensure that atmospheric pollutants are ejected above the stable boundary layer and widely dispersed. However, under unstable boundary layer conditions, the plumes can loop down to ground level, with high ground-level concentrations of gases being experienced close to the source (Turner, 1996).

At a local scale, there are a number of locations in the country where air pollution is perceived to be a problem (Annegarn, 1997). Coal burning for cooking and space heating is considered to be the greatest threat to human health in towns such as Soweto. Despite a countrywide electrification programme, and the threat to health and welfare, coal is still the preferred energy source for space heating (Sithole *et al.*, 1998). Industrial complexes are obvious pollution sources and include those in Durban south, Richards Bay, Secunda and the Vaal Triangle which comprises the towns of Vereeniging, Vanderbijlpark and Sasolburg. Windblown dust from goldmine dumps near Johannesburg was once considered a serious problem but, with the exception of a few specific sites, has largely been overcomed by revegetating and reprocessing of dumps.

Photochemical haze is an increasing phenomenon in the larger cities such as Johannesburg and Pretoria with vehicle emissions considered to be the major contributor (Annegarn, 1997). The brown haze prevalent in Cape Town is also attributed to vehicle emissions (De Villiers *et al.*, 1997). Besides these local incidences, a general haze is often experienced at a regional scale, particularly when the continental high pressure system prevails. Natural processes, such as veld burning and biogenic emissions, are the greatest contributors to the regional haze (Tyson, 1998).

Sulphur dioxide (SO_2)

Most emissions of SO_2 are from industrial sources (generally power stations) with emission stacks of 200 metres or more. The coal burned for electricity generation is considered to be low in sulphur, although the ash content is high compared with coals from elsewhere in the world (Wells *et al.*, 1996). Emissions information for South African industries is currently reported on a ¼ degree square basis and only for industries scheduled under the existing air pollution legislation. More detailed information on industrial emissions is not readily available.

At ground level, burning coal discard dumps were a considerable source of sulphur emissions in the past, although these have largely been brought under control since the 1980s. Coal burning for space heating and cooking is an additional source of SO_2, but emission estimates from such sources are also not readily available.

Table 2. Concentrations of SO_2 at various highveld and rural sites in South Africa.

	Maximum hourly mean concentrations[b] (ppb)	Maximum monthly mean concentrations[b] (ppb)	Two week average[c]	
			Summer (ppb)	Winter (ppb)
Bedworth Park	105	14.6	–	–
Elandsfontein	125	12.1	8.4	11.2
Makalu	131	12.5	–	–
Palmer	277	12.5	1.4	4.4
Verkykkop	63.8	5.8	–	–
Blyde[a]	–	–	0.9	2.1
Skukuza[a]	–	–	1.4	1.5
Ben MacDui[a]	–	–	0.98	0.90
DEAT guideline	300	50	NA	NA
Threshold value[d]	500–1000 (for visible damage to sensitive vegetation)	200 (for a reduction in yield)	NA	NA

Note: [a]Sites are considered rural
[b]Rorich and Galpin, 1998
[c]Zunckel, 1998
[d]Darrall, 1989

Although information detailing emissions is difficult to access, several ambient air quality studies have been conducted (Table 2). Monitored data are compared with the South African guidelines for different averaging periods. These guidelines are primarily aimed at protecting human health. Thresholds for reduced plant yield and visible damage are also presented. Consideration of the SO_2 data reported in Table 2 for various sites in South Africa reveals that none of the threshold concentrations for either human health or vegetation have been reached for either monthly or hourly averaging periods.

An annual mean concentration of 11.5 ppb SO_2 (30 $\mu g.m^{-3}$) is the European critical level or threshold at which a 5% yield reduction can be expected for agriculture (Ashmore and Marshall, 1998). Measurements by van Rensburg (1992) established that the annual mean SO_2 concentration over the central highveld over a five-year period was only 8.7 ppb (Table 3). More recently however, Marshall *et al.* (1998) reported that over the course of a single year, the critical level was exceeded over an area of 650 km^2 east of Johannesburg. Annegarn *et al.* (1996) also reported periods or annual means for various sites on the highveld where this critical level is exceeded, for example, Komati Airstrip (13.3 ppb), Soweto (17.6 ppb), Esther park (20 ppb) and Illiondale (14 ppb). Therefore, when annual means are considered for individual years, ambient SO_2 levels at a number of monitoring stations in the highveld region exceed the critical 11.5 ppb level for crop yield reduction.

Measurements conducted in other parts of the country such as Durban and Richard's Bay in KwaZulu-Natal are shown in Table 4. Local emissions of SO_2 in Durban are mostly produced by petrol refineries whereas smelters are the major sources in Richard's Bay.

Table 3. SO_2 levels over the industrial highveld of Mpumalanga over a five-year monitoring period (1981–1986).

	Hourly mean (ppb)	Daily mean (ppb)	Monthly mean (ppb)	Annual mean (ppb)	Five-year mean (ppb)
50% quantile	4	5.3	6.7	8.7	7.2
99% quantile	54.2	30	15.7		

Table 4. Observed concentrations of SO_2 in Durban and Richards Bay (KwaZulu Natal).

Monitoring site	Instantaneous concentration (ppb)	Three-minute concentration (ppb)	Hourly concentration (ppb)	Daily concentration (ppb)
Wentworth Hospital — Durban[a]	–	37	37	–
Southern sewage works — Durban[a]	–	19	19	–
Richards Bay[b]	163–672 (summer) 391–477 (winter)	–	91–138 (summer) 109–179 (winter)	16–50 (summer) 15–44 (winter)

Note: [a]Observed mean concentration of SO_2 at two monitoring stations in Durban (monitored from 14 June to 17 June, 1996) (Burger and Hurt, 1996).
[b]Range of maximum concentrations where summer is September 1997 to February 1998 and winter is June to August 1997 and March to May 1998 (Hurt *et al.*, 1998).

SO_2 emissions were also identified as a potential cause for concern at the Bothasig monitoring station in Cape Town. Hourly maximum SO_2 concentration peaked in the early winter months with the 40 ppb level being exceeded five times in April and 15 times in May (Botha *et al.*, 1990).

Nitrogen oxides (NO_x)

As with the SO_2 emissions, industrial emissions of NO_x are quantified on a ¼ degree square basis. Domestic coal burning is also a contributor of NO_x emissions. The amounts of NO_x generated in the process of coal combustion depend on both the amount of coal burned and on combustion conditions, as such, quantifying NO_x emissions from domestic coal burning is quite difficult. (Wells *et al.*, 1996).

Data for ambient NO_x concentrations are few, but existing studies show that the maximum daily mean concentration of NO_x reported on the highveld was only 131 ppb which is far below the threshold of 1000 ppb of NO_2 (over a 24-hour exposure period) at which sensitive plants show injury (Table 5).

Table 5. Ambient concentrations of NO_x on the highveld in 1994 (Rorich and Galpin, 1998).

	Maximum hourly mean concentrations (ppb)	Maximum daily mean concentrations (ppb)	Maximum monthly mean concentrations (ppb)
Bedworth Park	370	130.9	73.5
Elandsfontein	157	28.7	12.5
Makalu	150	56.4	25.3
Palmer	228	38.4	9.2
Verkykkop	87.7	15.3	6.3
DEAT guideline	800	400	300
Threshold value for vegetation damage[a]	–	1000	–

Note: [a]Darrall, (1989).

Ozone (O_3)

Global scale measurements of tropospheric O_3 have shown that levels for the northern hemisphere normally range between 50 to 60 ppb compared with 25 to 30 ppb in the southern hemisphere, reflecting the greater industrial influence in the north (Marenco *et al.*, 1990). However, during the dry season (December to March), O_3 levels over Africa range from 60 to 90 ppb. These enhanced levels of O_3 have been attributed to vegetation fires that occur over large areas of the continent. Smoke from savanna fires produces the precursors for tropospheric O_3 formation, as well as other trace gases and aerosols which accumulate over the tropical oceans surrounding Africa (Marenco *et al.*, 1990; Lindesay *et al.*, 1996).

Biogenic and anthropogenic sources are also major contributors to the tropospheric O_3 maximum although there is still uncertainty over the relative size of these contributions. Consideration of the seasonal patterns and total amounts of trace gases emitted does not allow any of the major sources to be eliminated. As such, the tropospheric O_3 maximum possibly results from a combination of the tail end of pyrogenic emissions (August to October) and the beginning of biogenic emissions (September to October). Scholes and Scholes (1998) argue that if such is the case, the tropospheric O_3 maximum over the southern Atlantic Ocean has probably recurred for millions of years and is thus a natural phenomenon. However, the addition of an increasing anthropogenic pollution load could result in damage thresholds being exceeded in the late winter and early spring with consequent damage to vegetation. Land use change could also have a considerable effect on the rates of biogenic emissions. As land is converted from one vegetation type to another, for example from savanna to farmland, the emission characteristics and vegetation architecture change. These changes will also determine whether damage thresholds are exceeded or not.

Monitored ambient concentrations of O_3 are shown in Table 6. Although data reported in Table 6 are the maximum hourly mean concentrations, they range from 76 to 110 ppb for various highveld sites and can be expected to fall within the 50 to 100 ppb (98–196 $\mu g.m^{-3}$) range reported to cause damage within two to four

Table 6. Concentrations of O_3 over the highveld in 1994 (Rorich and Galpin, 1998).

	Maximum hourly mean concentrations (ppb)	Maximum daily mean concentrations (ppb)	Maximum monthly mean concentrations (ppb)
Bedworth Park	95.1	31.9	22.3
Elandsfontein	110	48.8	33.3
Makalu	90.2	40.0	28.7
Palmer	85.3	63.6	40.9
Verkykkop	76.8	56.2	38.8
DEAT guideline	120	–	–

hours exposure (Lacasse and Treshow, 1976). The threshold dose of O_3 that causes injury varies tremendously between species and even cultivars of the same species. Visible symptoms of injury can occur at concentrations between 200 to 400 ppb for many economically important species (Heagle, 1989; Scholes *et al.*, 1996).

Particulate matter

The high ash content of South African coal results in high particulate emissions from coal combustion processes. Eskom has considerably reduced particulate emissions from its power stations with the installation of bag filters and SO_3 injection technology. In 1999, particulate emissions amounted to 67,080 tonnes compared with 122,000 tonnes in 1994 (Eskom Environmental Report, 1999; Rorich and Galpin, 1998).

Recent work in the Cape Town area has focused on the brown haze phenomenon which has been growing in intensity over recent years (De Villiers *et al.*,1997). The most important component of the brown haze was identified as primary particulate matter ($PM_{2.5}$), of which 48% is estimated to originate from diesel vehicles. $PM_{2.5}$ refers to particulate matter with an effective diameter $PM_{2.5}$ μm. Emissions of $PM_{2.5}$ are likely to increase by almost 50% over the next decade, thus the intensity of brown haze is likely to increase by a similar amount.

Groundlevel dust problems are also exacerbated by the aridity of the region, particularly in the dry winter months. Nuisance dust is most often caused by heavy traffic on unsurfaced roads and on windy days from dust tailings from gold and platinum mining and processing.

Fluoride

Aluminium smelters and fertiliser manufacturers are the major sources of fluoride emissions in Richards Bay. Monitoring of ambient concentrations is not performed, although dust fallout is measured and analysed for total fluoride content. Gaseous and particulate fluoride and the relative proportions of fluoride compounds cannot be distinguished by this monitoring method. In 1992, Dr. L.H. Weinstein, an acknowledged expert on plant responses to fluoride, visited the area and determined that the patterns of vegetation injury were consistent with the predicted distribution of gaseous fluoride in the ambient air (Bryszewski *et al.*, 1992). An annual average of 0.4 $\mu g\ m^{-3}$ has been set as a guideline to minimise plant injury in residential and agriculture areas of Richards Bay, whereas the guideline is 1.0 $\mu g\ m^{-3}$ for industrial and commercial areas.

1.2. *Vegetation*

South Africa is well recognised for the high level of diversity and endemism of its vegetation. Of the more than 23,000 species recorded, 8,500 are from the Cape Floral Kingdom and of these, 3,500 occur nowhere else in the world. The southwestern corner of South Africa has a Mediterranean climate (cool wet winters and hot dry summers), and supports a sclerophyllous shrubland known as "fynbos". The region is mountainous but the valleys are suitable for crops such as winter wheat, deciduous fruit and vines. The southern Cape coastal region east of Cape Town also has a temperate climate but rain falls throughout the year. Indigenous forest, and fynbos comprise the natural vegetation while crops include wheat, fruit, vegetables and timber. The west coast of South Africa is hot and dry with less than 100 mm of annual rainfall. This area supports the succulent karoo, a

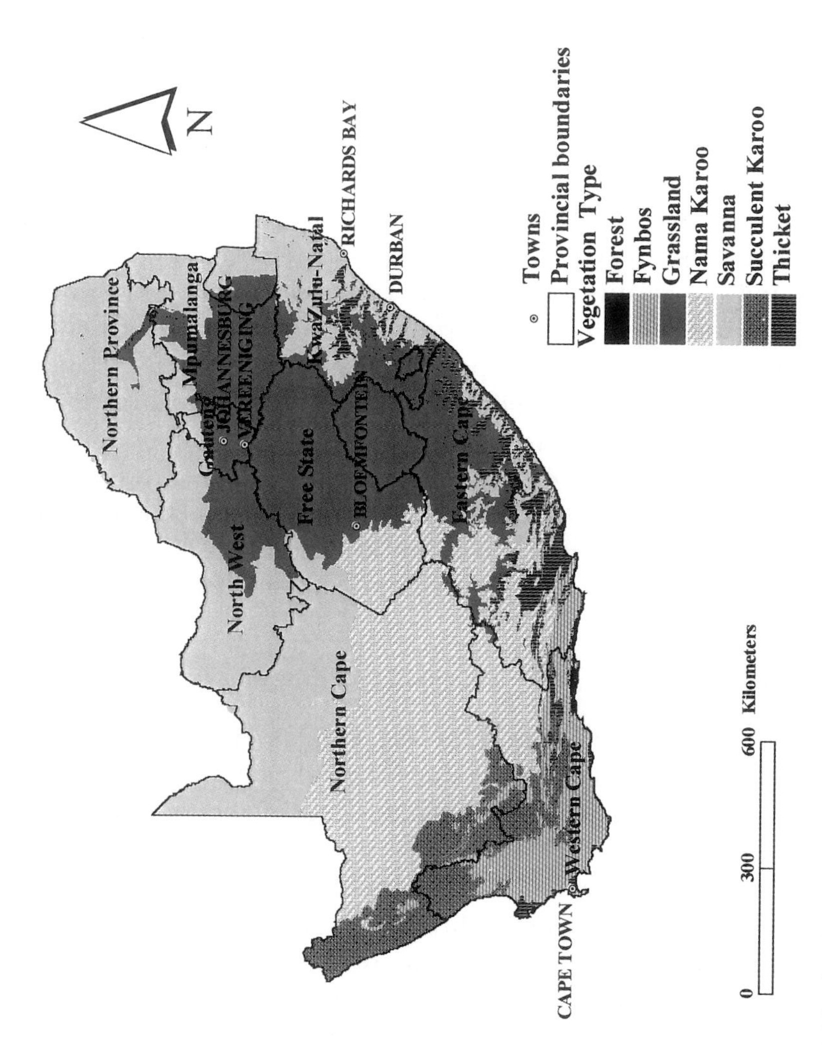

Fig. 1. Map of South Africa showing the major biomes (classification of the National Botanical Institute).

highly biodiverse flora from where 80% of the world's succulents originate.

The interior of South Africa is dominated by a plateau which slopes gradually to the west coast but forms a steep escarpment on the eastern seaboard. The western portion of the plateau hosts the Nama karoo (semi-desert shrub) which mostly supports sheep farming. Grassland replaces the karoo vegetation on the central interior of the country (highveld) which receives more rain and is largely devoted to the cultivation of maize, sorghum, sunflowers and potatoes. Most of South Africa's industrial activities are also located in this region, centred around the major cities of Johannesburg and Pretoria and extending eastwards. The northern-most part of South Africa is sub-tropical savannah where cattle- and wildlife ranching is the dominant land-use. The eastern escarpment experiences the highest rainfall (> 1000 mm/year) and hosts pockets of indigenous forest amongst commercial forestry plantations. Below the escarpment, the narrow coastal belt of the eastern seaboard is sub-tropical and frost free, allowing crops such as sugarcane, tobacco, citrus, banana and other sub-tropical fruit crops to be produced. The thickets of the east coast occur on fertile soils where the rainfall is too low to support forest. The thickets are evergreen, low-growing and species-rich. Figure 1 describes the major biomes of South Africa.

2. Impacts of Air Pollution

There are few South African studies on the impacts of air pollution on local vegetation. There is an inherent difficulty in adopting information for northern hemisphere species to suit local needs because of the considerably different environmental factors that are involved. The responses of plants to air pollutants in hot, dry climates may be influenced by water and temperature stresses. Patterns of temperature-dependent plant responses to air pollutants seem quite different in different species, so generalisations about the influence of exposure temperatures on resistance of plants to air pollutants are limited. Similarly, water stress may protect plants from air pollutants. Stomata close in response to severe water stress to reduce the loss of water by transpiration. If the stomata are closed, the uptake of air pollutants decreases and damage to plants is reduced (Schenone, 1993; Levitt, 1980). These examples suggest that species dose-responses to air pollutants that have been developed elsewhere should be applied with due care in South Africa.

The following sections provide more detail on the available studies that have been performed in South Africa.

2.1. *Field Evidence*

Indigenous vegetation

In the early 1990s, farmers in the vicinity of the industrial highveld of South Africa speculated that the deterioration of the grassland (veld) was attributable to air pollution. Breytenbach (van Rensburg, 1993) compared the floristic composition of grassland sites with that of the same sites sampled 50 years earlier. Vegetation changes were evident at both broad (1:1 500,000) and detailed (1:50,000) scales. Changes in the vegetation composition were greatest close to the pollution source. At the broad scale, a gradient of site degradation correspon-ding to distance downwind of the pollution source was found. However, these findings were based on only six sample plots and factors such as selective- and/or over-grazing, prevailing climatic conditions and fire management practices were thought to have a greater influence on the vegetation at the local scale than pollution impacts (van Rensburg, 1993).

Concern over the influence of air pollution on the Cape Floral Kingdom in the southwestern Cape, as well as on the surrounding vineyards, fruit and vegetable farms prompted an investigation into possible effects on these plant communities (Botha *et al.*, 1990). The genera *Gladiolus* and *Freesia* are both represented in the Cape Fynbos and were used as biomonitors in conjunction with air pollution monitoring instrumentation. The *Gladiolus* variety "Snow Princess" was used to detect atmospheric fluoride during summer, whereas in winter, a *Freesia* cultivar was used. Foliar fluoride levels correlated well with the ratings of visible injury on the leaves of *Freesia*, but no such correlation was evident for *Gladiolus*.

O'Connor *et al.* (1974; cited in Schenone, 1993) determined the relative susceptibility of 141 Australian tree and shrub species to acute SO_2 injury. The genera *Acacia* and *Eucalyptus* were found to be the most sensitive, suffering acute leaf injury after a three-hour exposure to 1000 nl l^{-1} SO_2 (approximately 1 ppm). Such high concentrations of SO_2 are unlikely to occur naturally. Nevertheless, the *Acacia* genus is widespread in Africa, and several *Eucalyptus* species are important commercial timber plants in southern Africa. The response of these species to long-term exposure of elevated SO_2 may not be immediate, but it would be prudent to monitor the trees for signs of damage.

Commercial forests

Commercial forests in the eastern and southeastern parts of Mpumalanga were first highlighted by Tyson *et al.* (1988) as a major resource at risk from air pollution impacts because of their downwind location from the industrial highveld. Thus several studies have focused on forestry species such as *Eucalyptus* and pines. Tyson *et al.* (1988) reviewed the existing literature on responses of forestry species relevant to South Africa to atmospheric pollution (Table 7). Australian species of Eucalypts were found to be particularly sensitive to SO_2 (40 to 100 ppb). Species grown in South Africa may be equally sensitive. However, the *Eucalyptus* plantations in South Africa are quite remote from SO_2 sources and are more likely to be exposed to the transformation products of particulate sulphate than to gaseous SO_2 itself. The transformation products of the sulphur and nitrogen oxides are acidic compounds which could detrimentally affect forests through soil acidification processes.

Foliar symptoms on *P. patula* were surveyed in five forestry regions in Mpumalanga and KwaZulu-Natal which corresponded to a gradient of potential pollution impact. Symptom types and severity were compared across regions revealing that most trees exhibited some type of chlorosis, although generally less than 5% of the needle was affected (Olbrich, 1990). One symptom, namely broad banding, appeared to be correlated with the potential pollution impact at the site. The symptom appeared as a broad chlorotic band (> 4 mm) that encircled the entire needle segment, with or without a necrotic

Table 7. Relative sensitivity to the major gaseous pollutants of certain pines and eucalypt species grown in South Africa (from Tyson *et al.*, 1988).

Species	Pollutant					
	SO_2	O_3	SO_2	PAN	Fl^-	Cl^-
Pines			Tolerant		Sensitive	Tolerant
P.elliotii	Intermediate	Intermediate				Sensitive
P. taeda	Sensitive	Sensitive				Sensitive
P. radiata		Sensitive		Sensitive		
Eucalypts	Sensitive					
E. robusta	Sensitive					
E.viminalis	Sensitive					

centre. Further analyses failed to identify any other site-, soil- or tree-related factor, other than distance from the highveld pollution source, that could account for the observed distribution of the symptom (Olbrich, 1992). Anatomical investigations showed that the chlorotic portions of the bands were not visibly damaged, but the mesophyll tissue showed some accumulation of a secondary cellular product. Concentrations of calcium and magnesium were also lower in the chlorotic bands than in undamaged green tissue (Whiffler, 1990). Although air pollution is suggested to cause the symptom, the exact mechanism and pollutants involved still need to be investigated.

Further monitoring of the symptom over time failed to reveal any other obvious causes (Olbrich, 1993), although a number of other potential factors have not yet been addressed. These include the incidence of fungal pathogens, viruses, insects and genetic abnormalities.

It is recognised that the analysis of foliar blemishes must be interpreted with caution as it is difficult to ascribe foliar symptoms to air pollution in the absence of experimentation. A specific pollutant or combination of pollutants must be identified that could cause the observed symptom, and the experiments must be repeatable (Scholes *et al.*, 1996; Winner, 1989).

Carlson (1994) conducted a study to determine the risk posed by ambient gaseous O_3 concentrations to the commercial forests of Mpumalanga. Records of O_3 concentrations were analysed to deter-

Table 8. Comparison of the number of days in the year when at least one hour during the day exceeded the various thresholds for ozone. In addition, the mean number of hours per day that exceeded the threshold are given, together with the cumulative doses (Carlson, 1994).

Site	Year	25 ppb threshold for O_3			40 ppb threshold for O_3			80 ppb Threshold for O_3		
		No. of days where at least 1 h exceeded threshold value	Mean No. of hours per day exceeding threshold	Cumulative dose (ppb.h)	No. of days where at least 1 h exceeded threshold value	Mean No. of hours per day exceeding threshold	Cumulative dose (ppb.h)	No. of days where at least 1 h exceeded threshold	Mean No. of hours per day exceeding threshold	Cumulative dose (ppb.h)
Palmer	1990	310	18.39	95,558	231	12.44	31,331	8	1.25	177
	1991	220	18.04	65,463	157	12.64	20,262	2	2.00	13
	1992	317	20.14	99,761	242	12.24	28,724	5	2.20	670
	1993[a]	225	18.38	56,069	141	10.85	14,643	6	1.83	527
Rivulets	1992[b]	147	12.83	22,806	93	6.81	5,326	1	1.00	10
	1993[a]	173	7.59	3239	60	3.67	1,470	0	0	0

Note: [a]Monitoring only begins in the second half of the year.
[b]Data only exists for the first half of the year.

mine the frequency with which potentially phytotoxic events occurred (Table 8). The critical level concept was adopted for the assessment. At the Palmer monitoring station, the cumulative dose over 40 ppb (AOT 40) exceeded the 10,000 ppb h threshold, which has been established to protect forestry in Europe, for every year of the study. O_3 concentrations were generally found to be relatively high by northern hemisphere standards, particularly on the highveld, with frequent potentially phytotoxic events.

Agricultural crops

Most of the economically important crops on the industrial highveld are annuals such as maize and sunflowers. Crops grown in this region include maize, sorghum, sunflower, Phaseolus bean, soybean and cotton. Maize is relatively tolerant to SO_2 and NO_x, but is sensitive to O_3. Sunflower is relatively sensitive to SO_2 and NO_x but tolerant of O_3. Dry beans are sensitive to all three pollutants (Tyson *et al.*, 1988). SO_2 concentrations are reported to exceed or approach the critical level of 30 mg m^{-3} annual mean established for agricultural crops in Europe. Crops are also suggested to be at risk from O_3 but over a greater area than SO_2 (Marshall *et al.*, 1998). Potatoes grown on the highveld, and tomatoes and tobacco grown in the lowveld, could be sensitive to high O_3 levels.

Based on the SO_2 concentrations measured during a five-year monitoring period (Table 3; van Rensburg, 1992), primary phytotoxic effects on crops were *not* expected to occur. The author concluded that if any damage was to occur, it was more likely to be from interactive effects of various pollutants and environmental factors than from SO_2 alone. It is appropriate to note that this conclusion was based on mean exposure values, whereas plants generally have a greater response to episodes of high concentration of pollutant gas, even if the episodes are infrequent.

2.2. *Experimental Research*

Few experimental studies on air pollution effects on vegetation have been conducted in South Africa.

Commercial forests

One of the earliest South African studies determined the response of three commercially important forest species (*Pinus patula, Pinus elliotii* and *Eucalyptus grandis*) to SO_2 at different concentrations and exposure times. These three species were fumigated in specially constructed fumigation chambers (1 m × 1 m × 2 m) which were housed in a greenhouse. Two-year-old plants were exposed to SO_2 at concentrations of 50, 100, 500 and 1000 ppb for one or two hours per day over 26 days. Visual damage to *E. grandis* was evident at short duration exposures to concentrations of 500 and 1000 ppb whereas *P. patula* was unaffected. *P. elliotii* showed symptoms of damage at 1000 ppb. However, due to insufficient replicates, conclusive results were not possible (Kelly, 1986).

Agricultural crops

Air pollution impact studies on South African crops have also investigated the use of various crop plants as bioindicators of air pollution. The bean cultivar "Lazy housewife" was used as an indicator for SO_2 and O_3 pollution in the Cape Town area (Botha *et al.*, 1990). The experimental results for O_3 injury were inconclusive, despite the ambient O_3 levels reaching concentrations that could have caused mild symptoms on vegetation. However, the measurements of dry mass and leaf surface area of beans grown in open plots were significantly less than those grown in either filtered on unfiltered open-top chambers, suggesting that exposure to ambient concentrations of SO_2 was reducing plant growth.

The tobacco variety Bel W-3 was also used as a bioindicator for ground level O_3 in a study in parts of KwaZulu-Natal and Mpumalanga (Blair, 1998). The greatest visible leaf injury was evident in winter, but there were no significant differences between the sites. Moreover, leaf injury was apparent at O_3 concentrations below the widely accepted 40 ppb threshold. Although the O_3 precursor, NO_2, was monitored with diffusive samplers, no measures of SO_2 were taken. Thus the possibility of synergistic effects could not be investigated further, but could account for the foliar injury observed below the 40 ppb threshold for O_3.

In an earlier study (Engelbrecht, 1987; Tyson *et al.*, 1990), bean varieties (*Phaseolus*) were exposed to varying SO_2 regimes in a greenhouse. Concentrations of SO_2 ranged from 50 ppb to 100 ppb and exposure times ranged from 30 minutes to two hours. The responses proved complex with reductions in biomass and pod yield at low exposures (50 ppb h^{-1}) but increases at doses of 100 ppb h^{-1}. An extreme dose of 3000 ppb for eight hours caused severe damage to leaves as well as a significant reduction in biomass.

3. Applicability of European and North American Air Quality Standards

Standards are assumed to be limits at which human health or vegetation are protected from harm, regardless of the costs or practical considerations necessary to achieve the standards.

Air Quality Standards developed for Europe and North America would not be directly applicable in South Africa for several reasons.

The first and most obvious reason is that South Africa's climatic conditions are very different to those in Europe and North America. South Africa is generally warmer, drier and experiences far more sunshine. There is also considerable climatic variation within South Africa's boundaries which is manifested in the variety of biomes found here, ranging from forest to desert scrub. Many of the species found in South Africa are endemic and the responses of these plants to air pollutants are unknown at this stage.

The environmental stresses to which vegetation in South Africa is exposed are also quite different to those in Europe or North America. Drought conditions are a common occurrence in South Africa, and background levels of pollutants such as O_3 may be naturally high. Whether these stresses will serve to increase or decrease vegetation susceptibility to air pollution is still unclear.

Finally, the stable whirlpool of air that recirculates over the southern African region in winter traps and accumulates pollutants, and serves to spread them across the sub-continent. Superimposed on this background air quality are numerous local pollution sources which may contribute to the pollutant mixture. Plant responses to these pollutant combinations are also unknown at present.

The pollutant levels at which South African vegetation is injured or damaged will be influenced by the above factors. If standards are based solely on the impacts measured, then other considerations may be ignored. Factors such as technical feasibility, practicality, costs, and social and cultural conditions will all influence whether or not the standards are achievable. The development expectations of South Africa must be considered on the same scale as protection of the unique diversity of vegetation.

4. Future Research

Two studies are underway which will consider air pollutant effects on the highveld biota (van Hamburg, pers. comm.).[1] One study will investigate the impacts of SO_2 pollution on major insect pests of key agricultural crops on the highveld. The study involves both field and laboratory components as crop plants which have been exposed to ambient air will be relocated into laboratories where they will be infected with insects. In the second study, a biomonitoring network has been established to investigate the ecotoxicological effect of airborne C2-chlorohydrocarbons and the deposition of their phytotoxic metabolites to the vegetation in the Johannesburg area. The collection of appropriate plant samples and the investigation of the photosynthetic activity of intact plants has already been initiated.

Early studies showed that ambient O_3 concentrations in the Cape Town region could possibly cause injury symptoms in sensitive plants (Botha *et al.*, 1990). More recent O_3 data are absent, but O_3 levels can be anticipated to rise in parallel with particulates from vehicle emissions. Whether increasing O_3 levels will affect the fynbos biome is presently speculative. Research on vegetation in the Mediterranean region, however, has shown that xerophytic plants are more resistant to O_3 and this is likely to be the same for xerophytic fynbos species (Inclan *et al.*, 1998).

[1]Professor H. van Hamburg, School of Environmental Sciences and Development, Potchefstroom University.

Concern has also been expressed about the potential impacts of particulate matter pollution on the fynbos vegetation surrounding Cape Town. The fynbos biome has evolved on nutritionally poor soils but is now receiving nutrient inputs through atmospheric deposition. The consequences of such inputs on plant response and ultimately species diversity are presently unknown, but should nevertheless, be considered (Stock, pers.comm.).[2]

Future research should be focused on improving our understanding of the biogeochemical cycles in the southern African context, and how these are affected by gaseous pollutants. The SAFARI 2000 experiment is an international initiative which aims to explore the linkages between land-atmosphere processes. Within the sub-Saharan region, the various emissions, their transport and transformations in the atmosphere, their influence on regional climate and meteorology and their eventual deposition will be studied and the effects on regional ecosystems explored.

5. Conclusions

Direct impacts of pollutants such as SO_2 are likely to be limited to areas around local sources and to the more heavily industrialised parts of the highveld. Crop production on the highveld is also likely to be affected to some extent.

At a regional scale, O_3 is emerging as the pollutant of major concern. Although high O_3 levels may be a natural feature of the region, anthropogenic contributions through land-use change and industrial and vehicle emissions could raise O_3 concentrations to levels that may affect crop production.

Acknowledgements

Grateful thanks are extended to K.A. Olbrich and Dr. R.J. Scholes for their valuable comments on earlier drafts of the chapter. Appreciation

[2]Professor W. Stock, Department of Botany, University of Cape Town.

is also extended to D. Vink and A.J. Singh for producing the map. The position of Professor Mary Scholes at the University of the Witwatersrand is sponsored by the South African Paper and Pulp Industry (sappi).

References

Annegarn H.J., Turner C.R., Helas G., Tosen G.R. and Rorich R.P. (1996) Gaseous pollutants. In *Air Pollution and Its Impacts on the South African Highveld*, eds. Held *et al.* Environmental Scientific Association, Cleveland, pp. 144

Annegarn H.J. (1997) A bird's eye-view of air pollution in South Africa. In *National Association for Clean Air — 28th Annual Conference Proceedings,* Vanderbijlpark, South Africa, November 1997 pp 110.

Ashmore M.R. and Marshall F.M. (1998) Direct impacts of pollutant gases on crops and forests. In *Regional Air Pollution in Developing Countries —* Background Document for Policy Dialogue, eds. Kuylenstierna J. and Hicks K., Harare, September 1998 pp. 21–33. Stockholm Environment Institute.

Blair S.A. (1998) *Monitoring Ground Level Ozone and Nitrogen Dioxide in Parts of KwaZulu-Natal and Mpumalanga (South Africa) By Means of Chemical and Biological Techniques.* M.Sc Thesis, Department of Biology, University of Natal, Durban, South Africa.

Botha A.T., Moore L.D. and Visser J.H. (1990) Gladiolus and bean plants as biomonitors of air pollution impacts in the Cape Town area. Presented at the *1st IUAPPA Regional Conference on Air Pollution: Towards the 21st Century.* Pretoria, 24–26 October 1990.

Bryszewski W., Wells R.B. and O'Beirne S. (1992) *Alusaf Expansion Project — Environmental Impact Assessment: Air Quality Impacts of Sites E, D and F.* CSIR report EMAP-C 92043.

Burger L.W. and Hurt Q.E. (1998) A tool for air quality management: real time atmospheric dispersion modelling in two large industrial regions — South Durban and Richards Bay. In *Proceedings of the 11th World Clean Air and Environment Congress.* Durban, South Africa, 13–18 September, 1998, Vol. 1, 3F–2.

Carlson C.A. (1994) *An Assessment of the Risk Posed by the Gaseous Pollutant Ozone to Commercial Forests in the Eastern Transvaal.* Report to the Department of Water Affairs and Forestry, Report FOR-DEA 754, CSIR, Pretoria.

Darrall N.M. (1989) The effect of air pollutants on physiological processes in plants. *Plant, Cell Env.* **12**, 1–30.

De Villiers M.G., Dutkiewicz R.K. and Wicking-Baird M.C. (1997) The Cape Town Brown Haze Study. *J. Energy Southern Africa*, November 1997, pp. 121–125.

Engelbrecht E. (1987) *Effects of Sulphur Dioxide on Bean Cultivar under Greenhouse Conditions.* Eskom Report TRR/N870/013, Cleveland, pp. 68.

Eskom Environmental Report (1999) Available from *Corporate Environmental Affairs Corporate Technology*, Eskom, PO Box 1091, Johannesburg, 2000.

Garstang M., Tyson P.D., Swap R., Edwards M., Kallberg P. and Lindesay J.A. (1996) Horizontal and vertical transport of air over southern Africa. *J. Geophys. Res.* **101**(D19), 23, 721–723 and 736.

Heagle A.S. (1989) Ozone and crop yield. *Ann. Rev. Phytopathol.* **27**, 397–423.

Hurt Q.E., Posollo P., Sullivan C. and Schroder H.H.E. (1998) Development, implementation and operation of an air pollution monitoring network in Richards Bay. In *Proceedings of the 11th World Clean Air and Environment Congress.* Durban, South Africa, 13–18 September, 1998, Vol. 6, 17E–3.

Inclan R., Ribas A., Peñuelas, J. and Gimeno B. (1998) On the relative sensitivity of different Mediterranean plant species to ozone exposure (Abstract). In *IUFRO 18th International Meeting for Specialists in Air Pollution Effects on Forest Ecosystems. Forest Growth Responses to the Pollution Climate of the 21st Century.* Heriot-Watt University, Edinburgh, UK, 21–23 September 1998.

Kelly B.I. (1986) *Visual Response of Pinus patula, Pinus elliottii and Eucalyptus grandis to Sulphur Dioxide Fumigations.* Eskom Specialist Environmental Investigations, Report No. TRR/N86/101, pp. 41.

Lacasse N.L. and Treshow M. (1976) *Diagnosing Vegetation Injury Caused by Air Pollution.* U.S. Environmental Protection Agency Handbook.

Levitt J. (1980) Miscellaneous stresses. In *Responses of Plants to Environmental Stresses. Volume II: Water, Radiation, Salt and Other Stresses*, ed. Levitt J. Academic Press, New York, Chapter 11, pp. 607.

Lindesay J.A., Andreae M.O., Goldammer J.G., Harris G., Annegarn H.J., Garstang M., Scholes R.J. and van Wilgen B.W. (1996) International geosphere- biosphere programme/international global atmospheric chemistry SAFARI-92 field experiment: background and overview. *J. Geophys. Res.* **101**(D19), 23, 521–523 and 530.

Marenco A., Medale J.C. and Prieur S. (1990) Study of tropospheric ozone in the tropical belt (Africa, America) from STRATOZ and TROPOZ campaigns. *Atmos. Env.* **24A**(11), 2, 823 and 834.

Marshall F.M., Ashmore M.R., Bell J.N.B. and Milne E. (1998) Air pollution impacts on crop yield in developing countries — a serious but poorly recognised problem? In *Proceedings of the 11th World Clean Air and Environment Congress.* Durban, South Africa, 13–18 September, 1998, Vol. 5, 13E–2.

O'Connor J.A., Parbery D.G. and Strauss W. (1974) The effects of phytotoxic gases on native Australian plant species: Part 1. Acute effects of sulphur dioxide. *Env. Pollut.* **48**, 197–211.

Olbrich K.A. (1990) *The Geographic Distribution of Possible Air Pollution Induced Symptoms in the Needles of Pinus patula.* CSIR Report FOR-1, pp. 39.

Olbrich K.A. (1992) *The Influence of Site, Soil and Foliar Characteristics on the Type and Distribution of Possible Air Pollution Symptoms on the Foliage of Pinus patula.* Report to the Department of water Affairs and Forestry, Deliverable 18-90325-2-0.

Olbrich K.A. (1993) *Patterns of Shoot Growth, Needle Growth and Foliar Symptom Development in Fertilised and Unfertilised P. patula.* Report to the Department of Water Affairs and Forestry, Report No 665.

Piketh S.J., Formenti P., Freiman M.T., Maenhout W., Annegarn H.J. and Tyson P.D. (1998) Industrial pollutants at a remote site in South Africa. In *Proceedings of the 11th World Clean Air and Environment Congress.* Durban, South Africa, 13–18 September, 1998. Vol. 6, 17E–5.

Rorich R.P. and Galpin J.S. (1998) Air quality in the Mpumalanga Highveld region, South Africa. *South African J. Sci.* **94**, 109–114.

Schenone G. (1993) Impact of air pollutants on plants in hot, dry climates. In *Interacting Stresses on Plants in a Changing Climate,* eds. Jackson M.B., and Black C.R. NATO ASI Series 1, Global Environmental Change Vol. 16, Springer-Verlag, Berlin.

Scholes M.C., Olbrich K.A. and van Rensburg E. (1996) The environmental impact of atmospheric pollution in the industrial highveld and adjacent regions: crops, indigenous vegetation and commercial forests. In *Air Pollution and Its Impacts on the South African Highveld,* eds. Held *et al.* Environmental Scientific Association, Cleveland.

Scholes R.J. and Scholes M.C. (1998) Natural and human-related sources of ozone-forming trace gases in southern Africa. *South African J. Sci.* **94**, 422–425.

Sithole J., Lethlage D., Mphati D., Annegarn H.J., Kneen M.A., Tapper A.M., Dhladhla S. and Mahlalela J. (1998) Soweto air monitoring: Project SAM — four year air pollution study. In *Proceedings of the 11th World Clean Air and Environment Congress.* Durban, South Africa, 13–18 September, 1998 Vol. 1, 3G–4.

Turner C.R. (1996) Dispersion modelling for the highveld atmosphere. In *Air Pollution and Its Impacts on the South African Highveld,* eds. Held *et al.* Environmental Scientific Association, Cleveland. pp. 144.

Tyson P.D., Kruger F.J. and Louw C.W. (1988) *Atmospheric Pollution and Its Implications in the Eastern Transvaal Highveld.* South African National Scientific Programmes, Report No 150, CSIR, Pretoria.

Tyson P.D. (1998) Regional transport of aerosols and trace gases over southern Africa and adjacent oceans: implications for the future. In *Proceedings of the 11th World Clean Air and Environment Congress.* Durban, South Africa, 13–18 September, 1998 Vol. 1, 2G–2.

van Rensburg E. (1992) *Ambient SO_2 Concentrations in the Eastern Transvaal Highveld and Their Potential to Impact on Crop Yields — A Review.* Eskom report TRR/S92/012.

van Rensburg E. (1993) *Assessment of Air Pollution on the Natural Grassland of the Eastern Transvaal Highveld.* Eskom report TRR/S93/144.

Wells R.B., Lloyd S.M. and Turner C.R. (1996) National air pollution source inventory (inventory of scheduled processes). In *Air Pollution and Its Impacts on the South African Highveld,* eds. Held *et al.* Environmental Scientific Association, Cleveland, pp. 144.

Whiffler J.J. (1990) The effect of acid rain on the morphology, anatomy and nutrient composition on the needles of *Pinus patula.* Honours Project, Department of Botany, University of the Witwatersrand, pp. 62.

Winner W.E. (1989) *Vegetation and Air Pollution — Perspectives in the Northern Hemisphere and South Africa.* Report prepared for the FRD.

Zunckel M. (1998) Dry deposition of sulphur in South Africa. In *Proceedings of the 11th World Clean Air and Environment Congress.* Durban, South Africa, 13–18 September, 1998 Vol. 6, 17A–3.

CHAPTER 12

AIR POLLUTION IMPACTS ON VEGETATION IN MEXICO

M.L. de Bauer

1. Introduction

Mexico covers an area of almost two million square kilometres and links the United States, probably the most developed country in the world, with Latin America, a region that is quite different, both in economic and cultural terms. In recent decades, Mexico has embarked on a rapid programme of modernisation; this has presented many problems for the country due to the population's strong desire to maintain Mexico's historical and traditional cultures. In general, it is the large and ever-increasing low-income proportion of Mexico's population that most fervently wishes to preserve the country's cultural heritage, making the transfer to a more modernised society even more difficult.

The progression towards modernisation has brought with it many advantages but also many disadvantages, which include increasing levels of environmental pollution and associated impacts. Of the different types of environmental contamination that occur, atmospheric pollution has now become one of the most serious problems affecting both human health and ecosystems in the Republic of Mexico. Much of the atmospheric pollution load in Mexico is due to emissions from the rapidly expanding urban areas located across the country and the oil-producing regions in the south and southeast of Mexico. A number of studies, many of them performed in the Valley of Mexico, have used indicator plants to provide evidence of damage caused by photooxidants resulting from urban emissions (de Bauer, 1972). Since the 1970s, more detailed experimental studies investigating the impact of air pollutants on vegetation have been perfromed in agricultural areas on crops such as beans (Laguette, 1985).

Forests located in the south and southwest of the Valley of Mexico have also shown signs of damage similar to that induced by exposure to photooxidants. Elevated ozone levels have been recorded in this region and it would seem that the damage observed may be due to the sensitivity of the forest trees to this pollutant. A number of studies are now being performed to further explore the relationship between damage and pollutant concentrations (de Bauer *et al.*, 2000). One of the reasons for linking this forest damage with ozone has been that the damage shown by affected species, such as *Pinus hartwegii*, resembles the symptoms and general features of injury which were exhibited by *P. ponderosa* trees in the San Bernardino Mountains of California, USA. For example, Krupa and de Bauer (1976) were able to attribute damage observed on pines in the southern forested area of El Ajusco Park, located downwind from Mexico City, to O_3 pollution due to the similarity of these injury responses with symptoms on pines affected in California. Collaborative studies have been initiated over the past ten years to investigate these relationships further (Miller *et al.*, 1994). As was the case in California, bark-beetle attack was very severe on weakened trees. This fact led to some investigators attributing to insect attack damage observed on pine species, and sacred fir tree mortality observed in later years. However, Alvarado *et al.* (1993) demonstrated that the presence of bark-beetles was not absolutely necessary for trees to die.

Sacred fir trees (*Abies religiosa*), a very important native species of Mexico, showed new unexplainable symptoms of decline (Alvarado *et al.*, 1993; de Bauer *et al.*, 1985; Hall *et al.*, 1996). The seriousness of the problem was emphasised by the fact that the sacred fir population started to die in the mountainous terrain that encircles the city of Mexico. Eventually, the massive decline of this species necessitated a large programme of sanitation salvage logging during the 1980s. The evidence to suggest that *Abies religiosa* was affected by ozone is based on the fact that the types of injury found on these plants were similar to those found in *P. hartwegii*, at higher altitudes in the same locality. *P. hartwegii* showed similar symptoms to *P. ponderosa*, the species studied in California.

Urban and industrial types of air pollution are predicted to cause serious environmental problems for the forseeable future. Emissions from the oil-producing areas will continue to increase, with emissions being greater in some years compared to others, in line with the demand for oil in the international market. In contrast, urban air pollution will increase steadily and therefore follow more obvious trends with emission levels rising in line with increases in the population. Further increases in urban emissions may be reduced at the end of the century as the urban population stabilises (Bravo and Torres, 2000).

The geographical locations of Mexico, the US and Canada and the similarity of the landscape, characterised by continuous coastal mountainous ranges that run along the western seaboard edge of these countries, mean that they tend to suffer similar yet not particularly well understood environmental problems. During the last few decades, three social phenomena have profoundly affected the social and economic structure of the country, these are: (i) intense centralisation, (ii) rapid industrialisation, and (iii) accelerated urbanisation due to rural migration to the cities.

The Mexican Republic is located at the confluence of two of the world's largest biogeographic regions, the Nearctic and the Neotropical regions. As such, Mexico experiences extremely variable conditions. Due to the varied topography of the country, there are large mountainous systems that generate a wide diversity of physiographic conditions together with distinct climatic zones along altitudinal gradients. These conditions have imposed selective pressures on living organisms and forced the process of co-evolution (de Bauer *et al.*, 2000). The natural resources of the country are extremely valuable. The forested areas of the country have an extremely high diversity with over 60 native *Pinus* species alone and, as such, are amongst the most valuable natural resources of the country. Many of these native forest tree species have been taken abroad and can now be found on all continents, e.g. *P. patula*.

In recent decades, Mexico has experienced very many environmental perturbations. For example, severe aquifer pollution has occurred in certain parts of the country and deforestation poses one of the most serious environmental problems. In addition, in the south and southeast and the Gulf of Mexico, dramatic flooding in coastal regions has occurred, which is thought to have been exacerbated by anthropogenic changes to the vegetation cover.

The overwhelming problems present in Mexico at the beginning of the 21st century are:

(1) Population increase
(2) Low level of agricultural productivity
(3) Risk of exhaustion of natural resources.

1.1. *Emissions*

By dividing the country into three main regions comprising the north, south and central areas, it is possible to gain a general overview of the main air pollution problems and their relation to native vegetation and crop damage.

The most advanced intensive agricultural area in Mexico is situated in the north and northwest of the country. Horticulture is one of the most economically important agricultural types in this region since crops are grown under irrigation for export. At Baja California, to the northwest of Mexico, air pollution damage to vegetation was first observed as early as the 1960s by Middleton and Haagen-Smit (1961). This is considered to be the first report of vegetation damage attributable to air pollution in Mexico, as well as the first example of the transboundary transport of air pollution from North America.

Since the early air pollution studies carried out in Mexico during the 1970s, it has become apparent that urban areas are the most important emission sources and subsequently the most polluted areas. This has been a result of the concentration of industry and the rapidly increasing numbers of motor vehicles present in urban regions. National control policies have been established in an attempt to reduce what are considered to be the six most important urban pollutants: ozone (O_3), sulphur dioxide (SO_2), carbon monoxide (CO), nitrogen dioxide (NO_2), particles and lead (INEGI, 1994).

Urban air pollution originating from medium size localities may have some effect on the horticultural areas mentioned above, although it is considered unlikely that impacts would be particularly significant. At present, it is perhaps much more important to consider the northeast region where the industrial city of Monterrey, Nuevo Leon, is located. This is the third largest city in Mexico and is located in an enclosed valley inland. Industrial complexes on the border between the US and Mexico may be causing damage to vegetation and crops. Some of the agriculture in this region is very important such as the citrus growing industry situated close to Monterrey. Under specific dispersion conditions, it is possible that the urban plume might be an important pollutant source for areas located downwind of this urban area. However, only limited evidence exists to link vegetation damage with the industrial emissions of this region.

Recent estimates have calculated that 25.4% of the entire Mexican population is concentrated in only three metropolitan areas: Monterrey, Guadalajara and Mexico City, this value is thought to be steadily increasing. Approximately 40% of the air pollution load produced across the entire country is generated in these three cities. More than 80% of the air quality monitoring equipment that exists in Mexico is concentrated in these three cities. The fourth largest city in Mexico, Puebla, is the location for an important industrial development and has a large and ever-increasing number of motor vehicles, both being sources of air pollutant emissions (Fig. 1).

The area experiencing probably the worst air pollution problems in Mexico is the Mexico City Metropolitan Area (MCMA). This is one of the biggest urban complexes in the world and the environmental situation in this urban area is considered critical (Pick and Butler, 1997). Mexico City has been described as an example of the worst air pollution problems occurring at a global level (United Nations, 1994).

The MCMA is located in an elevated valley (2250 m a.s.l.) in central Mexico (at latitude 20°N) and, with the exception of the northeastern part of the metropolitan area, is surrounded by mountains that extend up another 800 m. The central plains to the northeast of the city are semi-arid with a mean annual precipitation of only 450 mm/ year. Vegetation in this region is scarce and as a result, the eastern suburbs are subject to dust storms at the end of the dry season (February

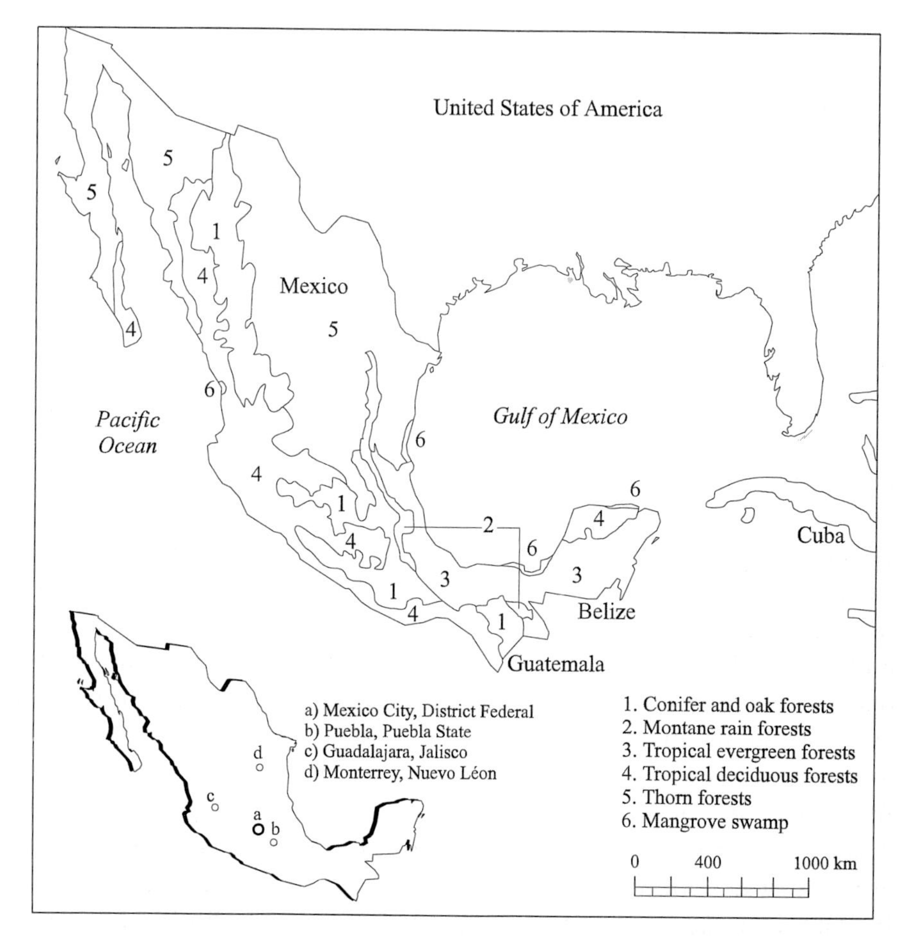

Fig. 1. Distribution of Mexican plant communities and the location of the four most important urban centres in Mexico.

to April) (Jauregui, 1989). At night and in the early morning (especially during the dry season), down-slope winds from the surrounding mountains converge with circulations induced by the heat island effect; these combine to restrict the lateral dispersion of pollutants over Mexico City. The vertical dilution of urban gases is further restricted by frequent surface inversions that occur for more than 70% of this part of the day. Over the course of the day, the stable layer is heated from below as a result of high insolation and the resulting convective turbulent mixing produces a dilution of the pollutants vertically. The regional winds that descend to the ground (usually originating from the north) transport this well mixed polluted boundary layer to the southern suburbs, where the highest levels of O_3 are frequently recorded (Jauregui, 1989; Bravo and Torres, 1995). This is a result of the occurrence of meteorological and topographical factors, combined with plentiful sunshine enhanced by ultraviolet radiation, which drives the atmospheric photochemistry leading to the formation of secondary pollutants such as O_3 and fine particulate matter (Streit and Guzman, 1986). Continuous inversions occur mainly during the winter and for up to 25 days each month (Bravo and Torres, 2000). Since 1950, the number of motor vehicles registered in the MCMA has grown almost 40 times with numbers currently estimated at 3.4×10^6, making the transport sector the main source of air pollutants. For example, during 1999 motor vehicles consumed an estimated 20 million litres of gasoline and diesel each day. The high level of emissions from transport has resulted not only from the large number of road vehicles in use, but also from poor vehicle maintenance, unfavourable fuel properties and localised traffic congestion. In addition, air quality has worsened many times as a consequence of the implementation of poorly thought out control and preventive measures, such as the introduction of low-leaded gasolines with a higher photochemical reactivity than leaded fuel (Benitez, 1992; Bravo and Torres, 2000).

The Mexican oil-producing areas located in the south of the country are the source of high SO_2 and other pollutant emissions. These industrial complexes may well as be causing serious damage to tropical vegetation located in this region as well as the many different crops grown in these parts of the country. Acid rain is also considered to be an important problem in this region. However, in spite of the possible risk that vegetation in this region is exposed to air pollution, almost no research has been done to investigate whether such impacts are occurring (Bravo and Torres, 2000).

Greenhouse and related gas emissions

The national inventory of greenhouse and related gaseous emissions was officially reviewed and updated with data collected in 1996 (SEMARNAP, 1997).

The following gases were included in this inventory: (i) those directly involved with the greenhouse effect, i.e. carbon dioxide (CO_2), methane (CH_4) and nitrous oxide (N_2O), and (ii) those mostly involved with O_3 formation: carbon monoxide (CO), nitrogen oxides (NO_x) and non-methane volatile organic compounds (NMVOC). In view of the results obtained by this inventory, the following conclusions were made:

(1) Emissions are produced predominantly as a result of the use of fossil fuels for energy production, changes in land-use and agriculture and from industrial processes.
(2) Of the gases considered, CO_2 comprises 96.42%, CH_4 0.79% and other gases 2.79% of total emissions.
(3) The energy sector is the most important anthropogenic source of CO_2 in Mexico.
(4) The transport sector contributes 32% of the emissions from fossil fuel combustion, with electricity generation and industry contributing 23% and 22%, respectively. The transport sector also represents the main source of NO_x, CH_4, N_2O and CO emissions.
(5) In 1990, almost 84% of energy and 62% of generated electricity were produced from fossil fuels.
(6) It was estimated that between 1987 and 1993 the CO_2 emissions per capita decreased 7.1%, i.e. from 3.75 to 3.48 tonnes.
(7) The country contributes less than 2% of global greenhouse gas emissions.

1.2. *Vegetation*

Leopold (1950) described the vegetation of Mexico. This description divided the country into natural biogeographical units and identified 12 different vegetation types. Four additional vegetation types were also identified, these included arctic-alpine meadows of elevated areas of major volcanoes; xeric vegetation of lava flows; littoral communities such as salt marshes, mangrove swamps and dune grasslands; and aquatic communities. Since these vegetation types had distributions restricted to only a small fraction of the total area of Mexico, they tended to be discussed separately from the 12 main categories. Leopold

Table 1. Twelve main vegetation types present in Mexico (Leopold, 1950).

Temperate	Tropical
Boreal forest	Cloud forest
Pine-oak forest	Rain forest
Chaparral	Tropical evergreen forest
Mesquite-grassland and desert	Savannah
	Tropical deciduous forest
	Thorn forest
	Arid tropical scrub

(1950) grouped these 12 classes as temperate or tropical vegetation types as described in Table 1.

There were several attempts to re-classify the Mexican vegetation in the years following 1950. Rzedowski (1978) identified ten Mexican vegetation types as follows: tropical perennial forest, tropical subcaducifolious forest, tropical caducifolious forest, thorn forest, scrub xerophytic forest, oak forest, conifer forest, mountain mesophytic forest, grassland and aquatic and subaquatic vegetation; to give a general view of the vegetation distribution of Mexico. Figure 1 has grouped these classes into six broad vegetation types. The Rzedowski (1978) classification system is now the generally accepted classification of Mexican vegetation.

2. Impacts of Air Pollution

Air pollution research in Mexico has tended to focus on oxidant effects on vegetation; this is largely due to the fact that early observations and experimental work involved this pollutant. For example, de Bauer (1972) demonstrated the importance of oxidant effects on vegetation from early investigations that involved exposing indicator plants in the metropolitan area of Mexico City. The description of O_3-induced damage on *Pinus* spp. in the southern areas of the Valley of Mexico further supported the case that O_3 is a pollutant capable of causing extensive injury (Krupa and de Bauer, 1976). Finally, observations made in 1980 of O_3-induced injury on certain varieties of beans grown upwind from Chapingo, confirmed O_3 to be a serious pollutant and resulted in the initiation of several studies to investigate the cause-effect relationship further (Laguette, 1985; Laguette *et al.*, 1986; de Bauer and Hernández, 1986).

In recent years, new information generated by a number of researchers has been collected. It is evident that suspended particulate matter (SPM) is also an important air pollutant in the Valley of Mexico, as well as in some other parts of the country. Unfortunately, no research has been done to assess the impacts of this air pollutant on vegetation in the Valley of Mexico or elsewhere in the country. Nevertheless, Bravo and Torres (1995) identify both SPM and O_3 as being the cause of the most serious air quality problems in the metropolitan area of Mexico City. Only a few studies have been carried out in other parts of the country besides the MCMA. These studies have tended to focus on instances of acute damage. For example, in 1985 severe injury to local crops was found to be caused by hydrofluoric acid (HF) emitted from a local fluorite processing plant located in the northeast of the country close to border with USA (Etchevers *et al.*, 1994).

2.1. *Field Evidence*

Instances of visible injury

As early as 1971, air pollutant-induced injury to sensitive plants was observed after exposure at different sites across the metropolitan area of Mexico City. Symptoms included bronzing of the underside of leaves of romaine lettuce and leaf-stippling on tobacco Bel W-3, characteristic O_3 injury symptoms (de Bauer, 1972). These results suggested the presence of phytotoxic concentrations of oxidants and their primary pollutant precursors at the sites investigated. In a study to determine the direction and deposition of the urban plume, Krupa and de Bauer (1976) observed photochemical oxidant symptoms on pine species, *Pinus hartwegii* Lindl. (shown in Fig. 2) and *P. leiophylla* Schl. et Cham. in the southern forested area of Mexico City at Ajusco, Distrito Federal (D.F.) located downwind from Mexico City.

These observations of oxidant type damage on pine trees initiated a series of research projects that were performed over several years,

Fig. 2. Chlorotic needles of *Pinus hartwegii* at Ajusco National Park in 1976 (courtesy of J. Galindo, Colegio de Postgraduados).

most of these studies were conducted at Ajusco, D.F. However, a massive decline of sacred fir (*Abies religiosa* H.B.K. Schl. et Cham.) was first observed in 1982 in the National Park "Desierto de los Leones" and caused alarm due to the severity and magnitude of the injuries observed (Fig. 3). The park has an area of 1529 ha and is located to the southwest of Mexico Valley at an elevation of between 2800 and 3800 m a.s.l. (see Fig. 1). The sacred fir forest is predominantly located between 2800 and 3200 m a.s.l. Precipitation is usually 1300 mm per year and annual mean temperatures range between 7 and 15°C. The soils of

Fig. 3. Massive decline of *Abies religiosa* at Desierto de los Leones National Park, 1985 (courtesy of T. Hernández, INIFAP).

the park are generally rich, with a pH between 5 and 7 (de Bauer *et al.*, 1985; Hernández and de Bauer, 1986; Vazquez, 1986).

Symptoms normally associated with damage resulting from exposure to peroxyacetyl nitrate (PAN) have also been observed in annual horticultural and ornamental crops at sites downwind of Mexico City such as the floating gardens of Xochimilco (de Bauer and Hernández, 1986; Fig. 4).

Fig. 4. Lower leaf surface bronzing of spinach, apparently induced by peroxyacetylnitrate (PAN) in the downwind area of Mexico city, Xochimilco (courtesy of T. Hernández, INIFAP).

To date, the most important and frequent visible foliar injury symptoms observed on different plant species have been related to O_3, although some cases of visible injury attributable to SO_2 pollution have been observed close to oil refineries (de Bauer and Hernandez, 1986).

An automated air quality monitoring network is currently in operation across the MCMA, this network has been financially supported by the World Bank and consists of 32 stations. Each of these stations measure several, if not all, of the following air pollutants: O_3, SO_2, NO_x, CO and SPM less than 10 μm (PM_{10}) (Bravo and Torres, 2000).

Field transect studies

A two-year field study was perfromed to continue efforts to follow the trajectory of the Mexican city urban plume. This study led to observations of vegetation injury on *P. hartwegii* and *P. montezumae* var. *lindleyi* at certain sites located along the mountain road from Mexico City to Cuernavaca. The extent of injury suffered by these two species clearly declined with distance from Mexico City, as observed up to approximately 56 km south of the city. This would suggest that a significant pollution gradient exists across this region (Hernandez and de Bauer, 1984).

The hypothesis that photochemical oxidant pollution is the cause of the damage observed on pine trees species, e.g. on *P. hartwegii*, and *P. montezumae* var. *lindleyi* Loud, was supported by the similar injury symptoms shown *in situ* on locally grown oats (*Avena sativa* L.) at Ajusco. Chlorotic mottling and banding was observed on current-year needles of *P. hartwegii*; in addition, premature senescence was observed on needles more than a year old. *P. montezumae* showed very similar but less severe symptoms. In both of these species of pine, the most severe damage was apparent at the end of the spring and the beginning of the summer season. Photochemical-induced injury to oat leaves showed chlorosis and single surface and bifacial necrotic bleaching (Hernandez, 1981). In addition, transverse leaf cuts in *P. hartwegii* showed histological changes including collapse and necrosis of the vein tissue. Cellular damage in oats was characterised by plasmolysis of the mesophyll cells and cell wall collapse. In addition, cells adjacent to the stomata also showed severe plasmolysis (Hernandez, 1981; de Bauer *et al.*, 2000).

Studies on *P. hartwegii* have shown this species, which is commonly found at elevations between 2850 and 3500 m a.s.l., to be highly sensitive to ambient O_3 concentrations. Of the nine native tree species found in the Valley of Mexico this would appear to be the most sensitive to O_3 (Hernández and de Bauer, 1984). O_3 injury has also been observed on many herbaceous plants at Ajusco and along the highway heading south towards Cuernavaca (Hernández, 1984). The scale of visible injury caused by oxidants on two species of pine tree, *P. hartwegii* and *P. montezumae*, was assessed using a method proposed by Miller (1973). From visible injury data collected over a two-year period, it was apparent that the foliar injury occurring on *P. hartwegii* was more severe than that observed on *P. montezumae.* In addition, foliar injury developed three months earlier on *P. hartwegii* compared to *P. montezumae* (Hernandez and de Bauer, 1984).

The sensitivity of *P. hartwegii* and the characteristic visible injury symptoms that occur on exposure to elevated O_3 levels have resulted in this species being used as a bioindicator for this pollutant. The characteristic symptoms include extensive yellow mottling and banding [as a result of the loss of chlorophyll "a" (Hernandez *et al.*, 1986)] and premature senescence of mature needles. This species has been studied in other areas such as the Desierto de los Leones National Park, where very severe damage has been observed, not only on this species, but also on sacred fir (Alvarado, 1989; Alvarado *et al.*, 1993). Several investigators who made observations within the Desierto de los Leones National Park have suggested a possible relationship between the severity of the O_3-induced symptoms observed on *P. hartwegii* and elevation, with greater injury occurring at higher altitudes (de Bauer *et al.*, 1985; Hernandez and de Bauer, 1986).

Recent studies have investigated the response of *P. maximartinezii* to photochemical oxidant air pollution at a tree nursery located in the south-central part of Mexico City. Observations of visible foliar injury indicated that this species was only moderately sensitive to photochemical oxidants, otherwise the plants appeared to be growing healthily across the study site. However, photooxidant exposure did appear to affect needle loss since those trees grown at the study site only maintained their needles for a three-year period. When this species is grown under native conditions, needles tend to be present for a whole five-year period (Hernandez and Nieto, 1996).

The sequence of visible injury that occurs on the species *Abies religiosa* has been documented for trees located in the Desierto de los Leones National Park, where the serious decline of this species has been attributed to photooxidant pollution. The first visible injury symptoms appear as the formation of small whitish lesions on the upper surfaces of older needles; these lesions turn reddish-brown in colour before the entire needle becomes necrotic and is shed from the tree (Fig. 5). Sacred fir trees in sample plots located to the south and southeast of the "Cementerio" ravine, were most affected, as shown by the mortality percentage obtained in the surveys conducted between 1986 and 1988 (Alvarado, 1989; Alvarado *et al.*, 1993). Histological examinations and chemical analyses of damaged and healthy needles of sacred fir were perfromed. Results clearly indicated a similarity in both the damage to the palisade tissue, as well as typical phenolic compound accumulations, that tend to be associated with O_3-induced injury (Alvarado, 1989; Alvarado *et al.*, 1993). Similar injury responses were found in the same species from the same National Park by Alvarez

Fig. 5. Advanced chlorotic and necrotic stippling and defoliation of *Abies religiosa* needles (courtesy of T. Hernández, INIFAP).

(1996) further supporting the role of O_3 in causing the injuries observed.

Extreme reductions in tree ring growth have been observed in sacred fir tree species since the beginning of the decline in this species in the 1970s (Alvarado, 1989; Alvarado *et al.*, 1993). Investigations using *A. religiosa* found that the formation of visible injury could be prevented by placing branches in charcoal-filtered chambers. Similarly, needles protected by an anti-transpirant maintained their typical dark green colour and were prevented from premature senescence and needle shedding (Alvarado, 1989; Alvarado *et al.*, 1993).

In addition to the O_3 sensitivity exhibited by *P. hartwegii*, investigations have identified synergistic interactions between biotic pathogens and photo-oxidant pollution. At Ajusco, *P. hartwegii* exhibited the characteristic symptoms associated with O_3 exposure (i.e. chlorotic banding and mottling of needles) whilst also being severely attacked by the leaf pathogen *Lophodermium* spp. The magnitude of the damage resulting from these two stressors could be directly related to the age of the needles and distance of the trees from the urban area (Alvarado and de Bauer, 1991). Observations also showed that the most affected pine trees tended to show signs of severe attack by dwarf mistletoe (*Arceuthobium* spp.) (de Bauer and Hernandez, 1986).

A standard procedure devised by Miller *et al.* (1994) was used to evaluate the O_3 injury exhibited by different pine species. This method compared the crown condition of damaged *P. hartwegii* at the Desierto de los Leones National Park with a mixture of *P. ponderosa* and *P. jeffreyi* at Barton Flats, California, USA. The O_3 injury index showed that the Mexican pine species suffered the greatest crown injury, with the Californian species exhibiting only moderate damage. Although the degree of damage varied between the Mexican and Californian trees, the types of injury evolved in a similar manner indicating that damage was caused by a common stressor. The patterns of growth of *P. hartwegii* were assessed using dendrochronological techniques and show a marked decline in growth since the early 1970s at both the Ajusco and Desierto de los Leones sites, located to the south-southwest of Mexico City. No signs of subsequent recovery have been observed to date (Alarcon *et al.*, 1995).

In a recent study, Fenn *et al.* (1999) measured deposition of inorganic nitrogen (N) and sulphur (S) over a one-year period at two locations; one downwind and the other in an eastern direction upwind of Desierto de los Leones National Park. The accumulation of S concentrations in the foliage of *P. hartwegii* was measured at both sites and was found not to be statistically different between sites. However, measurements of the accumulation of N compounds in the system supported the hypothesis that elevated N deposition at Desierto de los Leones increased the level of available N, and subsequently increased the status of *P. hartwegii* resulting in export of excess N as nitrate (NO_3^-) in streamwater.

Within the Mexico City airshed, broadleaf forest tree species also suffer from oxidant air pollution stress; symptoms include premature chlorosis and early leaf senescence indicative of season-long oxidant exposures (de Bauer and Krupa, 1990). In a recent study, Skelly *et al.* (1997) conducted surveys of foliar injury on "capulin" black cherry (*Prunus serotina* var. *capuli*) within the Desierto de los Leones National Park. During their surveys, typical adaxial leaf surface stipple was observed on approximately 41% of the indigenous "capulin" black cherry trees with injury increasing with elevation within the Park boundaries.

2.2. *Experimental Research*

Fumigation trials

An unusual case of visible foliar injury was observed during the summer of 1980, when typical ozone-induced foliar injuries were recorded on *Eucalyptus globulus* Labill. seedlings located in a newly forested urban area. To establish the possible cause-effect relationship of this damage, seedlings of *E. globulus* were exposed to 0.40 ppm O_3 for either two or four hours per day under environmentally controlled conditions. The plants subject to both treatments showed symptoms consisting of flecking and bleaching on the upper surface of intermediately aged leaves, with damage intensity being directly related to the duration of the exposure. Such symptoms resembled those observed under natural conditions in southern parts of Mexico City (Hernandez Tejeda *et al.*, 1981). In recent years, Martinez and Chacalo (1994) have observed pollutant injury in different tree species within the urban area of Mexico City. However, further research is needed to clearly establish cause-effect relationships in these and other cases.

In an attempt to demonstrate the occurrence of O_3 damage on *P. hartwegii*, Hernández (1981) developed a simple wind-activated branch exposure chamber to enclose pine branches at Ajusco, D.F. The chamber consisted of a clear plastic covered cylinder with 50%, or

100% activated charcoal in the end panels to allow varying amounts of ambient O_3 to pass through the chambers via wind movement. It was used to confirm that changes in the pigments of the pine needles were due to O_3 (Hernández *et al.*, 1986; Lefohn, 1992). The results showed clear prevention of pigment leaf changes on filtering of the ambient air through the chamber.

Field beans

During the summer growing season of 1980, symptoms resembling those caused by O_3 injury were first observed on certain varieties of bean. These varieties had been grown under field conditions in an experimental plot at the National Institute for Agricultural, Animal Husbandry and Forest Research (INFAP), in Chapingo, Mexico, a location upwind from Mexico City. The first varieties to show symptoms were Amarillo 153 and Amarillo 154. As a result of these initial observations of damage, numerous studies have been conducted over the last 20 years investigating several aspects of air pollution effects on field beans. These have included the behaviour of different varieties of *P. vulgaris* as well as other *Phaseolus* species (Laguette, 1985; Ortiz, 1988).

These field bean studies have been performed using open top chambers and the results have shown significant differences in sensitivity among the species and varieties investigated in terms of yield responses. A series of experimental trials were performed to compare two bean varieties. Control plants were treated with N-[2-{2-oxo-1-imidazolidinyl}ethyl]-n2 phenylurea (EDU) an O_3 protectant. The varieties used were Canario 107, a local bush-type variety and Pinto 111, a runner bean commonly used as a bioindicator for O_3. Measurements of several productivity parameters clearly showed that yield reductions occurred in both cultivars, with 4.5% and 40.7% yield reductions in Canario 107 and Pinto 111, respectively (Laguette *et al.*, 1986).

An experiment carried out in Montecillo, Mexico between 1994 and 1995 using the O_3-sensitive bush bean variety "Tempo", investigated the influence of three soil humidity levels in determining plant response to O_3 (control plants were again treated with EDU). The results clearly showed that plants grown under high levels of soil humidity sustained more damage than those grown under lower soil humidity. There appeared to be threshold at 40% soil humidity, only below this value were plants afforded some degree of protection against O_3-induced yield reduction (Rodriguez *et al.*, 1997).

3. Air Quality Standards

The Mexican regulations are based on the American Environmental Protection Agency (EPA) standards. An index called "IMECA" has been developed to inform the public of daily air quality (SEMARNAP, 1997). Ambient air pollution levels are measured by the Mexican Air Quality Standards (MAQS) which record concentrations for the main air pollutants.

4. Future Research

Air pollution problems in Mexico are closely related to the socioeconomic state of the country. Technological and scientific solutions should be implemented to improve air quality. These include: (i) effective emissions control; (ii) use of cleaner fuels; and (iii) implementation of reforestation projects. To ensure such solutions represent the most appropriate means of improving air quality, both scientific and technological research is required. Social aspects that affect air quality should also be considered when identifying appropriate air quality management strategies.

5. Conclusions

Given the seriousness of the air pollution problems in Mexico caused by recent rapid increases in industrialisation and urbanisation, improvements are urgently required in the development and implementation of appropriate technology to mitigate air pollution and its associated environmental impacts.

In order to achieve this, it is first considered necessary to educate the population with respect to air pollution and its consequences, and secondly to receive more governmental, as well as international support to mitigate air pollution and its effects. A priority should be to slow the increase in the country's population in order to reduce the environmental degradation that is currently occurring in Mexico.

In short, air pollution in Mexico is causing ecological, social and economic damage which requires careful consideration in terms of the scientific, governmental and public action that should be taken to improve air quality and reduce air pollution impacts.

Acknowledgements

Special recognition is gratefully expressed to Victor Perea for technical help in preparing this manuscript.

References

Alarcon M.A., de Bauer L.I., Jasso J., Segura G. and Zepeda E.M. (1995) Patron de crecimiento radial en arboles de *Pinus hartwegii* afectados por contaminación atmosferica en el Suroeste del Valle de Mexico. Agrociencia, Serie: *Recursos Naturales Renovables* **3**(3), 67–80.

Alvarado D. (1989) Declinacion y muerte del bosque de oyamel (*Abies religiosa*) en el Sur del Valle de México. Tesis de Maestria en Ciencias. Colegio de Postgraduados. Montecillo, Mex. Mexico, 78 pp .

Alvarado D., de Bauer L.I. (1991) Ataque de *Lophodermium* sp. en poblaciones naturales de *Pinus hartwegii* de "El Ajusco", México, bajo el efecto de gases oxidantes. *Micol. Neotrop. Apl.* **4**, 99–109.

Alvarado D., de Bauer L.I. and Galindo J. (1993) Decline of sacred fir (*Abies religiosa*) in a forest park south of Mexico City. *Env. Pollut.* **80**, 115–121.

Alvarez D.-E. (1996) *Determinacion de alteraciones a nivel citologico e histoquimico en Abies religiosa (H.B.K.) Schl. Et Cham. Del Desierto de los Leones del D.F., y su relación con el ozono.* Tesis de Maestria en Ciencias, Facultad de Ciencias, Universidad Nacional Autonoma de Mexico, 82 pp.

de Bauer L.I. (1972) Uso de plantas indicadoras de aeropolutos en la Ciudad de México. *Agrociencia, Serie* **D9**, 139–141.

de Bauer L.I., Hernandez Tejeda T. and Manning W.J. (1985) Ozone causes needle injury and tree decline in *Pinus hartwegii* at high altitudes in the mountains around Mexico City. *J. Air Pollut. Control Assoc.* **35**(8), 838.

de Bauer L.I. and Hernandez Tejeda T. (1986) *Contaminacion: Una Amenaza para la Vegetacion en Mexico.* Talleres Graficos del Colegio de Postgraduados, Chapingo, Mex. Mexico, 84 pp.

de Bauer L.I. and Krupa S.V. (1990) The Valley of Mexico: Summary of observational studies on its air quality and effects on vegetation. *Env. Pollut.* **65**, 109–118.

de Bauer L.I., Hernandez T.T. and Skelly J. (2000) Air pollution problems in the forested areas of Mexico and Central America. In *Air Pollution and the Forests of Developing and Rapidly Industrializing Countries*, eds. Innes J.L. and Haron A.H. CABI Publishing, Oxon, UK, pp. 35–61.

Benitez F. (1992) *Historia de la Ciudad de Mexico.* Salvat Mexicana de Ediciones, México D.F.

Bravo H.A. and Torres R.J. (1995) Trends of Air Pollution in the Metropolitan Area of Mexico City. Sección de Contaminacion Ambiental, Universidad Nacional Autónoma de México. México, D.F. 04510.

Bravo H.A. and Torres J.R. (2000) The usefulness of air quality monitoring and air quality impact studies before the introduction of reformulated gasolines in developing countries. Mexico City — A real case study. *Atmos. Env.* **34**, 499–506.

Etchevers J.D., Galvis S.A. Hernandez T. and de Bauer L.I. (1994) Water-soluble fluoride levels in plant species grown near a hydrofluoric acid production plant. *Terra* **12**(4), 383–392.

Fenn M.E., Bauer L.I., Quevedo Nolasco A. and Rodriguez Fraustro C. (1999) Nitrogen and sulfur deposition and forest nutrient status in the Valley of Mexico. *Water Air Soil Pollut.* **113**, 155–174.

Hall J.P., Magasi L., Carlson L., Stolte K.W., Niebla E., Bauer L.I., Gonzalez-Vicente C.E. and Hernandez-Tejeda T. (1996) *Health of North American Forests*/L'état de santé des forêts nord-americaines/Sanidad de los Bosques de América del Norte. Natural Resources Canada. Ottawa, Ontario, Canada, 66 pp.

Hernandez Tejeda T. (1981) *Reconocimiento Evaluación del Daño por Gases Oxidantes en Pinos Avena del Ajusco,* D.F. Tesis profesional, Universidad Autonoma Chapingo, 90 pp.

Hernandez Tejeda T., Krupa S.V., Pratt G.C. and de Bauer L.I. (1981) Sensibilidad de plantulas de eucalipto (*Eucalyptus globulus* Labill.) al ozono. *Agrociencia* **43**, 89–95.

Hernandez Tejeda T. (1984) *Efecto de los Gases Oxidantes Sobre Algunas Especies del Genero Pinus Nativas del Valle de Mexico.* Tesis de Maestria en Ciencias, Colegio de Postgraduados, Chapingo, Mex. Mexico, 109 pp.

Hernandez Tejeda T., de Bauer, L.I. (1984) Evolucion del daño por gases oxidantes en *Pinus hartwegii P. montezumae* var. Lindleyi en el Ajusco, D.F. *Agrociencia* **56**, 183–194.

Hernandez Tejeda T. and de Bauer, L.I. (1986) Photochemical oxidant damage on *Pinus hartwegii* at the Desierto de los Leones, D.F. *Phytopathology* **76**(3), 377.

Hernandez Tejeda T., de Bauer L.I. and Ortega M.L. (1986) Identificación determinacion de los principales pigmentos fotosinteticos de hojas de *Pinus hartwegii* afectadas por gases oxidantes. *Agrociencia* **66**, 71–82.

Hernandez Tejeda T. and Nieto C. (1996) Effects of oxidant air pollution on *Pinus maximartinezii* Rzedowsky in the Mexico City region. *Env. Pollut.* **92**(1), 79–83.

Instituto Nacional de Estadistica Geografia e Informatica (INEGI) (1994) *Estadisticas del Medio Ambiente, Mexico.* Mexico D.F, 447 pp.

Jauregui E. (1989) Meteorological and environmental aspects of dust storms in Northern Mexico. *Erdkunde* **43**, 141–147.

Krupa S.V. and de Bauer L.I. (1976) La ciudad daña los pinos del Ajusco. *Panagfa* **4**(31), 5–7.

Laguette R.H. (1985) *Impacto de Oxidantes Ambientales en el Cultivo de Frijol en Montecillo, Estado de México.* Tesis de Maestria, Colegio de Postgraduados, Montecillo, Mexico.

Laguette H.D., de Bauer L.I., Kohashi-Shibata J. and Ortega D. (1986) Impact of ambient oxidants on the bean crop at a locality within the Valley of Mexico. *Ann. Rep. Bean Improv.* **291**, 83–84.

Lefohn A.S. (1992) *Surface Level Ozone Exposures and Their Effects on Vegetation.* Lewis Publications Chelsea, Michigan, 366 pp.

Leopold S.A. (1950) Vegetation zones of Mexico. *Ecology* **31**(4), 507–518.

Martinez Gonzalez L. and Chacalo Hilu A. (1994) *Los Arboles de la Ciudad de Mexico,* ed. UAM/Unidad Azcapotzalco. Mexico, D.F, 351 pp.

Middleton J.T. and Haagen-Smit A.J. (1961) The occurrence, distribution and significance of photochemical air pollution in the United States, Canada and Mexico. *J. Air Pollut. Control Assoc.* **11**(3), 129–134.

Miller P.R. (1973) Oxidant-induced community change in a mixed conifer forest. In *Air Pollution Damage to Vegetation. Adv. Chem. Ser.* **122**, 101–117.

Miller P.R., de Bauer L.I., Quevedo A. and Hernandez Tejeda T. (1994) Comparison of ozone exposure characteristics in forested regions near Mexico City and Los Angeles. *Atmos. Env.* **28**(1), 141–148.

Ortiz Garcia F.C. (1988) *Efectos de los Oxidantes Ambientales en el Rendimiento Agronómico del Frijol (Phaseolus vulgaris L).* Tesis de Maestria, Colegio de Postgraduados Montecillo, Mexico.

Pick J.B. and Butler E.W. (1997) *Mexico Megacity.* Westview Press, Boulder, Colorado, 411 pp.

Rodriguez F.C., de Bauer L.I., Arteaga R.R. and Hernandez T.T. (1997) Behavior of dry beans (*Phaseolus vulgaris*) var. Tempo at three different soil humidity levels under the influence of ozone at Montecillo, México. In *Proceedings of the Dahlia Greidiner International Symposium on Fertilization and the Environment.* Haifa, Israel, pp. 89–99.

Rzedowski J. (1978) *Vegetación de México,* ed. Limusa. México, D.F, 432 pp.

Secretaría del Medio Ambiente, Recursos Naturales Pesca (SEMARNAP), México (1997) *Primera Comunicación Marco de las Naciones Unidad sobre el Cambio Climático.* Desarrollo Gráfico Editorial, México, D.F, 149 pp.

Skelly J.M., Savage J.E., de Bauer L.I. and Alvarado D. (1997) Observations of ozone-induced foliar injury on black cherry (*Prunus serotina, var. capuli*) within the Desierto de los Leones National Park, Mexico City. *Env. Pollut.* **95**(2), 155–158.

Streit G.E. and Guzmán F. (1986) Mexico City air quality: progress of an internal collaborative project to define air quality management options. *Atmos. Env.* **30**, 723–733.

United Nations (1994) *World Urbanization Prospect: The 1994 revision.* New York.

Vazquez J. (1986) *El Saneamiento la Limpia Forestales en el Desierto de los Leones.* COCODER-DDF, México, D.F, 24 pp.

CHAPTER 13

DISTURBANCES TO THE ATLANTIC RAINFOREST IN SOUTHEAST BRAZIL

M. Domingos, A. Klumpp and G. Klumpp

1. Introduction

The air pollution climate is extremely variable in Brazil as a result of the localised areas of intensive urban and industrial development that occur across the country. At a regional scale, very high levels of air pollution exist in the neighbourhood of large industrial complexes and megacities. For example, the industrial complex at Cubatão and the megacity of São Paulo both give rise to high pollutant emissions contributing to the poor air quality in the region of southeastern Brazil where they are located. Inadequate planning of the structure of industrial complexes and urban areas can lead to impaired air pollution dispersal that can also contribute to elevated levels of pollutants in these areas. Agricultural practices, vegetation burning and mining processes are additional factors that also contribute to the increasing pollution levels experienced in Brazil. However, despite the fact that regional air pollutant concentrations may exceed those observed in developed countries, continuous physicochemical air pollution monitoring stations are still scarce (Klumpp *et al.*, 2000a). According to Miguel (1992) and Kretzschmar (1994), progress in controlling air pollutants has mainly been limited to the large urban centres of Brazil situated in the southern and southeastern regions of the country.

Only a few detailed studies have been performed to investigate the atmospheric chemistry at remote locations in Brazil (e.g. the Amazon in northern Brazil). The results from studies that have been conducted suggest that concentrations of tropospheric ozone (O_3) are increasing in the Southern Hemisphere. This may be due to increases in O_3 precursor emissions resulting from changing agricultural practices, as well as the higher frequency of natural and man-made fires occurring over large tracts of natural vegetation in the region. In addition, transport of O_3 formed in urban centres to remote sites may also be contributing to increased O_3 levels in these remote regions (see references cited in Klumpp *et al.*, 2000a).

Until recently, the susceptibility of native vegetation or plant species to primary and secondary air pollutants was almost unknown, even in the more developed southern and southeastern regions of Brazil (Klumpp *et al.*, 2000a). Differential plant response to air pollution might be expected over the extensive area covered by the Brazilian territory (about 8.5 million km^2) due to the large differences in environmental conditions that occur across the region. These include variable climatic conditions (ranging from equatorial to subtropical climates), geomorphology and soil characteristics, all of which play a part in determining the establishment of several kinds of vegetation resulting in an extremely high diversity of plant species.

As urban and industrial development is more intense along the Atlantic coast, it is probable that the most severe disturbances are occurring to the Atlantic vegetation system found in this region. The most well-known example of Atlantic forest decline is found in the vicinity of the industrial complex of Cubatão, in the State of São Paulo, southeast Brazil. Since the beginning of the 1980s, extensive field investigations have been performed in this area (Domingos *et al.*, 1998; Domingos, *et al.*, 2000; Klumpp *et al.*, 2000a). As such, this review will focus on the environmental problems of this region.

However, there are other areas in Brazil where deteriorating environmental quality is also considered to be a serious problem. Extremely high pollutant emissions are associated with the petrochemical complex at Camaçari, located in the northeastern coastal region of Brazil. This is thought to be the largest industrial complex of South America and represents another potential source of stress to the Atlantic vegetation system. Unfortunately, emission data for this region are scarce and those that do exist are often considered unreliable. A biomonitoring programme has only recently been established in this region and as such, information relating to the extent of vegetation damage in this area is still lacking (Lima *et al.*, 2000). Other isolated studies that have been conducted in Brazil to assess the impacts of air pollution on vegetation are summarised by Klumpp *et al.* (2000a).

1.1. *The Disturbances to the Atlantic Vegetation System in Brazil*

The Atlantic vegetation system is located along the coast of the northeastern, southeastern and southern regions of Brazil. This extensive system comprises a diversified mosaic of ecosystems, including mangroves, forests on sandy soils ("restinga" forests), low-land and rain sub-montane/montane forests. They each have distinct structures and floristic compositions, which are the result of the high soil diversity and variable topography and climatic characteristics that exist across the region (Mello Filho, 1991/1992).

Disturbances to this vegetation system probably began around 500 years ago when the area was first colonised. This resulted in the territorial occupation of the region and the uncontrolled harvesting of the "pau-brasil" (*Caesalpina echinata*), a very common and attractive species of the Atlantic vegetation. Several other economic cycles based on agriculture (sugarcane, coffee, etc.), and exploitation of other forest resources and of minerals, greatly contributed to the disappearance of the Atlantic vegetation system in the country. The entire coastal area also has a high density of metropolitan conglomerations with more than 80 million inhabitants, a large number of ports and several industrial, chemical and petrochemical complexes. As a result, the Atlantic system has been identified as suffering from the greatest threat of extinction of all the tropical vegetation types (Mello Filho, 1991/1992).

In spite of the stresses that this Atlantic vegetation system has been placed under, there still exist secondary remnants of this vegetation type of extreme beauty, especially in the "Serra do Mar" mountain range in southeastern Brazil. This is where the Pluvial Tropical Forest, generically referred to as the Atlantic Rainforest, is located (Câmara, 1991/1992). In this region, the Atlantic Rainforest is represented by three different kinds of forest formation, namely the low-land, sub-montane and montane forests. These forest types are found at altitudes ranging from sea level to 900 metres a.s.l. and have distinct floristic compositions and physiognomies (Leitão Filho *et al.*, 1993). In general, the flora of these forests is extremely rich in evergreen tree species, palms, herbaceous plants and epiphytes, among them orchids and bromeliads (Peixoto, 1991/1992). The city of Cubatão was established within this region at the base of the "Serra do Mar" mountains, along the coast of the state of São Paulo (Leitão Filho *et al.*, 1993; Gutberlet, 1996).

Cubatão is located approximately 16 and 44 km from the cities of Santos and São Paulo, respectively (23°45' to 23°55'S; 46°15' to 46°30'W). The region has experienced rapid and uncontrolled industrial development between the 1950s and 1970s due to a number of factors that include its proximal location to the port of Santos, the availability of cheap energy and manpower, and the relatively flat terrain of the region. Currently Cubatão comprises a complex of 11 chemical/petrochemical plants, seven fertiliser factories, a non-metallic mineral plant, a paper mill, a cement factory, and a steel smelter which in total give rise to about 260 pollution sources (CETESB, 1999). A number of factors have contributed to the Atlantic Rainforest decline that has occurred in this region. These include the large number of different pollutants emitted, the local topography, the prevailing climatic conditions (characterised by high mean annual levels of precipitation, humidity and temperature) and the local air circulation patterns (Gutberlet, 1996). The latter consists of predominant daytime — air circulation from the sea towards the slopes of the "Serra do Mar" mountains. These patterns result in the transport of air pollutants through the narrow valleys that are located at the foot of these mountains; the Mogi river valley is one such valley that experiences very high levels of air pollution as a result of these circulation patterns.

The following sections briefly describe: (a) some of the characteristics of the air pollutant emissions from the industrial complex; (b) details of a programme to improve the environmental quality in

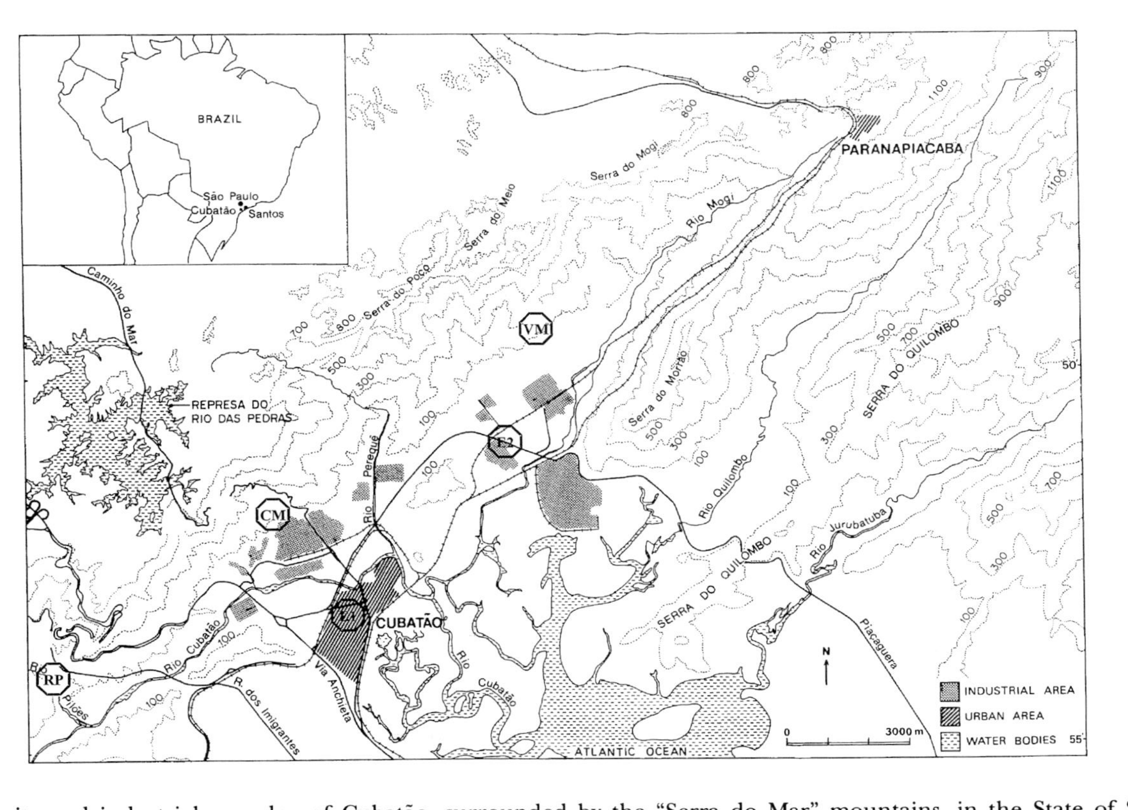

Fig. 1. The city and industrial complex of Cubatão, surrounded by the "Serra do Mar" mountains, in the State of São Paulo, southeast Brazil, and location of the study sites (RP, CM and VM) and the air monitoring stations of CETESB [station 1 (S1) and station 2 (S2)].

the region; and (c) the results of scientific studies performed over several years which clearly show the air pollution impact on the Atlantic Rainforest.

1.2. *Air Pollutant Emissions From the Industrial Complex of Cubatão*

The region of Cubatão is comprised of two distinct areas that each have a unique distribution of industry, air circulation and mass transport patterns. These characteristics have distinct influences on the air pollution found in these regions, as is apparent from studies made at two polluted sites situated in this area. The first site is located near to the centre of Cubatão, situated alongside the "Caminho do Mar" road (identified by "CM" in Fig. 1). The chemical/petrochemical complex is located close to this site resulting in high ambient levels of sulphur and nitrogen oxides (SO_x and NO_x), organic compounds (such as formaldehyde, toluene, benzene and other hydrocarbons), and other gaseous pollutants. The other site is situated in the Vila Parisi district at the Mogi river valley (identified by "VM" in Fig. 1) where several fertiliser, steel and cement plants are located. The presence of elevated concentrations of SO_x and NO_x, ammonia, fluorides and particulate matter characterise the atmosphere in this area (Alonso and Godinho, 1992).

The industrial complex at Cubatão has a history of emitting large quantities of pollutants; only in the last 15 years have these emissions been in any way controlled. In 1984, the Environmental Protection Agency of the State of São Paulo (CETESB) performed an inventory of the air pollution emissions originating from the industrial complex at Cubatão, the results of this inventory are described in Table 1.

In 1985, when environmental problems and forest decline were perceived to be at their worst, the Government of the State of São Paulo initiated several political and technical measures in an attempt to reduce environmental impacts. These were grouped within three main project areas: (i) the creation of a special commission for the recovery of the "Serra do Mar" mountains, (ii) the initiation of programmes to restore vegetation through hand planting and aerial sowing of seeds of native species, and (iii) the improvement of air quality through the establishment of a program to control emissions

Table 1. Estimates of total air pollutant emissions (in tons/year) from the industrial complex of Cubatão in 1984, 1990, 1994 and 1997, according to CETESB (1985, 1991, 1995, 1998 and 1999).

Pollutant	1984	1990	1994	1997	1998
Particulate matter	114,600	31,700	31,600	55,600	58,600
Hydrocarbons	32,800	4000	4000	4300	11,100
Sulphur oxides	28,600	18,100	17,000	40,600	35,300
Nitrogen oxides	22,300	17,400	17,300	7400	8400
Ammonia	3200	70	70	110	44
Fluorides	1000	70	70	110	28

(Alonso and Godinho, 1992; Gutberlet, 1996; Pompéia, 1997; CETESB, 1999). The initiation of these projects has lead to some improvement in the condition of the environment at Cubatão, for example, significant reductions in the levels of air pollutants have been observed, as illustrated by the air pollutant emission data estimates for 1990 shown in Table 1 (Alonso and Godinho, 1992). However, emission estimates made in subsequent years show that after 1994, levels of particulate matter, hydrocarbons and sulphur and nitrogen oxides have once again started to increase.

The State Environmental Agency (CETESB) maintains three fixed monitoring stations in the Cubatão region each linked to the automatic air monitoring network. These stations enable the continuous monitoring of air quality in the region providing information that can be used to impose actions to control pollutant emissions when necessary. These stations were established with the specific aim of collecting data that would be beneficial in improving the quality of life and health of the local population (Alonso and Godinho, 1992). The data generated at two of these monitoring stations, "Centro" (station 1 identified by "S1" in Fig. 1) and "Vila Parisi" (station 2 identified by "S2" in Fig. 1) are also useful in the evaluation of the potential stress caused by air pollution to the Atlantic Rainforest. This is despite the fact that both stations are situated about 2 km from the forest (Gutberlet, 1996). The air concentrations of total particulate matter, inhalable particulate matter (PM_{10}), SO_2, NO, NO_2, NO_x, O_3 and some organic compounds are continuously monitored at station 1, while only the concentrations of total particulate matter, inhalable particulate matter and SO_2 are measured at station 2 (CETESB, 1999).

Air quality can be evaluated by comparing the concentrations of air pollutants recorded at the monitoring stations to the national limits proposed by the Brazilian Institute of Environment (IBAMA). These national limits are regulated by the National Council of Environment (CONAMA) under resolution 03/90 of 28 June 1990 (www.mma.gov.br/port/CGMI/institu/index.html). Primary and secondary limit concentrations are defined for particulate matter, SO_2, O_3 and NO_2 as described in Table 2.

Primary limits determine maximum tolerable concentrations, which when exceeded can cause negative effects on the health of the population. These limits are very similar to those proposed by the Environmental Protection Agency of the United States of America (US EPA) and by the World Health Organization (WHO). Secondary limits define pollutant concentrations below which only minimum

Table 2. Primary and secondary limit concentrations ($\mu g\ m^{-3}$) of some air pollutants, regulated by the resolution CONAMA 03/90 and adopted by CETESB to determine air quality in the Cubatão region (CETESB, 1999).

Air pollutant	Averaging period	Primary limit	Secondary limit
Inhalable particulate matter (PM_{10})	24 hours* AAM‡	150 50	150 50
SO_2	24 hours* AAM‡	365 80	100 40
O_3	1 hour*	160	160
Total particulate matter	24 hours* AGM†	240 80	150 60
NO_2	1 hour* AAM‡	320 100	190 100

*Must not be exceeded more than once a year; †Annual geometric mean; ‡Annual arithmetic mean

Table 3. Annual arithmetic mean concentrations of inhalable particulate matter, SO_2 and O_3 and annual geometric mean concentrations of total particulate matter ($\mu g\ m^{-3}$) at the monitoring stations "Centro" (Station 1) and "Vila Parisi" (Station 2), in the Cubatão region between 1991 and 1998 (CETESB, 1992–1999).

	Inhalable particles (PM_{10})		SO_2		O_3 (Max. — 1 hour)		Total particulate matter	
Years	Station 1	Station 2	Station 1	Station 2	Station 1	Station 2	Station 1	Station 2
1991	67	147	15	10	296	—	82	201
1992	60	94	7*	29	266*	—	77	168
1993	47	129	6	18	181*	—	80	202
1994	18*	190	7*	12	66*	—	70	184
1995	57	160	#	15*	#	—	72	189
1996	44*	97*	19*	23*	197*	—	76*	197
1997	41	98	25*	14	329*	—	74	186
1998	39	95	16*	26*	227	—	64	199
1991–1995	50	144	9	17	202	—	76	189

*The criterion of sampling sufficiency not reached, according to the State Environmental Agency (CETESB); # Data not available; - Not measured in the station adverse effects on the human population, fauna, flora and the environment would generally be expected to occur. These secondary standards can be regarded as desirable levels of pollutant concentration, and as such should represent target levels for policy to aim towards for the long-term prevention of air quality degradation. The CONAMA resolution 03/90 stipulates that before applying either of the standards, the national territory should be divided into different classes of land use. This has not been performed and consequently only primary limits have been applied to date (CETESB, 1999).

It is interesting to compare the primary limit values described in Table 2 with the annual concentrations of air pollutants recorded between 1991 and 1998 at the CETESB monitoring stations described in Table 3. Measured values of SO_2 concentrations do not exceed either the primary or secondary limit values indicating that emissions of this pollutant are appropriately controlled from the industrial complex. However, concentrations of particulate matter (both inhalable and total) at station 2 (located in the Mogi valley (VM)) greatly exceeded the primary limit values established for this pollutant. Similarly, limit values for O_3 were exceeded for most years at station 1.

Concentrations of atmospheric fluorides are not continuously monitored in the Cubatão region. However, the clustering of fertiliser and steel industries at the entrance of the Mogi river valley (VM) is likely to lead to high emissions and atmospheric concentrations of this

Table 4. Mean concentrations of gaseous fluorides ($\mu g\ m^{-3}$) recorded during monitoring campaigns conducted over one month at a location in the Mogi river valley (VM), between 1985 and 1997 (Pompéia, 1997; Lopes, 1999).

Years	Mean	Max. daily value
1985	4.4	10.3
1987	2.4	5.2
1991	0.9	1.5
1992	0.6	5.6
1998	1.4	3.6

pollutant in the vicinity. To evaluate the effect of these industries on local air quality, monitoring campaigns of one-month duration were performed in the area. The results of these campaigns are presented in Table 4 and show large reductions in the levels of gaseous fluorides from 1985 to 1998. However, based on critical levels suggested by "Verein Deutscher Ingenieure" (VDI), gaseous fluoride concentrations are still considered high enough to threaten the Atlantic Rainforest in the Mogi valley (Lopes, 1999). In summary, the values given in Tables 3 and 4 show that since 1990, there has been little change in the concentration levels of all the pollutants monitored.

2. Air Pollution Impacts on the Atlantic Rainforest Around the Industrial Complex of Cubatão

2.1. *Field Evidence*

Air pollution impacts on the Atlantic Rainforest were first observed at both the landscape and physiognomic level. Increasing degradation to this vegetation system was most obvious between 1960 and 1985 when the highest emissions were recorded (Table 1). Damage covers over an area of 60 km^2 and has been observed over a wide altitudinal range (from sea level to 800 metres a.s.l. at the Paranapiacaba plateau), but has been most severe on exposed slopes (Pompéia, 1997). The upper tree stratum showed intense injury and a high mortality rate, resulting in a reduction in species diversity and in disturbances in the ecological capacity to maintain the water table level and soil stability. As a consequence, a number of large-scale landslides have occurred which have threatened both populated areas and factory complexes (Gutberlet, 1996; CETESB, 1998). Disturbances to nutrient cycling have also been detected together with structural changes in parts of the forest situated on the Paranapiacaba plateau. Confirmation that air pollution was the cause of the stress which this forest was experiencing was found through changes in the nutrient fluxes through litterfall and in the litter layer on forest floor (Domingos *et al.*, 2000).

Bragança *et al.* (1987) compared aerial photographs of secondary vegetation taken in 1962, 1977 and 1985. They found that more than 50% of the 70 km^2 study area was dominated by portions of forest with clear symptoms of degradation, which occurred more frequently on the steep slopes of the Mogi valley and along the "Caminho do Mar" road. Symptoms observed included increased tree mortality, higher frequencies of the occurrence of gaps and isolated trees and the absence of a closed forest canopy.

Phytosociological studies showed that remnant vegetation found at the lower altitudes of the Mogi valley was composed of a dense herb stratum, which comprised several invader species and a few tree species. High tree mortality and a large heterogeneity in the tree canopy was also observed in the forest near "Caminho do Mar" (Leitão Filho *et al.*, 1993; Pompéia, 1997). Pompéia (1997) identified 70 species, 53 genera and 30 families in the tree stratum of one hectare located in the Mogi valley, and 88 species, 67 genera and 39 families in an area of the same size located at "Caminho do Mar". In comparison, one hectare of forest located at an unpolluted site in the Pilões river valley (identified by RP in Fig. 1), was found to have a higher number of tree species (96), genera (69) and families (38) than either the "Caminho do Mar" or the Mogi valley sites.

2.2. *Experimental Research*

Objectives and methods

An integrated research project between German and Brazilian institutions was established in 1989 to investigate the dispersion, transformation and deposition of air pollutants released from the industrial complex at Cubatão. This study also investigated the direct and indirect effects of deposition on the soils and forests covering the slopes of the "Serra do Mar" mountains. The following sections briefly summarise the main conclusions from this project that are of relevance to the impacts of air pollution on the Atlantic Rainforest.

Investigations were performed using both active and passive biomonitoring techniques, which represent two established methodologies that have been used extensively in Europe and North America. Both biomonitoring studies were conducted at the polluted sites of "Caminho do Mar" (CM) and the Mogi river valley (VM), situated at altitudes of 80 metres and 20 metres a.s.l., respectively

(Fig. 1). The Pilões river valley (RP) was used as a reference site because of its location away from the path of the prevailing winds from Cubatão and due to the protection from polluted air masses afforded by the surrounding geographical barriers. This valley is located 40 metres a.s.l., and the vegetation in this region shows no visible signs of degradation induced by air pollution. The secondary vegetation at this location is dense, with large trees, well-structured canopies and a high species diversity (Leitão Filho *et al.*, 1993).

Standard active biomonitoring methods were employed during the first series of studies in which indicator plants were exposed to the environment. Two types of indicator plant were used, those with known sensitivity to specific air pollutants (e.g. *Nicotiana tabacum* cv. Bel W3, an O_3 indicator species and *Urtica urens,* a PAN indicator species) and those with the potential to accumulate pollutants (e.g. *Lolium multiflorum* ssp. *italicum* cv. Lema, a fluoride, sulphur and metal accumulator). Visible foliar symptoms in sensitive plants and changes in foliar element concentrations in accumulator plants were then monitored in order to obtain information on the current air pollution situation and on the spatial and temporal distribution of air pollutants in the region. In the second phase of the project, some native tree species of the Atlantic Rainforest, and others of tropical origin, were introduced into the studies, among them *Tibouchina pulchra* and *Psidium guajava.* Leaves of these plants were submitted to several biochemical and chemical analyses, which included determination of peroxidase activity and ascorbate, fluoride, sulphur and nitrogen contents, in addition to growth parameter observations. This information was used to draw conclusions on the causes of vegetation damage, to improve transferability of results to native vegetation and to facilitate risk prognosis. Between 1990 and 1995, traditional bioindicator species were exposed at each site for consecutive periods of 28 days, over one to three years depending on the species. The saplings of *T. pulchra* and *P. guajava* were exposed for consecutive periods of 16 weeks, between 1992 and 1996.

Passive biomonitoring studies were performed to evaluate the present state of the vegetation, the effects of air pollution on plant vitality and to identify possible mechanisms of air pollution resistance. These studies used native tree species that are present in large

Table 5. Annual mean pollutant concentrations measured at Pilões valley (RP), "Caminho do Mar" (CM) and the Mogi valley (VM) and regional environmental parameters recorded between 1991 and 1995.

	Pilões valley (RP)	Caminho do Mar (CM)	Mogi valley (VM)
SO_2 ($\mu g\ m^{-3}$)	6.7	25.6	18.1
O_3 ($\mu g\ m^{-3}$)	29.8	78.2	27.1
NO_2 ($\mu g\ m^{-3}$)	7.5	19.4	24.9
Temperature (°C)		22.1	
Relative humidity (%)		85	
Precipitation (mm)		3000	

numbers in the Atlantic Rainforest such as *Tibouchina pulchra* and *Miconia pyrifolia* (Melastomataceae) (Leitão Filho *et al.*, 1993; Pompéia, 1997). Data detailing a wide range of biochemical (peroxidase activity, ascorbate concentrations, pH of cell sap and buffer capacity index) and chemical parameters (mineral contents) were collected. These data were used to indicate the extent of "hidden injuries", to aid identification of repair and detoxification mechanisms and to assess the accumulation of toxic substances and disturbance in the plant nutrient balance. At each site, leaves of adult individuals of each species were taken in Jun/91 (winter time) and in Mar/92 (summer time).

During the period of study (1991 to 1995), additional air quality monitoring to that carried out by the CETESB (Table 3) was made by both the German and Brazilian Project. These measurements recorded annual mean pollutant concentrations of SO_2, O_3 and NO_2 along with other environmental parameters as shown in Table 5.

Main results

Extremely high foliar accumulations of fluorine were recorded in both the active and passive bioindicator species (Tables 6 and 7) at the Mogi valley (VM) site indicating that this pollutant may pose a significant risk to the health of the Atlantic Rainforest at this location.

Table 6. Results of the active biomonitoring studies developed between 1991 and 1996 at the control site in Pilões river valley (RP) and at the polluted sites in "Caminho do Mar" (CM) and Mogi river valley (VM).

Bioindicator responses	**Bioindicator species**	RP	CM	VM	Reference
Foliar F contents (μg/g DW)	*L. multiflorum* *T. pulchra*	5 56	30* 112	247* 427*	Klumpp *et al.* (1996) Klumpp *et al.* (1998)
Foliar S contents (mg/g DW)	*L. multiflorum* *T. pulchra*	4.1 5.3	5.7* 90*	6.3* 9.1*	Klumpp *et al.* (1996) Klumpp *et al.* (1998)
Foliar N contents (mg/g DW)	*L. multiflorum* *T. pulchra*	38 13	44 17	44 23*	Domingos *et al.* (1998) Klumpp *et al.* (1998)
N/Ca ratio	*T. pulchra*	0.9	0.9	1.2*	Domingos (1998)
S/Ca ratio	*T. pulchra*	0.3	0.5*	0.5*	Domingos (1998)
Foliar O_3-induced damage (%)	*N. tabacum*	3.7	17.6*	5.6	Klumpp *et al.* (1996)
Foliar PAN-induced damage (%)	*U. urens*	3	21*	7	Klumpp *et al.* (1996)
Foliar ascorbate concentrations (mg/g DW)	*T. pulchra* *P. guajava*	3.1 4.0	17* 3.4	1.7* 2.4*	Klumpp *et al.* (2000b) Klumpp *et al.* (1998)
Peroxidase activity (ΔE/min/g DW)	*T. pulchra* *P. guajava*	41 79	125* 117	121* 264*	Klumpp *et al.* (2000b) Klumpp *et al.* (1998)
Shoot-root ratio	*P. guajava*	1.4	1.7	2.3*	Klumpp *et al.* (1998)

*Significantly different from the mean value observed in RP ($p < 0.05$; multiple comparison tests).

Table 7. Mean results of the passive biomonitoring studies developed in Jun'91 and Mar'92, at the control site at Pilões river valley (RP) and the polluted sites at "Caminho do Mar" (CM) and Mogi river valley (VM).

Bioindicator responses	**Bioindicator species**	RP	CM	VM	Reference
Foliar fluoride contents (μg/g DW)	*T. pulchra*	28	45	221*	Klumpp *et al.* (1998)
	M. pyrifolia	5	27	67*	
Foliar sulphur contents (mg/g DW)	*T. pulchra*	6.2	6.9	7.4	
	M. pyrifolia	3.2	5.2+	5.5+	
Foliar nitrogen contents (mg/g DW)	*T. pulchra*	23	24	25	
	M. pyrifolia	20	23	22	
N/Ca ratio	*T. pulchra*	1.7	2.3	2.6*	Domingos (1998)
	M. pyrifolia	3.6	4.6*	4.4*	
S/Ca ratio	*T. pulchra*	0.5	0.8*	0.8*	
	M. pyrifolia	0.5	1.1*	1.1*	
Foliar ascorbate concentrations (mg/g DW)	*T. pulchra*	2.6	3.1	2.1	Klumpp *et al.* (1997) Klumpp *et al.* (2000b)
Peroxidase activity (ΔE/min/g DW)	*T. pulchra*	46	64	64	
pH of cell sap	*T. pulchra*	3.5	3.2*	3.4	
	M. pyrifolia	3.9	3.6*	3.6*	
Buffer capacity index	*T. pulchra*	7.5	5.6*	6.8	
	M. pyrifolia	10.6	7.7*	8.5	

*Significantly different from the mean value observed in RP ($p < 0.05$; multiple comparison tests).
+Significantly different from the mean value observed in RP ($p < 0.10$; multiple comparison tests).

The increase in the foliar concentrations of sulphur, especially in *L. multiflorum* and in saplings of *T. pulchra* (Table 6), suggest that vegetation may be at risk from damage by sulphurous compounds at both polluted sites. In contrast, nitrogen was only significantly accumulated by young plants of *T. pulchra* when exposed at the Mogi valley site, indicating a potential risk to the Atlantic Rainforest from nitrogen-containing pollutants at this location (Table 6). Despite the fact that both minerals are essential macronutrients for plants, an increase in their uptake may lead to adverse effects. These include loss of other macronutrients from the soil and nutritional disturbances resulting from the function of the mineral compounds SO_2, NO_x and NH_4^+ as acidic and secondary pollutant precursors. For example, Mayer *et al.* (2000) observed the development of soil acidification and cation leaching at the Mogi valley. These processes might be the causes of the nutritional imbalances detected at this site which were illustrated by the estimates of N/Ca and S/Ca ratios for young and adult trees of native species (Tables 6 and 7).

Severe O_3 and PAN-induced injuries were observed on the leaves of *N. tabacum* and *U. urens*, with damage most frequently being observed at elevated sites along the "Caminho do Mar" road (CM). It should be emphasised that phytotoxic O_3 concentrations occur across the entire "Serra do Mar" region, representing a risk of damage by this pollutant to O_3 sensitive species (Table 6).

Peroxidase activity increased and ascorbate levels decreased in *P. guajava* exposed at the Mogi valley site (Table 6). Similar trends were observed from the passive biomonitoring studies (Table 7). These biochemical indicators are part of the endogenous protective mechanisms of plant cells against oxidative stress, indicating that oxidant pollutants may have been causing "hidden" injuries in these plants. Table 7 shows that the pH of leaf tissue and buffer capacity index of trees decreased at sites with high pollution loads (e.g. the "Caminho do Mar" site) indicating a reduced ability to compensate for increased acidic deposition. Exposure of *P. guajava* to ambient air at the Mogi valley site caused a strong effect on certain growth characteristics. For example, shoot/root ratio was significantly increased (Table 6) as a result of a significant reduction in root growth.

3. Conclusions

The results obtained from these biomonitoring studies clearly indicate that, despite the introduction of emission control measures and subsequent reductions in air pollutant emissions (Alonso and Godinho, 1992), the complex mixture of air pollutants originating from the industrial complex may still be causing damage to the Atlantic Rainforest. Sulphur compounds and secondary pollutants are probably most responsible for the forest decline observed in the "Caminho do Mar" region. Fluorides, sulphur and nitrogen compounds are involved in the forest disturbance that is occurring in the Mogi valley region. However, it is not possible to clearly attribute symptoms seen at these sites to any specific air pollutant, since the studies described have been conducted under field conditions where a range of pollutants and other environmental factors may be responsible for the plant responses observed. Synergistic or additive effects of air pollutants are well documented and highly likely to occur at these sites where a number of different pollutants are present at elevated concentrations.

Similarly, the risk from air pollution to the Atlantic Rainforest as a whole cannot be determined from studies indicating susceptibility of only a few individual plant species. Among the hundreds of species occurring in this particular part of the Atlantic Rainforest, only a few local species, *Cecropia glazioui* (Cecropiaceae), *Psidium cattleyanum* (Myrtaceae), *Tibouchina pulchra*, *Miconia cabucu* and *Miconia pyrifolia* (Melastomataceae), were included in the biomonitoring studies. The pioneer tree species *T. pulchra* occurs across the entire region and as such, the distribution of this species does not appear to be affected by ambient pollution levels. Therefore, it would appear that this species is relatively insensitive to air pollution, a conclusion that is supported by the data presented in Tables 6 and 7 as well as by results from other more recent studies not yet published. This tree species does however show chemical, biochemical and physiological responses to elevated pollution levels. Some of these responses are endogenous mechanisms of resistance against acidic and oxidative stresses and are probably ensuring the survival of plants at the polluted sites. Due to these characteristics, this species is increasingly being used in biomonitoring studies in the region (Klumpp *et al.*, 1997; Domingos *et al.*, 1998).

Finally, values obtained from the automatic air quality monitoring conducted by CETESB were compared with those made within the Brazilian and German Project (as presented in Table 5) which were located closer to the emission sources and adjacent to the vegetation biomonitoring sites. Based on the annual mean SO_2 concentrations, the only equivalent air pollution index measured by both organisations at the different sites, it would appear that the CETESB monitoring is under-estimating the level of air pollution to which vegetation is exposed.

4. Applicability of European and North American Air Quality Standards

The primary air quality standards proposed by the Brazilian Institute of the Environment, used to indicate the level of emission control necessary to improve air quality within the State of São Paulo, are similar to those proposed by the US EPA and WHO. However, for SO_2 and NOx, the limit values are much higher than those adopted in Europe for the protection of vegetation.

In order to offer more protection to the Atlantic Rainforest located close to the industrial complex of Cubatão, it would be desirable to improve air quality in the short term by applying the national secondary standards. However, the introduction of lower threshold concentration values similar to those established in Europe to protect vegetation should become a long-term goal. Additional monitoring stations should also be linked to the automatic monitoring network of CETESB, aiming to improve the evaluation of risk that the Atlantic Rainforest is under from ambient levels of air pollution. In addition, continuous monitoring of air quality should be extended to cover all the national territory. Monitoring stations should be located so as to provide information that can be used to develop policies to protect not only the human population, but also Brazilian vegetation, especially in areas and regions likely to be under threat from industrial or urban air pollution.

5. Future Research

The environmental problems caused by the great variety of air pollutants that have been emitted in the region near the city of Cubatão over the past 30 years and more, are complex and difficult to fully understand. Many extensive and multidisciplinary studies are still needed. Emphasis should be given to increasing understanding of the degree of sensitivity or resistance of the native plant species to air pollution stress, with reference to the response at different levels of biological organisation (cells, tissues, organs, organisms, populations and communities of plants). Further investigations need to be performed to more accurately assess the influence that specific air pollutants and air pollutant mixtures are having on the vegetation of the study area (e.g. the impact of organic compounds on vegetative systems) and to establish cause-effect relationships by conducting appropriate fumigation experiments.

On a national scale, far more research is needed since the extent of air pollution impacts on other types of Brazilian vegetation are still unknown. The experience acquired from conducting studies in the Cubatão region should be applied to study air pollution impacts in other high-risk areas of Brazil so as to indicate the extent and magnitude of the national air pollution problem with respect to vegetation.

Acknowledgements

We gratefully acknowledge the help given to us by the Chemistry section of the German and Brazilian Project in providing information regarding the air quality of the Cubatão region.

References

Alonso C.D. and Godinho R. (1992) A evolução da qualidade do ar em Cubatão. *Química Nova* **15**, 126–136.

Bragança C.F., Kono E.C., Aguiar L.S.J. and Santos R.P. (1987) Avaliação da degradação da Serra do Mar. *Ambiente* **1**, 77–85.

Câmara I.G. (1991/1992) Conservação da Floresta Atlântica. In *Floresta Atlântica/textos científicos* (coordinated by Monteiro S. and Kaz L. Edições Alumbramento, Rio de Janeiro, pp. 23–31.

CETESB. (1985) *Relatório de qualidade do ar no Estado de São Paulo — 1984.* Série relatórios.

CETESB. (1991) *Relatório de qualidade do ar no Estado de São Paulo — 1990.* Série relatórios.

CETESB. (1992) *Relatório de qualidade do ar no Estado de São Paulo — 1991.* Série relatórios.

CETESB. (1993) *Relatório de qualidade do ar no Estado de São Paulo — 1992.* Série relatórios.

CETESB. (1994) *Relatório de qualidade do ar no Estado de São Paulo — 1993.* Série relatórios.

CETESB. (1995) *Relatório de qualidade do ar no Estado de São Paulo — 1994.* Série relatórios.

CETESB. (1996) *Relatório de qualidade do ar no Estado de São Paulo — 1995.* Série relatórios.

CETESB. (1997) *Relatório de qualidade do ar no Estado de São Paulo — 1996.* Série relatórios.

CETESB. (1998) *Relatório de qualidade do ar no Estado de São Paulo — 1997.* Série relatórios.

CETESB. (1999) *Relatório de qualidade do ar no Estado de São Paulo - 1998.* Série relatórios.

Domingos M. (1998) *Biomonitoramento da fitotoxicidade da poluição aérea e da contaminação do solo na região do complexo industrial de Cubatão, São Paulo, utilizando Tibouchina pulchra Cogn. como espécie indicadora.* PhD Thesis, Universidade de São Paulo, São Paulo.

Domingos M., Klumpp A. and Klumpp G. (1998) Air pollution impact on the Atlantic Forest at the Cubatão region, SP, Brazil. *Ciência and Cultura* **50**, 230–236.

Domingos M., Lopes M.I.M.S. and De Vuono Y.S. (2000) Nutrient cycling disturbance in Atlantic Forest sites affected by air pollution coming from the industrial complex of Cubtaão, Southeast Brazil. *Revista Brasileira de Botânica* **23**, 77–85.

Gutberlet J. (1996) *Cubatão: desenvolvimento, exclusão social e degradação ambiental.* Editora Universidade de São Paulo, São Paulo.

Klumpp A., Klumpp G. and Domingos M. (1996) Bio-indication of air pollution in the tropics. The active monitoring programme near Cubatão (Brazil). *Gefahrstoffe — Reinhaltung der Luft* **56**, 27–31.

Klumpp A., Domingos M., Klumpp G. and Guderian R. (resp.). (1997) Vegetation module. In *Air Pollution and Vegetation Damage in the Tropics — The Serra do Mar as an Example: — Final Report 1990–1996*, eds. Klockow D., Targa H.T. and Vautz W., German/Brazilian Cooperation in Environmental Research and Technology, Studies on Human Impact on Forests and Floodplains in the Tropics (SHIFT) Programme. GKSS — Forschungszentrum Geesthacht GmbH, Geesthacht, pp. V1–V47.

Klumpp A., Domingos M., Moraes R.M. and Klumpp G. (1998) Effects of complex air pollution on tree species of the Atlantic Rainforest near Cubatão, Brazil. *Chemosphere* **36**, 989–994.

Klumpp A., Domingos M. and Pignata M.L. (2000a) Air pollution and vegetation damage in South America — state of knowledge and perspectives. In *Environmental Pollution and Plant Responses*, eds. Agrawal S.B. and Agrawal M. Lewis Publishers, Boca Raton, pp. 111–136.

Klumpp G., Furlan C.M., Domingos M. and Klumpp A. (2000b) Response of stress indicators and growth parameters of *Tibouchina pulchra* Cogn. exposed to air and soil pollution near the industrial complex of Cubatão, Brazil. *Sci. Total Env.* **246**, 79–91.

Kretzschmar J.G. (1994) Particulate matter levels and trends in Mexico City, São Paulo, Buenos Aires and Rio de Janeiro. *Atmos. Env.* **28**, 3181–3191.

Leitão Filho H.F., Pagano S.N., Cesar O., Timoni J.L. and Rueda J.J. (1993) *Ecologia da Mata Atlântica em Cubatão.* Editora UNESP, Rio Claro.

Lima J.S., Fernandes E.B. and Fawcett W.N. (2000) *Mangifera indica* and *Phaseolus vulgaris* in the bioindication of air pollution in Brazil. *Ecotoxicol. Env. Safety*, in press.

Lopes C.F.F. (1999) *Fluoretos na atmosfera de Cubatão.* Relatório Técnico, CETESB.

Mayer R., Liess S., Lopes M.I.M.S. and Kreutzer K. (2000) Atmospheric pollution in a tropical rain forest: Effects of deposition upon biosphere and hydrosphere. I. Concentrations of chemicals. *Water, Air Soil Pollut.* **121,** 59–78.

Mello Filho L.E. (1991/1992) A Floresta Atlântica. In *Floresta Atlântica/textos científicos* coordinated by Monteiro S. and Kaz L. Edições Alumbramento, Rio de Janeiro, pp. 17–21.

Miguel A.H. (1992) Poluição atmosférica urbana no Brasil: Uma visão geral. *Química nova* **15**, 118–125.

Peixoto A.L. (1991/1992) A vegetação da Costa Atlântica. In *Floresta Atlântica/textos científicos* coordinated by Monteiro S. and Kaz L. Edições Alumbramento, Rio de Janeiro, pp. 33–42.

Pompéia S.L. (1997) *Sucessão secundária da Mata Atlântica em áreas afetadas pela poluição atmosférica Cubatão, SP.* PhD Thesis, Universidade de São Paulo, São Paulo.

CHAPTER 14

ASSESSING THE EXTENT OF AIR POLLUTION IMPACTS IN DEVELOPING COUNTRY REGIONS

L. Emberson, J. Kuylenstierna and M. Ashmore

1. Introduction

Chapters 6 to 13 have described some of the key observational and experimental studies providing evidence of air pollutant impacts on vegetation in rapidly industrialising regions. These studies suggest that in many of these regions, particularly in parts of Asia, crop yields and the health and productivity of forests are being severely affected by local ambient air pollutant concentrations. However, these studies are relatively few in number, and not nearly as extensive in terms of the range of species, cultivars and pollutant concentrations covered as studies performed in Europe or North America (as discussed in Chaps. 2 and 3). As such, it is currently not possible to accurately assess the full regional extent of air pollution damage to vegetation in Asia, Latin America and Africa. However, to inform the implementation of emission reduction policy initiatives there is a real need to provide some indication of the potential risk to which agricultural crops and forest trees are exposed under current-day air pollution concentrations, and the increased risk which may arise from future increases in emissions. There are a number of different tools that have been developed in Europe and North America to estimate the risks of air pollution impacts to vegetation. This chapter examines these tools and considers the extent to which they can be applied in developing countries, concentrating on sulphur dioxide (SO_2) and ozone (O_3).

2. Tools for Assessing Air Pollution Impacts

Two major types of assessment method can be identified. The first is to set some form of air quality guideline, below which there is no risk of damage to vegetation or, alternatively, there is an "acceptable" risk of damage. Policy assessment may then be aimed at eliminating, or minimising, the area over which this guideline is exceeded. The second is to define exposure-response relationships, which may be used to assess the change in the degree or risk of damage for a given change in pollutant exposure. This information may then be used in socio-economic impact analyses or cost-benefit analyses which estimate the environmental benefits of investment in measures to reduce pollutant emissions. These two tools are in fact complementary. Thus, exposure-response relationships can be used to assess the size and importance of impacts on vegetation in areas where defined air quality guidelines are exceeded. They can also be used to identify threshold exposures for impacts on vegetation and thus in setting air quality guidelines. In this section, we consider each assessment method in turn.

2.1. *Exposure-Response Relationships*

There are a limited number of exposure-response relationships that have been derived for SO_2 impacts. These include the response of perennial ryegrass from UK studies (Fig. 1) as well as those relationships illustrated in Chap. 5 (from Australia) and Chap. 6 (from China).

Most of the recent work to develop exposure-response relationships has been carried out for O_3. The UN/ECE Integrated Cooperative Programme on Vegetation and Crops has collected and pooled published data describing exposure-response relationships (relative yield against 7-hour mean ozone concentrations) for a number of crop species (Fig. 2). These indicate, for example, that spring wheat

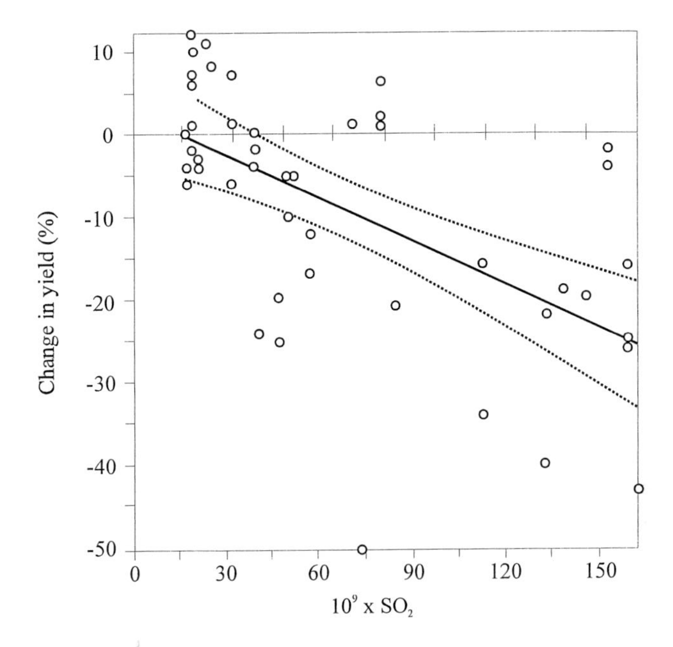

Fig. 1. Linear regression of SO_2 concentration and percentage yield reduction for chamber exposures of *Lolium perenne* (ryegrass) for more than 20 days at less than 200 $\mu g\ m^{-3}$. The regression equation was: yield loss (%) = $+2.75-0.18\ SO_2\ (/10^9)$ ($p < 0.001$). The dashed lines show the 95% confidence limits (Roberts, 1984).

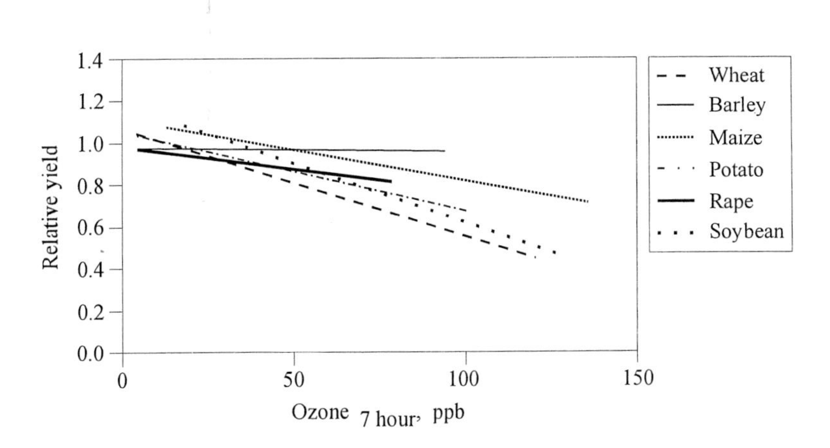

Fig. 2. Yield of various crops in response to O_3, showing regression lines for the response functions. The length of the lines reflects the range of data available (source: Mills *et al.*, 2000).

is the most sensitive arable crop for which data have been collected (Mills *et al.*, 2000).

These relationships can be compared to those derived for crops during the National Crop Loss Assessment Network (NCLAN) experiments in North America. Figure 3 shows yield relationships for (a) five crops plotted against 12-hour means over the growing season, and (b) four crops plotted against 7-hour means over the growing season. Data from the NCLAN exposure-response regression analyses (as reported by Percy, this vol.) indicated that at least 50% of species/cultivars tested exhibited a 10% yield loss at 7-hour seasonal mean O_3 concentrations of less than 50 ppb The NCLAN data suggest that soybean and cotton are both significantly more sensitive than winter wheat for 12-hour averages, but care should be exercised in comparisons between the European and US data, due to differences in the cultivars used in these experiments.

Such exposure-response relationships have been applied in North America and Western Europe to estimate crop yield reductions. For example, the OECD (1981) used exposure-response relationships (such as that in Fig. 1) to estimate the economic costs of yield loss due to SO_2. In the US, the relationships illustrated in Fig. 3 were combined with monitored or modelled O_3 concentration data and databases on crop distributions to estimate the impact of O_3 on crop production both nationally and regionally. For example, Adams *et al.* (1988) estimated that a national decrease in O_3 concentrations of 40% would provide a net annual economic benefit of US$3000 million, or about 3% of national production of agricultural crops. Olszyk *et al.* (1988) estimated yield losses in California due to O_3 to be 20–25% for cotton, beans and onion. In Europe, a recent evaluation in the context of EU emission control policy estimated the annual benefit of reducing O_3 exposure by 30% as being about 2000 million Euro (EC, 1998). Mills *et al.* (2000) estimated that economic loss in terms of wheat yield alone due to current O_3 exposures in the EU was about 1500 million Euro.

However, the uses of such relationships to estimate yield reductions may not be appropriate if there are fundamental assumptions that are invalid, either in the methods used to produce the exposure-response

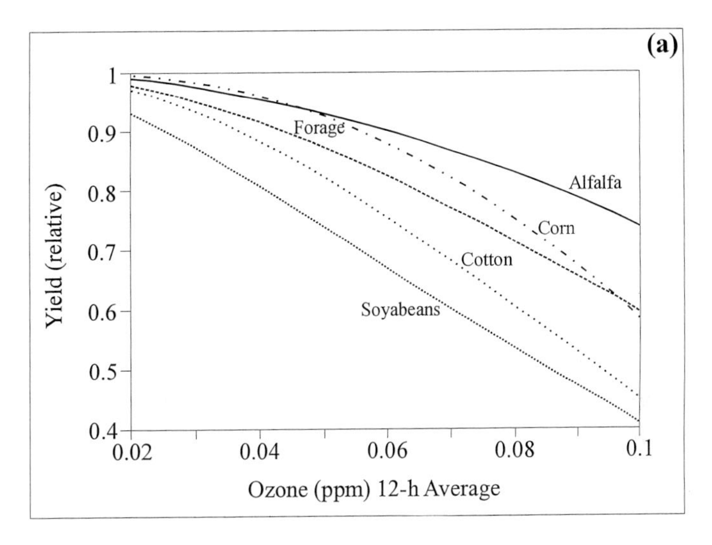

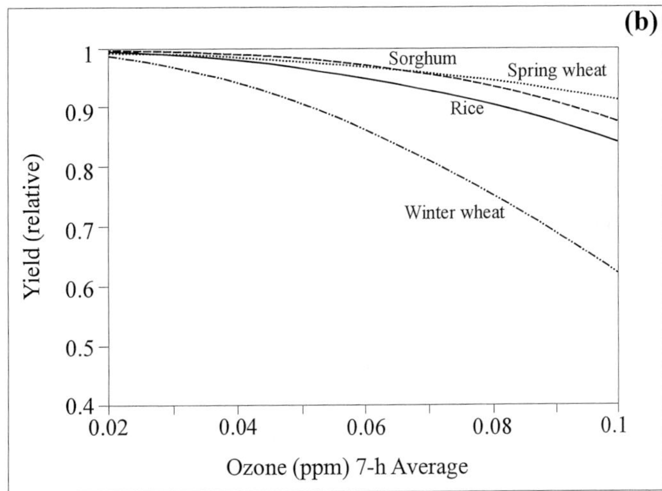

Fig. 3. Weibull dose-response functions for each crop used in an economic of crop yield losses in the United States. The ozone exposure was defined as (a) 12 hr mean and (b) 7 hr mean concentration (source: Adams *et al.*, 1988).

relationship or in applying the relationship to estimate yield reductions. For example, the model used by OECD (1981) to estimate crop yield losses was derived from relatively short-term experiments and assumed that there was no threshold for adverse effects. In fact, more recent data (CLAG, 1996; Bell *et al.*, 1993) suggested that below a certain concentration, the effect of SO_2 is minimal, or variable (i.e. either positive or negative), because of factors such as soil sulphur availability and the effects of SO_2 on the prevalence of fungal diseases and insect pests. This is important because the cumulative economic impacts of small modelled yield losses over large areas with low or moderate SO_2 concentrations may be considerable if it is assumed that there is no threshold for damage.

In the case of O_3, the value of exposure-response relationships, such as those in Figs. 2 and 3, has been questioned. This is because they are based on the external concentration, rather than the flux of O_3 into the leaves, which is known to be more closely related to effects (e.g. Pleijel *et al.*, 2000). The exposure-response relationships were derived from experiments using open-top chambers (OTCs), under fumigation patterns that frequently apply high O_3 concentrations under microclimatic conditions leading to a high flux into the leaves and hence O_3 damage. In contrast, in the field high O_3 levels are typically associated with environmental conditions (such as low wind speeds and high vapour pressure deficits) leading to a low O_3 flux. Therefore, the absorbed O_3 flux may be significantly greater in fumigation treatments than under actual field conditions, at the same O_3 concentration. Model simulations across Europe have shown that there are large regional differences in the spatial distribution of O_3 exposure compared to O_3 flux (Emberson *et al.*, 2000). Therefore, the use of the concentration-based indices, such as those in Figs. 2 and 3, may be a poor basis for estimating large-scale yield losses.

2.2. *Air Quality Guidelines and Critical Levels*

The "critical levels" approach uses air quality guidelines formulated from the definition that they represent:

> "the concentrations of pollutants in the atmosphere above which direct adverse effects on receptors, such as plants, ecosystems or materials, may occur according to present knowledge" (UN/ECE, 1988).

The critical level, and the associated critical load, concept, as discussed in Chap. 2, have proved important policy tools for "effects based" emission reduction negotiations within the UN/ECE Convention on Long-Range Transboundary Air Pollution (LRTAP). Critical levels are designed to protect vegetation (categorised as forests, agricultural crops or semi-natural vegetation) and can be used to identify areas which are potentially at risk from air pollutant damage. Identifying these exceedance areas, both for current day and future pollutant concentrations, enables the derivation of optimised emission abatement policies through their application in integrated assessment models.

Different criteria and methods have been used to define critical levels for different pollutants and receptors, as described in Chap. 3. For SO_2, the critical level was originally established from an analysis made by Jäger and Schulze (1988) during the UN/ECE Critical Levels Workshop in Bad Harzburg. They compiled data from field and chamber experiments in Europe and North America, recorded the lowest value at which responses were observed and then set the critical level below the damage threshold values documented. On the basis of the data available at that time, they established that most horticultural and agricultural plant species do not show adverse effects at mean SO_2 concentrations below 30 $\mu g\ m^{-3}$ over the growing season. They also concluded that particularly sensitive species of trees, mosses and lichens would be affected below this concentration and the critical level for these was set at 20 $\mu g\ m^{-3}$ as an annual value, which was expected to provide for a margin of safety. A subsequent workshop, using the analysis of Bell (1994), recommended a lower limit for lichens which was adopted in the UN/ECE Mapping Manual (Umweltbundesamt, 1996) where the critical level was reduced to 10 $\mu g\ m^{-3}$. It also recommended that critical levels be reduced to 15 $\mu g\ m^{-3}$ for forests in areas experiencing low effective temperature sum on the basis of the analysis of field observations performed in Czechoslovakia by Mäkelä *et al.* (1987).

For O_3, the critical level in Europe is based upon an accumulated dose above a threshold concentration of 40 ppb (AOT40) (Fuhrer *et al.*, 1997). Current critical levels for O_3 in Europe were derived from relationships between AOT40 and yield (e.g. Fig. 4, Chap. 3) for species considered sensitive to O_3 (wheat and beech), using statistical analysis which determined the level of yield reduction that could be discerned with 99% confidence (Pleijel, 1996).

Although the methods by which the critical levels have been set differ between the pollutants, they all aim to identify the concentration representing a "no-significant-effect" level. It is also clear that there are difficulties in defining thresholds, partly because a large volume of data is required to determine "no-significant effect" levels, partly due to difficulties in interpretation of that data and the influences of different biotic and abiotic factors (such as climate), and partly due to the lack of experimental data close to the proposed threshold.

Critical levels may be compared to monitored or modelled concentration data to determine where they are exceeded. In areas where exceedance occurs, it is assumed that there is a likelihood of adverse impacts on sensitive species. In order to use critical levels for regional assessments, various data are required (Fig. 4). Regional mapped concentration data may be derived from interpolated monitoring data or from atmospheric transfer models. Care needs to be taken in the interpolation; for example, for O_3 there are distinct spatial relationships, with AOT40 increasing with altitude and decreasing with increasing concentrations of NOx, which need to be included in the interpolation (PORG, 1997). Modelled concentrations require a detailed emission inventory, meteorological data and an appropriate atmospheric transfer model. Once concentrations have been characterised, receptor information is required to show the species location, preferably as a map with an associated database. Overlaying these data will give maps showing locations where concentrations exceed critical levels and hence where there may be a risk of yield reductions for sensitive crops.

In Europe, national-scale assessments have been carried out of exceedance of critical levels for SO_2, such as that performed for the

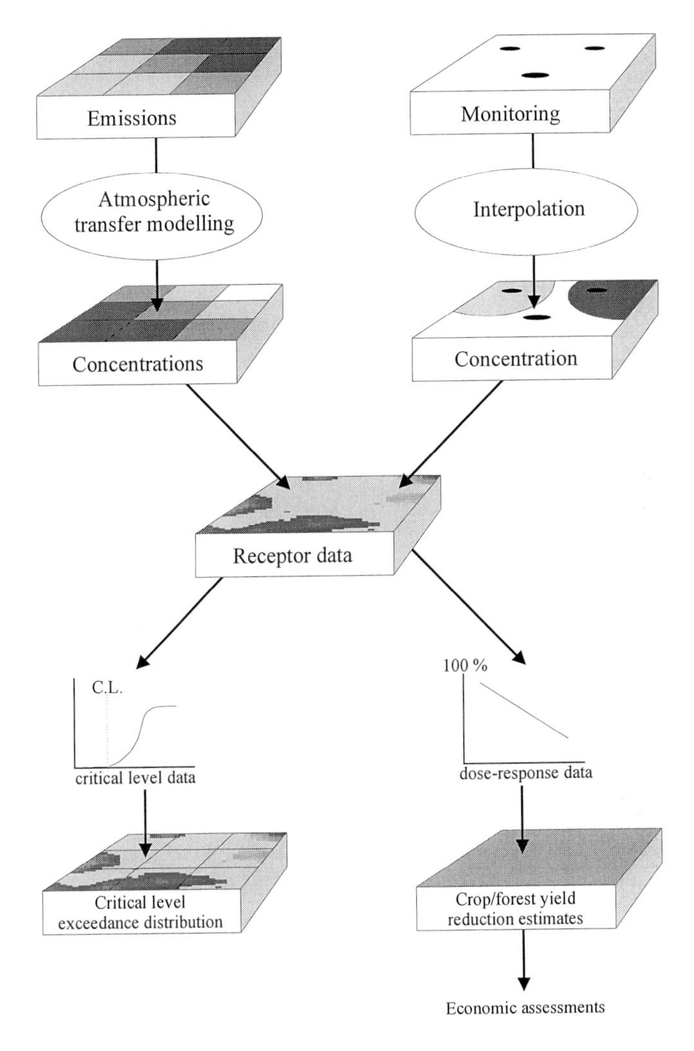

Fig. 4. Datasets required for regional application of the critical level or dose-response approaches

UK (CLAG, 1996). UK SO_2 mapping has been carried out at scales of 5 × 5 km and 20 × 20 km. The distribution of SO_2 at these scales gives useful information when compared to critical level values, showing considerable local exceedance in northern and central England. However, at the European scale, the 150 × 150 km model simulations for SO_2, only show exceedances in central Europe with modelled values greater than 20 μg m^{-3}. Therefore, the scale at which the concentration data are used and presented is important when using critical levels. The 150 × 150 km grids are too large for useful assessment of local areas at risk from SO_2.

The impacts to forest tree health and productivity, agricultural crop yields and semi-natural vegetation composition caused by exposure to ozone (Fuhrer *et al.*, 1997) make it perhaps the most important gaseous pollutant in Europe at the regional scale. Critical level exceedance studies have been carried out using the predictions of the EMEP photochemical model (Simpson, 1995) at a scale of 150 × 150 km. This exercise has shown that large areas of Europe exceed the critical level at which the yield of sensitive crops might be at risk from O_3 concentrations. As O_3 is a regional-scale pollutant, assessments at this scale are considered useful and have been used to negotiate emission reduction strategies within the UN/ECE.

3. Can the Different Tools be Applied in Developing Countries?

Exposure-response relationships, critical levels and air quality guidelines could be useful tools in developing countries for providing relevant information to policy-makers concerned with the extent of air pollution impacts to crops and forests. However, the key problem areas are the transferability of exposure-response relationships and air quality guidelines originally developed in Europe and North America, and the capacity to model or measure pollutant concentrations at the level of detail required for regional-scale assessments.

To expand, the two main issues are:

(1) The applicability of exposure-response functions and critical levels to developing countries.

Local species or cultivars may well respond differently to gaseous pollutants when compared to European and North American species, because of genetic differences. Furthermore, the same cultivar could respond quite differently in different countries due, for example, to variable pollutant concentrations and mixes, different climatic conditions (including precipitation patterns, number of sunshine hours, high temperatures, different atmospheric and soil humidities) and the different agronomic practices and forest growth patterns common to these regions. Questionnaires were sent to the authors of Chaps. 6 to 13 enquiring about the potential problems related to transferring and developing air quality guidelines and critical levels to their specific developing country region. The issues raised included a number of regional factors that may modify the response of species to pollutant concentrations such as cultivar type, agronomic practices (irrigation, soil fertilisation and use of pesticides) and prevailing climatic conditions.

(2) The application of these assessment tools at local and regional scales in these countries.

In order to make broad impact assessments, monitoring networks are required to monitor concentrations around, as well as within, cities, and these are not well established in all countries. Atmospheric transfer models at appropriate scales need to be used, in addition to monitoring data, to further model the distribution of pollutant concentrations and the impact of different proposed control measures.

In the following text, the evidence presented in this volume from Asia, Africa and Latin America is reviewed in the light of the development and application of critical levels and exposure-response relationships, with discussion of the availability of concentration data for regional or local assessments.

3.1. *SO_2 Impacts and Exposure-Response Data Outside Europe and North America*

Figure 5 describes SO_2 response data collected from different locations around the world and compares the response, in terms of relative yield, to 8-h or 7-h growing season mean SO_2 concentrations of different species and cultivars local to each region. The Chinese data were collected from field-based OTC studies assessing the effects of a range of SO_2 concentrations (clean air, 132, 264, 396 and 666 $\mu g\ m^{-3}$; 7-h day^{-1}) on the growth and yield of several crop species from seedling to final harvest (Feng *et al.*, 1999). Similarly, the Indian

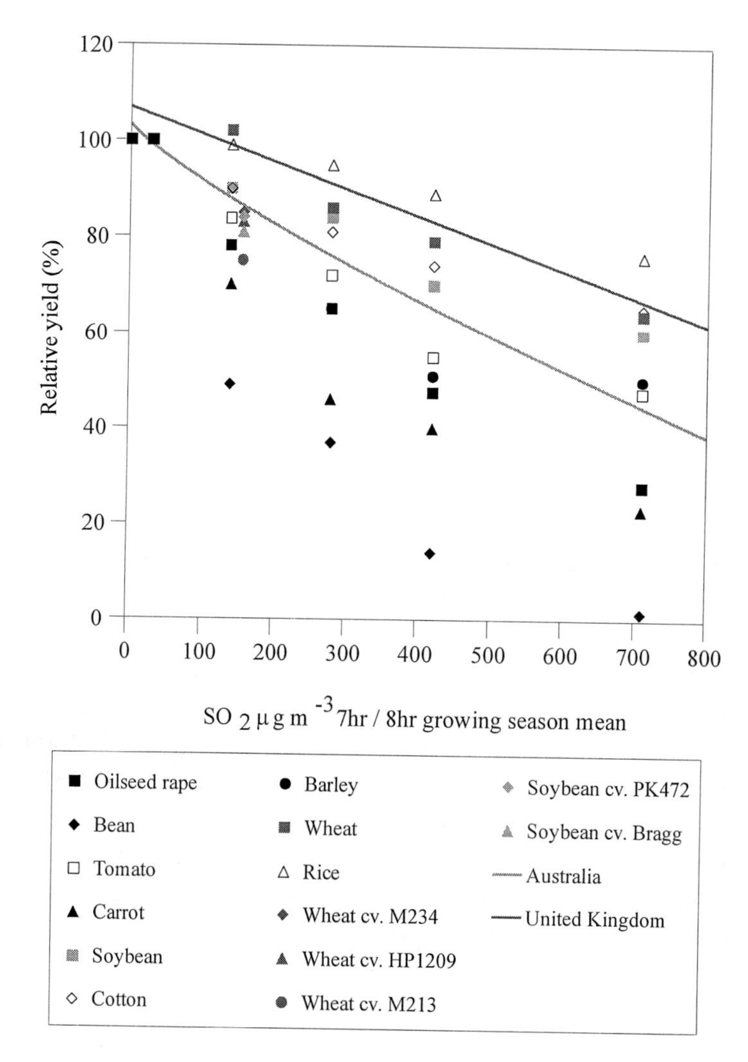

Fig. 5. SO_2 dose-response relationship pooling data from China (Feng *et al.*, 1999), India (Agrawal, this vol.), Australia (Murray, this vol.) and the UK (Roberts, 1984) for different species and cultivars.

data were from OTC studies recording relative yield of wheat and soybean cultivars exposed to SO_2 concentrations of 157 $\mu g\ m^{-3}$; 8-h day^{-1} for 70 days. The Australian crop species exposure-response relationship is a generalised relationship pooled from experimental data relating mean SO_2 concentrations predicted from continuous exposure for 8-h day^{-1}, 4-h day^{-1}, 2.5-h day^{-1} and 3 days week^{-1}. The bulk of the data came from experiments conducted with an exposure regime of 4-h day^{-1} over four or five months. The UK exposure-response relationship is based on a linear regression analysis of data pooled from chamber experiments relating exposures of *Lolium perenne* (ryegrass) to SO_2 concentrations of less than 200 μg m^{-3} for more than 20 days (Roberts, 1984).

From Fig. 5, it is apparent that there is a large variability in the SO_2 sensitivity of different species and cultivars. The UK data were predominantly for cereal crops and grasses; the Indian wheat data and Chinese rice data show a similar response, indicating a consistency in the SO_2 sensitivity of graminaceous species between different countries and regions. The Australian dose-response relationship included a larger proportion of data from non-graminaceous species, and this may explain why the Australian exposure-response relationship is more sensitive than the UK relationship. This interpretation is further supported by the Chinese data which suggests that the non-graminaceous species — bean, carrot and oilseed rape — all have a significantly greater SO_2 sensitivity than that defined by either the Australian or UK dose-response relationships.

However, this comparison is based on chamber studies only, and not on field responses. There are a number of regional factors that may modify the response of species to pollutant concentrations such as experimental conditions, cultivar type, agronomic practices (irrigation, soil fertilisation and use of pesticides) and prevailing climatic conditions. Climate can have a twofold effect by influencing pollutant uptake through the stomata (Emberson *et al.*, 2000), and also by determining the seasonal pattern of air pollution formation of secondary air pollutants like O_3. Vegetation response also varies with exposure to different air pollutant combinations, which may vary between regions. This is especially important when considering pollutant interactions, for example the long-established synergistic responses to SO_2 and NOx.

Comparing threshold concentrations or exposures for adverse effects is more difficult as studies in developing countries have focussed on the higher concentration ranges of relevance to field conditions. For example, the recent data collected by Feng *et al.* (1999), using OTCs to assess the effects of a range of SO_2 concentrations on the growth and yield of several crop plants, was then used to establish exposure-response relationships to estimate yield losses attributable to SO_2 in seven Chinese provinces. The lowest concentration used was 132 μg m^{-3}; this caused a 25% reduction in yield of the most sensitive species (bean) but had no effect on the yield of wheat. This concentration is well above the concentrations used to set SO_2 critical levels in Europe and therefore makes comparison of the "no-significant effect level" impossible. Similar problems arise when interpreting the available data on responses to SO_2 from India.

Feng *et al.* (1999) calculated the threshold concentration that would give a 5% yield reduction using a linear regression through the data points describing relative yield against these five different SO_2 concentrations. For the most sensitive crop (bean), that threshold was 72 μg m^{-3} SO_2. This is more than double the UN/ECE threshold of 30 μg m^{-3} and the Chinese authors suggest that crops in China may have been selected for SO_2 tolerance. However, such a statistical approach may not be sound, and it is essential that experiments be carried out in countries such as China and India at lower concentrations, before an appropriate air quality guideline for SO_2 impacts on crops can be reliably defined.

3.2. *Ozone Effects and Exposure-Response Relationships*

No exposure-response relationships for crop responses to O_3 in developing countries were found in our case studies. However, data from Japan could be used to compare the response of European and East Asian tree species. Figure 6 shows Japanese O_3 exposure-response data for 17 tree species, given as the relative increment in total dry weight per plant (TDW) at different values of AOT40 calculated over six months. Figure 6(a) shows exposure-response data for deciduous tree species. Also shown is the exposure-response relationship for beech (*Fagus sylvatica*) derived by pooling data from several European studies (Fuhrer *et al.*, 1997). This relationship was used to define the European critical level for forest trees (Ashmore, this vol.). Figure 6(a) shows that there is considerable variability in the response of different Japanese deciduous tree species to O_3 exposure.

Comparison with the European beech dose-response relationship, which is recognised as one of the most O_3 sensitive European forest tree species suggests that the most sensitive species in Japan are responding in a similar manner to O_3 to European beech. The Japanese species of maple and zelkova appear to have very similar O_3 sensitivities to that of the European beech, whilst the Japanese poplar species would seem to be somewhat more sensitive.

Figure 6(b) shows data for eight coniferous Japanese tree species. These data are compared with European data for three different coniferous species: Norway spruce (*Picea abies*), Scots pine (*Pinus sylvestris*) and silver fir (*Abies alba*). The data suggest that the most sensitive European species have a similar O_3 sensitivity to the most sensitive Japanese species. Overall, these data suggest a broad consistency in exposure-response relationships for sensitive species in Japan and in Europe. As with the SO_2 studies discussed above, this conclusion is based on chamber studies only, and field responses may be quite different.

Experimental data from a single O_3 exposure level may give a tentative indication of the relevance of exposure-response data from North America or Europe to the field situation elsewhere in the world. For example, data from Pakistan (Wahid, this vol.) shows 23–47% yield reductions for several crops (wheat, rice, soybean, mungbean and chickpea) at 6-h day^{-1} seasonal mean O_3 concentrations of between 33–85 ppb. Comparison of the Pakistan dose-response data with the dose-response relationships derived from the NCLAN experiments (Fig. 4) indicates that, of the species investigated in both these studies, the Pakistan varieties have far higher O_3 sensitivity. The highest yield loss estimated using the NCLAN dose-response relationships was 27% for winter wheat at a seven-hour mean O_3 concentration of 85 ppb. In contrast, the highest yield loss recorded in the Pakistan study was 47% at a maximum 6-h day^{-1} seasonal mean O_3 concentration of 85 ppb. Since the diurnal pattern of O_3 concentrations in Pakistan is unknown, and since the pollutant was only measured with a non-specific chemical method, such interpretations should be treated with caution. However, Maggs and Ashmore (1998) exposed these Pakistani cultivars of wheat and rice in a European fumigation facility, and found them to be highly sensitive to O_3.

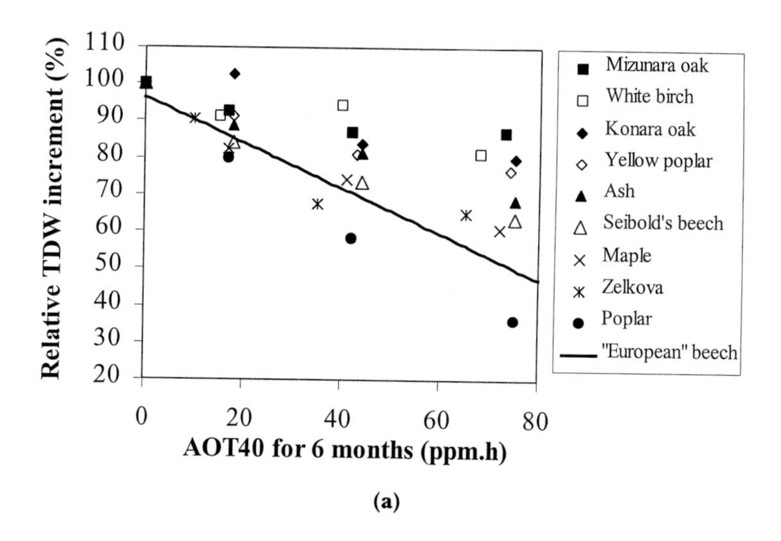

(a)

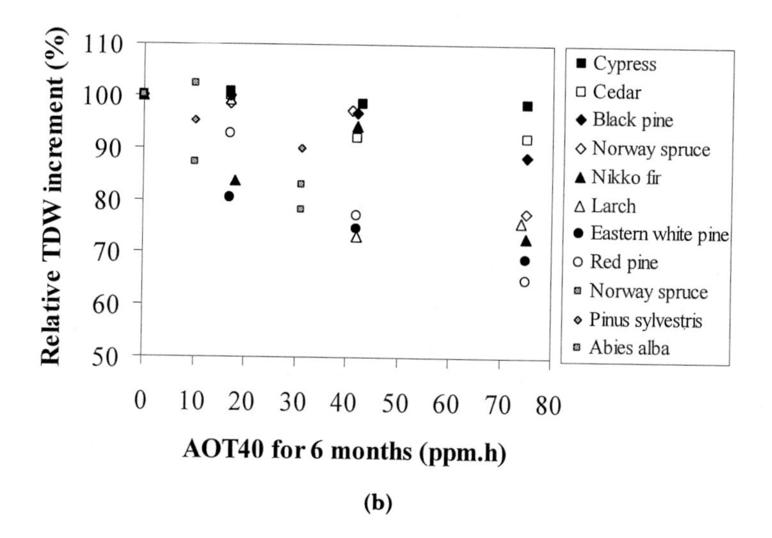

(b)

Fig. 6. The relationships between the AOT40 of O_3 for six months and relative increment of TDW for 17 Japanese tree species (source: Izuta and Matsumura, 1997; Matsumura and Kohno, 1999). Graph (a) compares relationships for nine Japanese deciduous tree species with the European beech dose-response relationship (shown as continuous line). Graph (b) compares eight Japanese coniferous tree species with three European species data (shown as grey data points) (source: Küppers *et al.*, 1994).

More data are urgently needed to define the O_3 levels at which significant impacts on the yield of local crops might occur in Asia, Africa and Latin America (Ashmore and Marshall, 1999).

4. Assessing the Scale of Air Pollution Impacts in Developing Countries

There is a strong desire from policy-makers in developing countries for data indicating the severity and scale of air pollutant impacts on crop yield and quality, and also on the growth of forests, and their socio-economic implications. It is therefore desirable to be able to map areas where high concentrations of gaseous pollutants may occur and also to assess the degree of change occurring in these areas. As has been discussed above, there is a lack of data concerning "no-significant effect" levels and exposure-response relationships relevant to developing countries and real concerns exist about the transferability of relationships and values derived elsewhere in the world. Therefore, it is not possible to provide any quantitative assessment of the current or future impacts of air pollution on crops and forests in developing countries.

However, it is still of interest to consider the possible relative scale of impact of SO_2 and O_3 concentrations in different parts of the world, while remembering these very significant uncertainties. In this section, we consider the implications of modelled global pollutant distributions, both current and future. It is important to note that these distributions cannot provide any realistic quantitative assessment of potential crop losses; rather they can indicate the relative likelihood of pollutant impacts in different parts of the world, and how this might change over the coming decades.

4.1. *Sulphur Dioxide*

The major problems with interpreting global models of SO_2 concentrations are connected with the fact that it is considered a local pollutant which causes most damage close to sources of emission. A relatively high resolution scale is required to make any sensible assessments of yield losses or critical level exceedance. This scale of concentration data, especially outside cities or around point sources, is not available for most developing countries. There are certainly data from urban areas in China indicating extremely high ambient concentrations of SO_2. For example, Zheng and Shimitzu (this vol.) found that in 1995, 52% of northern and 38% of southern Chinese cities recorded annual daily mean SO_2 concentrations exceeding 60 $\mu g\ m^{-3}$. In 1997, very high mean annual SO_2 concentrations were recorded in the cities of Taijuan (248 $\mu g\ m^{-3}$); Jinan (173 $\mu g\ m^{-3}$); Yibin (216 $\mu g\ m^{-3}$) and Chongqing (208 $\mu g\ m^{-3}$). Such concentrations are liable to cause acute impacts on vegetation located within the cities but concentrations are likely to decrease rapidly with distance from the city into agricultural areas.

SO_2 concentration data exist from modelling studies at a scale of $1° \times 1°$ (approximately 100×100 km) which is too coarse a scale for application with dose-response relationships. Figures 7 and 8 are based upon the MATCH model, which is an Eulerian multi-layer model modified for Asian conditions (Engardt, 2000). Emissions data have been calculated for the year 1995 (Vallack *et al.*, in press) and are spatially allocated to $1° \times 1°$ grids according to country GDP, population density and the distribution of large point sources (Olivier *et al.*, 1998). The projections for sulphur emissions are based upon Global Scenario Group projections (Raskin *et al.*, 1998) for activity variables together with appropriate emission factors (Vallack *et al.*, in press). These model runs are based upon scenarios which try to develop one possible picture of the future according to a consistent set of assumptions; they are not predictions of the future. The projections are based upon a "no further control of emissions beyond 1995 levels" variant of the reference scenario, which is therefore considered a high emission scenario. Both the emission inventory and the model have uncertainties associated with them, which need to be considered when interpreting the maps. The maps showing the concentrations in 2025 show increasing concentrations of SO_2 to very high levels in widespread parts of China and levels in India increasing to 1995 Chinese levels by 2025.

The calculated SO_2 concentrations for 1995 (Fig. 7) indicate that a large part of Asia experiences concentrations of between 0–5 $\mu g\ m^{-3}$ which is comparable to current concentrations in rural

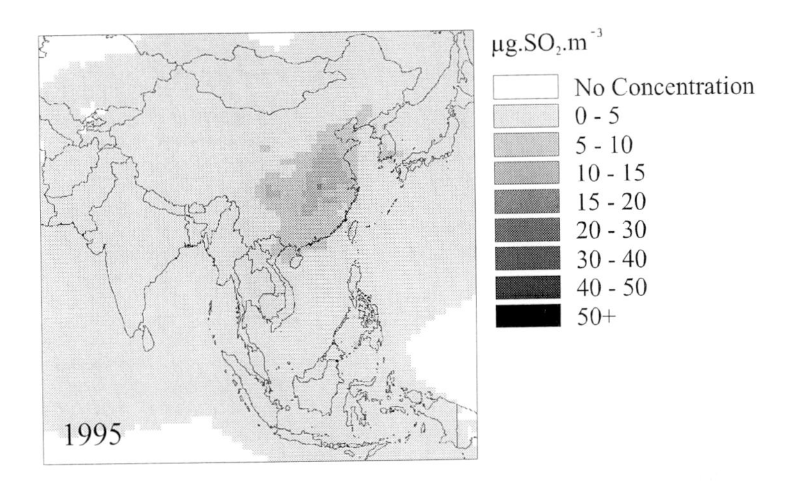

Fig. 7. Calculated SO_2 concentrations in Asia in 1995 according to runs of the MATCH atmospheric transfer model at a scale of 1° × 1° (Engardt, 2000) using 1995 emissions calculated from historical energy data (Vallack *et al.*, in press).

non-industrial parts of Europe. In China, Fig. 8 shows large areas with higher concentrations of 5–10 and 10–15 μg m^{-3}. At this scale, this equates with more heavily industrialised parts of Europe, where localised damage to vegetation has occurred. As reported in Chap. 6, impacts of SO_2 have been observed in China. Due to the extremely high SO_2 concentrations, trees in some cities have had to be replaced many times; 50% of 2000 ha of Masson Pine trees in Nanshan, southeast of Chongqing have died (attributed to SO_2 from the city); 90% of 6000 ha of *Pinus armentii* died due to extremely high SO_2 concentrations form a point source in Sichuan province (see Zheng and Shimitzu, this vol.); and studies comparing fruit trees around power stations to those at clean sites showed a 50% reduction in fruit yield (see Zheng and Shimitzu, this vol.). These studies show that SO_2 concentrations in different regions of China are already high enough to cause deleterious impacts at regional scales, even though the concentrations from the MATCH modelling at 1° × 1° scale are, in the main, below critical levels. The SO_2 concentrations in 2025, according to Fig. 8, will increase significantly in China. There are also increases in SO_2 concentrations on the Korean peninsula and in parts of

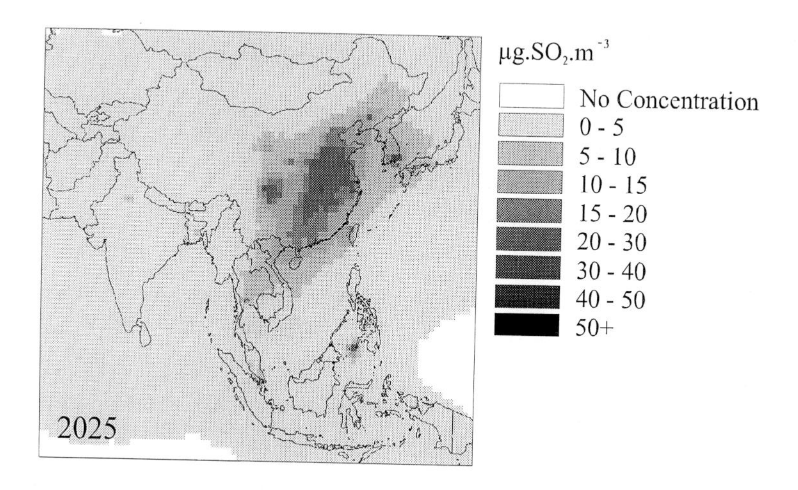

Fig. 8. Calculated SO_2 concentrations in Asia in 2025 according to runs of the MATCH atmospheric transfer model at a scale of 1° × 1° (Engardt, 2000) using a projection for 2025 emissions calculated from a Global Scenario Group reference scenario (Vallack *et al.*, in press).

Southeast Asia. This indicates that SO_2 concentrations, known to be causing damage in parts of Asia, are set to increase further in the absence of further emission prevention or control measures thus causing greater impacts.

4.2. *Ozone*

The 3D STOCHEM model (Collins *et al.*, 1997; Collins *et al.*, 2000) provides projected global O_3 concentrations for both the present day (1990) and the future (2030). Future O_3 concentrations are predicted according to emissions based on the IPCC (1992) "business as usual" scenario. O_3 concentrations are described at a spatial scale of 5° × 5° showing maximum growing season (defined as three months) means for each grid square. It should be noted that these O_3 concentrations represent values within the well-mixed planetary boundary layer that extends to approximately 50 metres above the surface. As such, concentrations at plant canopy heights will be lower than the 50 metres value due to the effect of atmospheric mixing. In addition, the coarse spatial scale of the model conceals large within-square variability in O_3 concentrations that result from the occurrences of local emissions and subsequent O_3 formation, transport and removal process operating on spatial scales much smaller than 5° × 5°.

Figure 9 shows that during 1990 there were five main "hot-spots" (North America, Europe, northeast Asia, southwest Asia and central Africa) where three-month mean concentrations are modelled to reach levels between 60 and 70 ppb over large parts of the region. The enhanced O_3 levels occurring over mid-Africa during the dry season (December to March) have been attributed to pyrogenic, biogenic and anthropogenic source emissions although their relative contributions to ozone maxima in each region are unknown. The leaf symbols identify areas where damage to vegetation has been observed and attributed to O_3 pollution in the eight developing country reports given in Chaps. 6 to 13. Of these eight countries, only those in Asia are experiencing the highest levels of mean three-monthly O_3 concentrations predicted by the STOCHEM model. However, as summarised below, evidence of O_3 damage collated in the expert

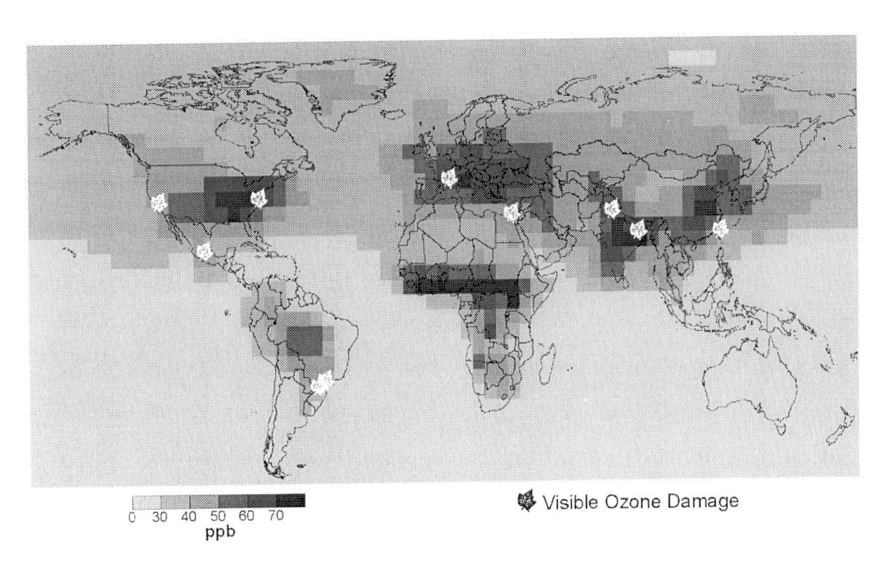

Fig. 9. Mean maximum growing season O_3 concentrations for 1990 and site-specific visible O_3 damage to vegetation observed in the eight selected developing countries.

reports is not limited to those areas where the highest O_3 concentrations are predicted globally. This has important implications when considering the likely extent of damage to vegetation caused by O_3 concentrations.

Even with the high O_3 concentrations modelled and measured in China, no observations of visible O_3 injury have been documented, which is probably due China's focus on SO_2 as the most serious gaseous pollutant. For example, in Taiwan, where visible O_3 injuries have been documented on agricultural crops (e.g. sweet potato, cucumber and musk melon) associated O_3 concentrations were similar to those in rural parts of China, with peak values of 120 ppb being recorded (Sheu and Liu, this vol.). In India, evidence of O_3 injury to potato crop has been observed in the Punjab (Bambawale, 1986) and OTC experimental investigations have shown that O_3 exposures similar to ambient levels can reduce yield and biomass accumulation in Indian cultivars of wheat and soybean plants. Experimental research in Lahore using OTCs showed that ambient O_3 concentrations (ranging from 35 to 85 ppb 6-h day^{-1} seasonal mean) can cause damage to a number of different crop species (wheat, rice, chickpea, mungbean and soybean).

Evidence of O_3 impacts to vegetation have also been documented in Egypt, Mexico and Brazil (Chaps. 10, 12 and 13, this vol.). However, the STOCHEM model predicts comparatively lower O_3 concentrations for these regions. In Egypt, the limited cultivatable region (accounting for 3.7% of the total area of Egypt) is predominantly located along the Nile since this represents the sole source of water for irrigation. Visible injury symptoms to local varieties of radish have been documented at a suburban location near Alexandria where ambient mean O_3 concentrations were only 55 ppb (6-h day^{-1}). Controlled fumigation experiments in OTCs have shown that O_3 exposures similar to ambient concentrations occurring in Egypt can cause significant yield reductions and associated visible injury to Egyptian varieties of turnip, radish, wheat, tomato and broad bean. In Mexico, the most important and frequent foliar injury shown by different plant species has been related to tropospheric O_3 pollution. The sudden decline in sacred fir trees observed in the "Desierto de los Leones" national park located to the southwest of the Mexico valley, is considered to be caused by O_3 pollution. In Brazil, active and passive biomonitoring has been performed close to the industrial complex of Cubatão in the state of São Paulo, finding severe O_3 injury to tobacco, increasing at higher elevations.

These instances of O_3-induced injury emphasise two points. Firstly, that current day O_3 concentrations are causing impacts on crop yields and forest health across the world, even in those areas that global O_3 models do not identify as being the worst polluted. Secondly, that the size of the areas experiencing high O_3 concentrations predicted by the global models would suggest that vegetation damage by O_3 is a much more serious problem than the limited site-specific observations of O_3 impacts collated in this report might at first indicate.

Figure 10 shows predicted future global O_3 concentrations assuming a "business as usual" scenario. Elevated O_3 concentrations are still centred around the five main hot-spots but the area covered by concentrations between 60 and 70 ppb is greatly increased. For example, Asia, the whole of India and eastern China are predicted to experience mean three-monthly O_3 concentrations between 60 and 70 ppb. This indicates that if nothing is done to curb O_3 precursor

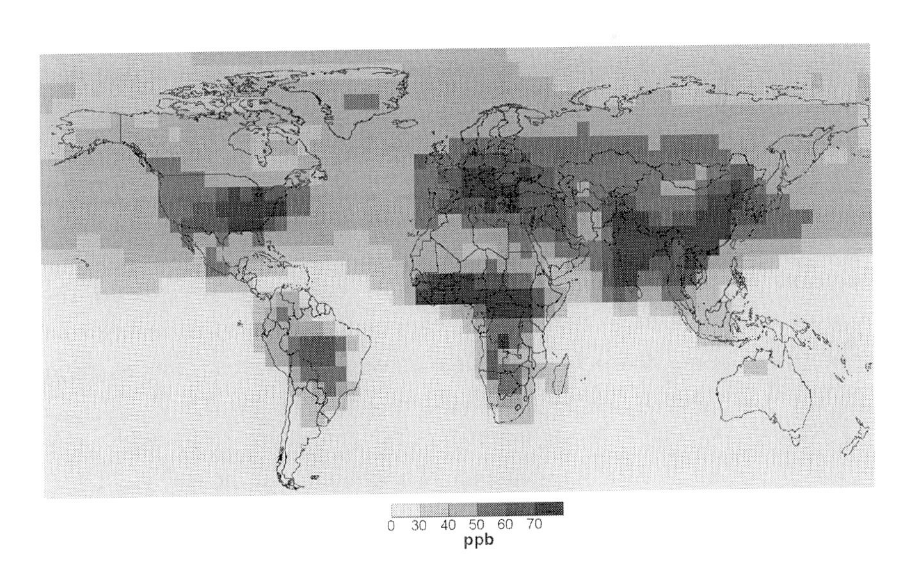

Fig. 10. Mean maximum growing season O_3 concentrations for 2030.

emissions, O_3 pollution and associated impacts to vegetation will become an even more significant problem in the future.

5. Conclusions

It is apparent, from the observational and experimental data described in Chaps. 6 to 13, that ambient concentrations of air pollutants have the potential to cause significant impacts to vegetation in the form of reduced crop yield and quality, and reduced forest health and productivity. It is also apparent that SO_2 and O_3 tend to be the most important air pollutants in developing countries. More information exists for Asia, compared to Africa or Latin America, reflecting the higher emissions, pollutant concentrations and impacts associated with the Asian region. Research also tends to be biased towards agricultural crops rather than forest trees. This is probably a reflection of the importance of crops as a food commodity. Agriculture also tends to be located close to urban or industrial/residential areas that experience higher pollution levels.

However, these data by no means provide a comprehensive picture of the extent and magnitude of local or regional impacts. For such comprehensive assessments to be possible requires the availability of a number of different types of information. Firstly, knowledge of the range, duration and spatial coverage of ambient pollutant concentrations is vital, both to define realistic exposure patterns for use in experimental exposure-response studies and to enable the application of impact assessment tools. At present, the limited monitoring stations that do exist are located within urban centres to address the risk of air pollution to human health. There is a real need to extend these monitoring activities to rural areas, especially when considering ozone pollution, which tends to be characterised by higher concentrations downwind from the source of its precursor pollutants.

There is very clearly an urgent need for more field and experimental research in these countries to establish the actual yield losses in areas known to, or suspected of, having elevated concentrations of SO_2 and O_3. Further development of modelling methods and monitoring networks can help to identify those areas which most urgently need more in-depth research. Currently, the uncertainties in the data are too large to estimate actual yield loss on a regional scale. Nevertheless, continued improvement in assessments of the potential scale of impacts, based on different exposure-yield loss scenarios, could, for example, be used to provide estimates of the possible scale of regional impacts on crops and forests. This would provide at least initial estimates for policy-makers' questions as to the potential impact of pollutant gases in Asia, Africa and Latin America.

References

Adams R.M., Glyer J.D. and McCarl B.A. (1988) The NCLAN economic assessment: approach, findings and implications. In *Assessment of Crop Loss from Air Pollution,* eds. Heck W.H., Taylor O.C. and Tingey D.T. Elsevier, London, pp. 473–504.

Ashmore M.R. and Marshall F.M. (1999) Ozone impacts on agriculture: an issue of global concern. *Adv. Bot. Res.* **29**, 32–52.

Bamawale O. (1986) Evidences of ozone injury to a crop plant in India. *Atmos. Env.* **20**(7), 1501–1503.

Bell J.N.B., McNeill S., Houlden G., Brown V.C. and Mansfield P.J. (1993) Atmospheric change — Effects on plant pests and diseases. *Parasitology* **106**, S11–S24.

Bell J.N.B. (1994) A reassessment of critical levels for SO_2. In *Critical Levels of Air Pollutants for Europe*, ed. Ashmore M.R. and Wilson R.B., UK Department of the Environment, London, pp. 6–19.

CLAG (1996) *Critical Levels of Air Pollutants for the United Kingdom.* Critical Loads Advisory Group, sub-group report on critical loads. Institute of Terrestrial Ecology, Penicuik.

Collins W.J., Stevenson D.S., Johnson C.E. and Derwent R.G. (1997) Tropospheric ozone in a global-scale three-dimensional Langrangian model and its response to NOx emissions controls. *J. Atmos. Chem.* **26**, 223–274.

Collins W.J., Stevenson D.S., Johnson C.E. and Derwent R.G. (2000) The European regional distribution and its links with the global scale for the years 1992 and 2015. *Atmos. Env.* **34**, 255–267.

EC (1998) *Economic Evaluation of Air Quality Targets for Tropospheric Ozone.* European Commission DG XI, http://europa.eu.int/comm/environment/enveco/studies2.htm#1.

Emberson L.D., Ashmore M.R., Cambridge H.M., Simpson D., and Tuovinen J.-P. (2000) Modelling stomatal ozone flux across Europe. *Env. Pollut.* **109**, 403–413.

Engardt M. (2000) Sulphur simulations for East Asia using the MATCH model with meteorological data from ECMWF. Swedish Meteorological Institute, RMK No. 88.

Feng Z., Cao H. and Zhou S. (1999) *Effects of Acid Deposition on Ecosystems and Recovery Study of Acid Deposition Damaged Forest.* China Environmental Science Press, Beijing (in Chinese).

Fuhrer J., Skärby L., Ashmore M.R. (1997) Critical levels for ozone effects on vegetation in Europe. *Env. Pollut.* **97**, 91–106.

IPCC (1992) Climate Change (1992) *Intergovernmental Panel on Climate Change.* Cambridge University Press, Cambridge, UK.

Izuta T. and Matsumura H. (1997) Critical levels of tropospheric ozone for protecting plants. *J. Japan Soc. Atmos. Env.* **32**, A73–A81 (in Japanese).

Jäger H.-J. and Schulze E. (1988) *Critical Levels for Effects of SO_2.* Final Draft Report of the UN/ECE Critical Levels Workshop, 14–18 March 1988, Bad Harzburg, pp. 15–50.

Küppers K., Boomers J., Hestermann C., Hanstein S. and Guderian R. (1994) Reaction of forest trees to different exposure profiles of ozone-dominated air pollution mixtures. In *Critical Levels for Ozone.* eds. Fuhrer J. and Achermann B. UN/ECE Workshop Report, Swiss Federal Research Station for Agricultural Chemistry and Environmental Hygiene CH-3097, Liebefeld-Bern, Switzerland, pp 98–110.

Maggs R. and Ashmore M.R. (1998) Growth and yield responses of Pakistan rice (*Oryza sativa* L.) cultivars to O_3 and NO_2. *Env. Pollut.* **103**, 159–170.

Mäkelä A., Materna J. and Schöpp W. (1987) *Direct Effects of Sulfur on Forests in Europe — A Regional Model of Risk,* Working Paper 87-57. IIASA, Laxenburg.

Matsumura H. and Kohno Y. (1999) Impact of O_3 and/or SO_2 on the growth of young trees of 17 species: an open top chamber study conducted in Japan. In *Critical Levels for Ozone — Level II,* eds. Fuhrer J. and Achermann B. UN/ECE Workshop Report, Swiss Agency for the Environment, Forests and Landscape (SAEFL), Berne.

Mills G., Hayes F., Buse A. and Reynolds B. (2000) *Air Pollution and Vegetation: UN/ECE ICP Vegetation Annual Report 1999/2000.* UN/ECE Working Group on Effects, Centre for Ecology and Hydrology (CEH), Bangor.

OECD (1981) *The Costs and Benefits of Sulphur Control.* Organisation of Economic Cooperation and Development, Paris.

Olivier J.G.J., Bouwman A.F., van der Hoek K.W.M. and Berdowski J.J.M. (1998) Global air emission inventories for anthropogenic sources of NO_x, NH_3 and N_2O in 1990. *Env. Pollut.* **102**, 135–148.

Olszyk D.M., Thompson C.R. and Poe M.P. (1988) Crop loss assessment for California — modelling losses with different ozone standard scenarios. *Env. Pollut.* **53** (1–4), 303–311.

Pleijel H. (1996) Statistical aspects of critical levels for ozone based on yield reductions in crops. In: *Critical levels for ozone in Europe: Testing and Finalizing the Concepts,* eds. Kärenlampi L. and Skärby L. UN/ECE Workshop Report, University of Kuopio, Department of Ecology and Environmental Science, Kuopio, Finland.

Pleijel H., Danielsson H., Karlsson G.P., Gelang J., Karlsson P.E. and Selldén G. (2000) An ozone flux-response relationship for wheat. *Env. Pollut.* **109**, 453–462.

PORG (1997) *Ozone in the United Kingdom.* Report of the DETR United Kingdom Photochemical Oxidants review group, Department of the Environment Transport and the Regions, London.

Raskin P., Gallopin G., Gutman P., Hammond A. and Stewart R. (1998) *Bending the Curve: Toward Global Sustainability.* Stockholm Environment Institute-Boston, MA, http://www.seib.org/publications/bendingthecurve.pdf.

Roberts T.M. (1984) Long-term effects of sulphur dioxide on crops: an analysis of dose-response relations. *Philo. Trans. Roy. Soc. London* **B305**, 299–316.

Simpson D. (1995) Biogenic emissions in Europe 2: implications for ozone control strategies. *J. Geophys. Res.* **100**(D11), 22891–22906.

Umweltbundesamt (1996) *Manual on Methodologies and Criteria for Mapping Critical Levels/Loads and Geographical Areas Where They Are Exceeded.* UN/ECE Task Force on Mapping, UN/ECE Convention on LRTAP, Umweltbundesamt, Berlin.

UN/ECE (1988) *UN/ECE Critical Levels Workshop Report.* Bad Harzburg, FRG.

Vallack H.W., Cinderby S., Kuylenstierna J.C.I. and Heaps C. Emission inventories for SO_2 and NOx in developing country regions in 1995 with projected emissions for 2025 according to two scenarios. *Water Air Soil Pollut.* (in press).

CHAPTER 15

SOCIAL AND ECONOMIC POLICY IMPLICATIONS OF AIR POLLUTION IMPACTS ON VEGETATION IN DEVELOPING COUNTRIES

F. Marshall and Z. Wildig

1. Introduction

Chapters 6 to 13 of this book documented key field and experimental observations proving that current day levels of air pollution are capable of causing considerable damage to vegetation at many site-specific locations in the developing world. Chapter 14 described the tools developed in Europe and North America that can be used to translate information collected from such site-specific experimental work into regional and local assessments of the spatial extent and magnitude of air pollution impacts. These methods rely heavily on the establishment of robust dose-response relationships developed using appropriate pollutant exposures in fumigation experiments. However, a number of difficulties exist in the application of North American and European dose-response relationships to developing country situations. Of these, perhaps the most important is related to the differing local species and cultivar types found in developing country regions, with other factors such as different pollutant concentrations and mixes; climatic conditions and agronomic practices also likely to modify plant response.

The question of biological applicability of dose-response relationships is however not the only problem in identifying areas where vegetation may be especially at risk from air pollution in developing countries. The application of these tools require additional information, namely appropriate spatial and temporal scale pollutant concentration and land-cover (receptor) data. Suitable pollutant concentration data are especially difficult to obtain in many developing country regions since pollution monitors have tended to be located in urban areas and global or regional models tend to be at such coarse spatial scales that local scale emission and deposition patterns are not represented.

In spite of these difficulties, Chap. 14 indicated areas where vegetation may be at significant risk from damage by air pollution for both sulphur dioxide (SO_2) and ozone (O_3). Global/regional modelled pollutant concentration data were discussed in relation to dose-response relationships and observations of damage on the ground. The data showed that the spatial extent of significant pollution impacts on vegetation may be considerable and that the situation may further deteriorate in the future if appropriate air quality management strategies are not implemented.

The previous chapters of this book have concentrated on the biological impacts of air pollution either in terms of visible injury or physiologically related growth reductions. In order to understand the broader costs of air pollution, these biological responses need to be supplemented by assessments of the effects on both economic and social systems.

This chapter discusses methods that have been developed in order to make this transition from biological to socio-economic impacts through the quantification of the benefits of air pollution abatement in socio-economic terms. Once again, this includes a discussion of the suitability of techniques developed in Europe and North America to the developing world. This assessment focuses on the inability of traditional market-based methods to fully quantify pollution costs in developing country situations where effects on non-marketable environmental goods and services are particularly crucial to specific sectors of society. The need for alternative approaches that provide a more inclusive and equitable assessment of the benefits of improvements in air quality for developing country situations are discussed.

An alternative approach is presented as a case study describing research conducted in peri-urban areas in India. Here, an innovative interdisciplinary approach is taken to assess the social and economic policy implications of air pollution damage, with particular emphasis on the poor.

2. Assessing Policy Implications in Industrialised Countries

Towards the end of the last century, recognition of the widespread and complex nature of environmental impacts became an issue of local and global political significance. This coincided with a major shift from social to market-based philosophy and the development of techniques that resulted in the command and control methods, which had previously been used in addressing these impacts, being replaced by decision-making tools allowing stakeholders to take a market-based approach (Garrod and Willis, 1999).

In terms of air pollution policy, the key issue lies in determining what is an appropriate level of pollution abatement. Cost-benefit analysis (CBA) is one of the most established and widely used econometric techniques used for this form of assessment. CBA essentially attempts to compare the costs of abatement (e.g. policy implementation) with the benefits (e.g. increased yields and quality of product) generated by this abatement, by attributing monetary figures to both. The valuation of benefits figure is obtained through the Willingness to Pay (WTP) for improvements or the Willingness to Accept (WTA) the negative impacts of pollution damage (Lanian, 1993). These methodologies (which may be carried out in a variety of ways) require that people are fully informed about and aware of the damage under examination, in order that the valuations given fully reflect the true societal costs of the environmental damage. The optimum level of pollution control is identified where the net present value of investment expenditure (abatement options) is equal to or less than the net present value of the social benefits generated (the direct and indirect value of the pollution damages which are avoided). One of the particular strengths of this method is that, when carried out properly, it ensures that the decision-makers are fully aware of all factors in the argument and of their relative import.

The use of economic assessments of the impact of air pollution on crops received particular interest within industrialised nations in the 1980s, largely through government-funded programmes in the United States, and later in Europe (Spash, 1997). Here the direct impacts of air pollution on crop yield have been assessed using dose-response relationships to determine the effect of a pollutant on crop output, and a market value has then attributed to the loss of produce/output. Perhaps the best known and most comprehensive example of this approach is the National Crop Loss Assessment Programme of the United States (1980–1987); this involved detailed field studies to define dose-response information for ten major annual crops, which were used in economic assessments of crop losses due to tropospheric ozone (see details in Chap. 14). Yield reductions due to ozone were estimated to be about 5% of national production, with the economic benefit of reducing ozone concentrations by 40% estimated to be about three billion dollars annually in terms of producers surplus (Heck *et al.*, 1988).

Whilst assessments of this type are clearly valuable, monetary figures of crop losses due to air pollution require careful interpretation, particularly when making comparisons of economic loss assessment between studies and geographic areas. Spash (1997) reviewed the methods used in estimating the economic damage from the impacts of tropospheric ozone on crops and the problems and issues that arise. Where the regions of study are generally the same, a number of factors affect comparability of economic damage assessments such as different dose-response models (as discussed in Chap. 14). These include: the type and number of crops grown; the nature of the supply shift in response to air pollution impacts on crops; the inclusion of cross-crop substitution in response to air pollution impacts and the type of benefit affected (i.e. producers' and/or consumers' surplus). For example, a number of early studies carried out prior to the early 1980s tended to adopt the traditional model to estimate the cost of losses due to air pollution (Spash, 1997). Here, increased yields as a result of pollution abatement are simply multiplied by current market price to give the producers benefit. The benefits of pollution abatement to the producer can be overestimated because any changes in price or alteration in crop mix in response to pollution losses are not taken into account. Some studies have adopted more complex approaches such as quadratic programming to derive the effects of regional production changes in national markets (e.g. Adams and McCarl, 1985) and linear programming models to assess cross-crop substitution in response to ozone damage (Brown and Smith, 1984).

There are a number of other factors that will tend to result in an underestimate of the benefits of pollution abatement. One issue of great significance here is that the majority of studies carried out in industrialised countries have focussed on economic damage to commercial crop harvest and have ignored other factors such as social benefits, food for home consumption/bartering and protection of soil and water and bio-conservation values.

The value of social benefits is of particular significance for forestry. Forestry worldwide is a multiple output production system and the assessment of pollution impacts can raise much wider issues of non-market valuation than agricultural crop loss in both the industrialised and developing world. As a result, forest damages are poorly represented in purely market-related models as applied to agricultural crop loss (Spash, 1997). As with other types of vegetation, the key factor in costing air pollution damage to forests is in dose-response relationships (possibly the loss of needles and the loss of increment). Although forest conditions generally depend on soil, tree age, climate, pests and diseases and other natural stressors; air pollution loads add to, or may interact with, these factors (Aamlida *et al.*, 2000). Studies show that one of the first responses of a stressed tree is reduction in annual increment (Innes, 1993), which suggests important long-term effects may need to be taken into account. This may result in damage to the local infrastructure, where forestry is a major source of income or where there is a decline in forest tourism. There is also a threat to amenity use, biodiversity and intrinsic existence value of the forests (in terms of conservation and heritage). This could be expanded e.g. to include the biodiversity/conservation of forests in National Parks, Nature Reserves and as local National or World Heritage values. Thus, whilst a few studies from industrialised countries have included value-added in the wood processing industry and non-timber social benefits such as tourism (Mackenzie and El Ashry, 1989), it is clear that current approaches to costing air pollution impacts will generally result in a significant under valuation of risks to forests.

In addition to a focus on marketable produce, economic loss assessments to date have concentrated on the direct impacts of air pollution on yield, and have not taken into account effects on crop quality or the indirect impacts on yield. Air pollution-induced crop damage not only affects the welfare of producers and consumers by changing costs, implying a supply response, but also by affecting demand. A supply response occurs where there is an impact of air pollution on yield whereas an impact on demand occurs when the attributes or perceptible quality (size, appearance or taste) of an agricultural product changes, affecting consumers demand for the product. Thus, reductions in income for vegetable producers and suppliers arise from visible damage to the edible portion of the crop, and by the possibility of increased susceptibility to post-harvest pest and disease damage (Bell *et al.*, 1993; Davies, 2001). In addition, there are other potential non-visible impacts of air pollution such as reductions in nutritional quality or accumulation of heavy metals, with important implications for consumers, particularly the poor (Marshall *et al.*, 1997; Ashmore and Marshall, 1999).

3. The Applicability of Market-Based Assessments to Developing Countries

Clearly, reliable estimates of the fiscal costs of air pollution impacts on an economy will, where appropriate, support and encourage the implementation of policies that lead to pollution abatement. However, it should also be recognised that applying detailed economic analysis within the context of a developing country may compound the difficulties associated with introducing and implementing abatement policies. Also, whilst in industrialised countries the markets are at least a measure of the effectiveness of policy implementation, in developing countries there are less quantifiable livelihood issues to consider. Some key issues in making an assessment of the feasibility and relevance of applying a CBA approach in developing countries are as follows:

(1) *Technical data for assessing costs and benefits*

A fundamental requirement for CBA are reliable dose-response relationships relating crop yield and air pollution exposure damage. However, in many developing countries even the preliminary steps required to provide this information, such as the systematic measurements of concentrations of air pollution have not yet been taken. Whilst dose-response data for major crops are available from studies elsewhere, there are considerable difficulties in transferring these across, or even within, regions of the globe (see Chap. 14 of this volume). There is a clear demand for field studies in developing countries to establish dose-response relationships specific to local tree and crop species and varieties and the climatic and physical conditions under which they are grown. In order to achieve this within the resource constraints of many developing countries, simple cost-effective techniques must be established and promoted (Bell and Marshall, 1999).

Whilst dose-response data are an essential starting point for CBAs, many of the shortfalls in monetary assessments of the benefits of pollutant abatement based on these data, as described in Sec. 2, are greatly magnified in developing countries. This can have a particular significance for the poor, who tend to live and work in the most polluted areas. For example, the omission of data concerning the possible indirect impacts of air pollution on crop yields through increased damage from fungal pathogens and insect pests, may be particularly significant under the climatic conditions of much of the developing world with the cost of disease and pest control tending to discriminate against the poor. Similarly, if air pollution reduces crop nutritional quality, this may have a disproportionate impact on the health of the poor.

(2) *Equitable assessments of costs and benefits*

Section 2 described how the evaluation of the protection of non-marketed environmental goods and services is particularly difficult and often overlooked in CBAs carried out in industrialised countries. This shortfall is much more important when CBAs are carried out in developing countries where the informal economy has a major role to play. The role of agriculture in purely market-based economies is relatively easy to assess, and the losses incurred as a result of air pollution can be represented reasonably with dose-response data. However, there is a lack of data to assess the wider role that agriculture plays in the developing world, as for example, a major source of food for home consumption or exchange for other goods and services. Notably, there is an extreme shortage of information relating to the role of agriculture in urban and peri-urban livelihoods, areas where air pollution is likely to have significant adverse impacts.

Forest resources are also relied on more heavily in developing countries, particularly as a source of fuel wood, shelter and food for the poor and a wide range of non-tree forest products. This represents a vast increase in forest value that is not represented within markets and therefore often inadequately addressed in policy agendas.

The above examples indicate how historically, the data pertaining to the benefits of abatement in CBAs tends to be biased towards the requirements of the richer sectors of society. This trend is also apparent in data relating to the costs of abatement measures. For example, decisions in Europe and the US are often related to national policies of diffuse or widely dispersed (e.g. through tall stacks) pollutants. In costing air pollution abatement, a top down strategy is generally used, in which the total air pollution damage to a particular area (often a country) is divided by the total emissions giving an average cost per unit of pollutant emitted. As such, no priority is given to the regeneration of areas where pollution is at its worst. If it were cheaper to reduce emissions in areas where there are low emissions this would also be reflected and it may be the preferred policy (Lanian, 1993). In developing countries, where pollution tends to be localised around particular industrial areas and cities, an approach based on assessing or simulating damage around specific sources may be more appropriate.

There are also difficulties in assessing the cost of abatement options. Until recently, most developing countries had little formal regulation of air pollution. These difficulties can be compounded by other factors such as literacy rates that affect the ability of stakeholders to comprehend the information that is available. Furthermore the validity of WTP/WTA figures are affected by high levels of poverty, which result in low environmental values in comparison with other priorities. It may be that environmental values and abatement will be favoured more highly in rich rather than in poor areas. This represents a particular injustice especially as poorer communities are more susceptible to health impacts of air pollution due to their generally poorer health status, a result of poor health, nutrition and medical facilities (Lanian, 1993).

(3) *Alternative approaches*

The above examples illustrate that whilst CBA is a valuable tool, it is essential to target risk areas effectively, and to develop innovative methods of benefit assessment in order to achieve an understanding of the full implications of reducing pollution impacts on crops and forests.

In order to meet the requirement for more inclusive and equitable assessments of the benefits of pollution abatement, there is a need to improve channels of communication between those most affected by environmental issues and scientists, planners and policy-makers in developing countries. This requires a more holistic approach than the methodologies adopted in industrialised countries can generally offer, involving a wider group of stakeholders in the research and reporting of activities. In this way, the overall capacity to link micro and macro level assessments is improved, and the process of developing, implementing and monitoring policy initiatives to address environmental issues becomes more effective. The success of this approach relies on establishing links with other relevant initiatives. For example, in India, as in many developing countries, there are already active campaigns for the abatement of air pollution on the basis of the impacts on human health and architectural heritage. Their case may be strengthened by information on other impacts, such as those on agriculture. Thus, there is potential to combine efforts into interdisciplinary initiatives, aimed at integrated assessments of the benefits of air pollution control, in terms of livelihoods and local and national economies (Marshall *et al.*, 1997).

4. Alternative Methodologies for Assessing Social and Economic Impacts: A Case Study from India

The following case study introduces an alternative methodology for air pollution impact assessment. The approach emphasises non-market valuation and the inclusion of the poor, whilst developing strategies to link the micro and macro level perspectives on pollution impacts and to influence ongoing policy debates (Marshall *et al.*, 2000). The case study described here considers agricultural crops, however it would be possible to apply the same principles to a forestry study.

Any threat to agricultural production in India is a matter of great concern. Agriculture accounts for a little over one-quarter of India's GDP, and nearly 80% of the population rely on agriculture as part of their livelihood. However, food security is still considered a luxury by many and 80% of the urban population spend 70% of their income on food (Mougeot, 1994). Whilst attention tends to focus on rural areas, urban and peri-urban agriculture makes an important contribution to the growing urban food demand, particularly for supplies of perishable produce.

Preliminary risk assessments indicated that significant reductions in crop yields may be occurring as a result of air pollution in India, and that urban and peri-urban agriculture, in addition to rural agriculture in close proximity to industries, was likely to be at greatest risk. However, there were limited data to support the preliminary risk assessments in terms of field studies at ambient air pollution levels using Indian crops and cultivars grown under local conditions, and no information on the implications of these air pollution losses to livelihoods or the economy (Marshall *et al.*, 1997). This led to further research to assess the impacts and policy implications of air pollution on crops in urban and peri-urban areas of India, focussing on two case study cities, Delhi and Varanasi.

This interdisciplinary study was carried out by British researchers in partnership with local and national governmental bodies, non-governmental organisations, research institutions and local farming communities in India. There were three major strands to the research programme as indicated in Fig. 1. Scientific data concerning the impacts of air pollution on crops of importance to the poor (see

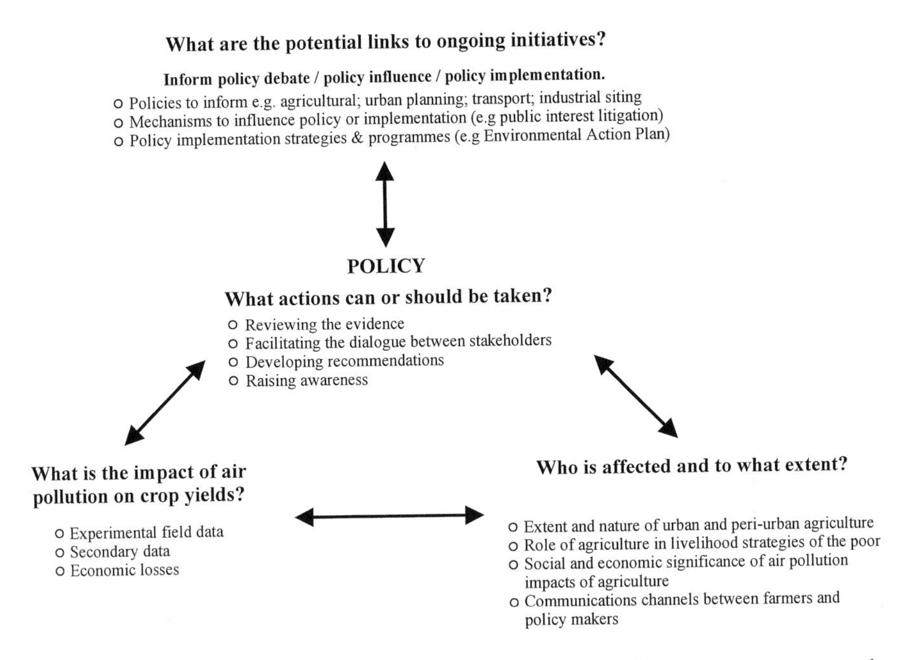

Fig. 1. Major research areas investigated in an interdisciplinary study to assess the impacts and policy implications of air pollution on crops in urban and periurban areas of India.

Box 1) was linked with information on the significance of these impacts to livelihoods of farming communities, to other vulnerable stakeholders and to the cities economy. The policy and scientific research aspects of the work were carried out in parallel, with fora created to allow affected communities, researchers and policy-makers to exchange views whilst consumers groups, polluting industries and other interested parties were also consulted. In addition, a policy group was developed to assess how the emerging research findings could be most effectively utilised within the existing policy agenda.

This programme also has a wider agenda, to improve information exchange between those most affected by air pollution impacts on agriculture in urban and peri-urban areas and the relevant researchers, policy-makers and planners. In this way, research, policy-making and policy implementation become more appropriate, efficient and cost-effective. The benefits of this model, and the cross-sectoral linkages that it develops, extend to a much wider range of environment and development issues (Marshall *et al.*, 2000).

There are two key processes involved, which take place in parallel to the scientific investigation, and make the approach to this study both distinct and particularly suited to addressing air pollution impacts.

(1) *The incorporation of participatory approaches into a holistic evaluation of social and economic impacts of air pollution*

In order to establish the socio-economic significance of air pollution as a constraint on agricultural production in developing countries, baseline information is needed on the nature and extent of agriculture in polluted areas, the economic value of the agricultural produce and a non-market valuation in terms of the importance of agriculture to the livelihoods of vulnerable groups.

The general nature and extent of urban and peri-urban agriculture (UPUA) was assessed using secondary data and was supported by field transect surveys. The results showed that UPUA is an important, if little recognised activity largely practised by landless and small-scale farmers. Wheat and rice, grown for subsistence use, predominate the land area, whilst vegetables are often also grown for sale.

Marketing surveys were also undertaken to determine the proportion of the cities food supply that was sourced from urban and peri-urban areas, and the economic value thereof. In fact, the majority of highly perishable products, including many vegetables that are consumed in Indian cities are produced in peri-urban areas. For example, 50–70% of cauliflower and 70–90% of spinach that is marketed in Azadpur market in Delhi (the largest fruit and vegetable market in Asia) is produced in Delhi itself or the six surrounding peri-urban districts (Marshall *et al.*, 2000).

Data from the experimental studies were used to make preliminary assessments of the economic costs of air pollution damage. These were limited to an illustration based on current market value, because little is known about responses (such as price fluctuation and crop substitution) to yield and quality constraints in urban and peri-urban agricultural produce. This issue is being pursued in further studies, but it is important to note that many of the possible scenarios suggest that it is the poor that will suffer the most. For example, should prices increase because a commodity is limited and/or is being transported from elsewhere it is the poor consumers who will suffer. In addition, price spread information from the study indicated that poor producers who are forced to farm in highly polluted areas would not only have less produce to sell, but are likely to reap little of the benefit from any increase in the price charged to the consumer.

The analysis described above provided an indication of the monetary value of crop losses as a result of air pollution, but does not include any costs that are not represented in the traditional economy. In order to address this, a more detailed assessment of the nature of agriculture and its role in the livelihood strategies of both individuals and communities in selected case study villages was developed through an intensive programme of participatory research activities (see Box 2). The study provided evidence of the many ways in which reductions in crop yield can threaten livelihoods in these areas. Amongst small-scale farmers in the survey areas around Delhi and Varanasi, the majority of crops are relied on for home consumption, and the crop residues maintain livestock and fuel supplies. Agriculture is also an important source of seasonal employment.

Whilst much of the participatory work focussed on poor producers, complementary activities examined the potential impact of air pollution on other stakeholders in the production-consumption chain. Once again, the poor were seen to be the most vulnerable to constraints on production in polluted areas. Street vendors for example, are exclusively involved in the marketing of highly perishable vegetable produce within a limited geographic area, and have limited scope to change to other commodities or areas in response to environmental threats.

(2) *Links with the existing policy agenda*

The second key process is based on the recognition that benefits of the participatory level methodologies are only fully realised by utilising the information gained from the analysis within the context of wider policies and programmes. Micro and macro level analysis within an iterative framework enables researchers to focus on effective means to inform the relevant policy debates, influence the policy-making process and improve policy implementation.

The first stage was to identify target policy fields and actors. These included legislators for emission standards, pollution control measures, land-use and industrial siting. One example of an area to target is the recent Supreme Court judgement in India that has forced polluting industries to relocate away from Delhi to the surrounding peri-urban areas. This is likely to be of benefit in terms of direct health impacts of air pollution, but impacts on agriculture could also be significant and there is case for monitoring them. There is also a requirement for conducting environmental impact assessments before new industrial installations are approved, but they do not currently include potential damage to agricultural crops. These are examples of where awareness of impacts on agricultural crops may help to support the case for more stringent emission controls.

Formal and informal channels for influencing environmental policy were then examined. The stakeholders involved included: central and state government departments for environment, agriculture, power/energy, industry, transport; pollution control boards; NGOs concerned with environment and public health; and industrial organisations and consumer groups. Policy influence may take place through direct actions such as lobbying senior government officials (the Ministry of Environment and Forests is a high priority here) or indirectly through, intermediate institutions or systems such as the democratic, judicial and market systems. Some high profile institutions through which direct interaction takes place, are still developing and provide an excellent opportunity for policy influence. These include the Indian Centre for the Promotion of Clean Technology which includes academic and NGO members, and various expert groups associated with the Central Pollution Control Board. Other major players in India include an extremely active environmental movement led by non-governmental organisations and the judiciary that have been pivotal to many recent changes in environmental policy in India during the past decade.

Recognising that different stakeholder groups have unequal access to policy-makers and implementers, communication channels between those most affected by air pollution and the authorities charged with controlling it were explored. The objective was to consider how the concerns expressed by farming communities affected by pollution issues (as expressed through the participatory research) are currently and could potentially be addressed by the policy and regulatory frameworks.

Following this, the translation of environmental policy statements to specific strategies, instruments and initiatives were reviewed and successes and failures recorded respective of their success in improving air quality particularly in urban and peri-urban areas. This included a review of problems that needed to be overcome to support a higher level of achievement of policy objectives. For example, despite recommendations and justification for extensive use of economic, voluntary and social tools, command and control is still widely used for environmental pollution issues. However, the financial penalties are often considered too low to act as a real deterrent. In addition, whilst there is a strong rhetoric for greater involvement of NGOs and more widespread use of voluntary and social tools for pollution control, this is hindered by poor public access to information which limits accountability of responsible agencies towards the public and electorate at large.

Finally, specific links with ongoing initiatives to raise environmental awareness and improve implementation of environmental policies were identified and facilitated. These included involvement in school-based environmental monitoring programmes and organisation of field based farmer/scientist/government extension workshops.

5. Conclusions

Whilst regional assessments of crop losses and macro level economic analysis offer a pollution control decision-making tool which is useful in market-based economies, in developing countries the methodologies should be adapted and augmented to allow for policy priorities which must, by necessity, be very different. Here, increases in population size, industrial production, energy demands and motor vehicle traffic use are causing rising levels of pollution and there may be a strong case for a reduction in air pollution to enhance agricultural production, maintain food security and the livelihoods of a large proportion of the population. However, it is particularly important to assess and demonstrate the potential of abatement options, which are argued for in the face of limited resources and a general desire to promote industrial development.

The above case study demonstrates that there can be considerable advantages in developing a scientific field research programme in parallel with community-based research and in maintaining a policy dialogue at all levels. It is important to note that the case study was carried out at a local level, and most traditional CBAs are carried out at national or regional level. However, whilst not all aspects of the case study will translate to wider scale assessments, some general principles do apply. This type of approach not only complements market-based assessments, but through the involvement of additional stakeholders improves access to the basic data required to carry them out. Unlike traditional CBAs, the non-market values of goods and services can be included, providing a comprehensive assessment of the possible social and economic impacts of air pollution on the livelihoods of vulnerable communities. In addition, the focus on specific stakeholder groups (in addition to the local or national economy) provides data to target further scientific research into the effects of air pollution on crops and forests to focus on commodities and field studies that are appropriate to a particular priority group.

Furthermore, a participatory approach coupled with links to policy-makers and the policy implementation process, helps to determine how policies and programmes aimed at controlling environmental pollution and supporting agricultural production can most effectively address the needs and priorities of stakeholders who are not traditionally involved in the decision and planning process, usually the poor.

Box 1 Assessing the Impacts of Air Pollution on the Yield and Quality of Urban and Peri-Urban Crops

A series of field studies were designed to address the general lack of primary data on the impacts of ambient levels of air pollutants and pollutant mixtures on Indian crops, and the ▶

recognition that measured and predicted levels of phytotoxic pollutants in urban and peri-urban areas were at damaging levels.

Target crops (spinach (*Spinacea oleracea*), mustard (*Brassica campestris*), wheat (*Triticum aestivum*) and mungbean (*Vigna radiata*) were selected largely on the basis of their importance to poor producers and their significance to the local economy.

An experimental programme was designed to meet the priority geographic areas, crops and target groups. Standard cultures of selected crops were exposed outdoors at agricultural locations with different levels of air pollution, identified on the basis of limited published data and the presence of industry, level of motor traffic and proximity to city centres. SO_2, NO_2 or O_3 were monitored and crop growth, development, yield and nutritional quality were assessed, whilst all other environmental parameters and crop management which might also influence yield and thus confound the effects of air pollution were maintained as constant as possible. Two parallel experiments were carried out concurrently in the two cities, over two separate growing seasons. In addition to the experimental data, the existing literature was surveyed to identify published studies, using these and other experimental techniques, which may assist in defining threshold concentrations for effects on visible injury, crop yield and crop quality in the different crops. These threshold concentrations were utilised in illustrative economic loss assessments.

Box 2 What is Participatory Research?

Participatory methods allow local people (rural or urban) to undertake their own appraisals, analysis, monitoring and evaluation, and to plan organise and act. There are a wide range of established techniques which are useful for natural ▶

resource research: mapping, well-being ranking to define distinct groups, historical and seasonal analysis to describe agricultural practices, evaluation matrices of crops, problems and solutions, causal linking and diagramming, and semi-structured interviews with both groups and individuals. Participatory methodologies also allow researchers to explore the following factors within communities or households (adapted from Hunter, 1978):

- socio-political constraints
- economic risk-aversion
- religious and cultural constraints
- relationships with institutional organisations and information services
- degree of change occurring within the community.

It is imperative to understand the above factors in order to appreciate both the impact and significance of the constraint (in this case, air pollution) under consideration but also in the development of coping strategies (in this case, possibly through improved links with support groups to develop agricultural measures to ameliorate the impacts of air pollution, or providing assistance to monitor and report on damaging pollution levels).

Acknowledgements

The Indian case study material presented here is an output from research projects funded by the United Kingdom Department for International Development (DFID) for the benefit of developing countries. The views expressed are not necessarily those of DFID. R6289 and R6992 Environment Research Programme.

We acknowledge the contribution of many colleagues and friends to the case study in India including: Dolf te Lintelo, Nigel Bell, Mike Ashmore, Simon Croxton, Madhoolika Agrawal, Neela Mukherjee (the participatory research team and farming communities in 28 villages around Delhi and Varanasi), D.S. Bhupal, Rana P.B. Singh, C. Chandra Sen, Ravi Agrawal, Anshu Sharma and SEEDS, Sandhya Chatterjee, C.K. Varshney, Madhushree Mazumdar and RMSI Limited.

Conversion Factors

Units for converting mass per unit volume to volume mixing ratios (at 20°C and 1013 mbar pressure)

SO_2: 1 μg (SO_2 as S) m^{-3} = 2 μg SO_2 m^{-3}
= 0.75 ppb

NO_2: 1 μg (NO_2 as N) m^{-3} = 2.1 μg NO_2 m^{-3}
= 1.72 ppb

O_3: 1 μg O_3 m^{-3} = 0.5 ppb

HF: 1 μg F m^{-3} = 1.2 ppb

1 ppb is equivalent to 1 nl l^{-1}

References

Aamlida D., Tørsethb K., Venna K., Stuanesc A.O., Solberga S., Hylend G., Christophersene N. and Framstadf E. (2000) Changes of forest health in Norwegian boreal forests during 15 years. *Forest Ecol. Manag.* **127** (1/3), 103–118.

Adams R.M. and McCarl B.A. (1985) Assessing the benefits of alternative ozone standards on agriculture: the role of response information. *J. Env. Econ. Manag.* **12**, 264–276.

Ashmore M.R. and Marshall F.M. (1999) Ozone impacts on agriculture: an issue of global concern. *Adv. Bot. Res.* **29**, 32–52.

Bell J.N.B. and Marshall F.M. (1999) Field studies of impacts of air pollution on agricultural crops. In *Environmental Pollution and Plant Responses*, eds. Agrawal M. and Krizek C. CRC Press/Lewis Publishers, Florida, pp. 99–110.

Bell J.N.B., McNeill S., Houlden G., Brown V.C. and Mansfield P.J. (1993) Atmospheric change: effect on plant pests and diseases. *Parasitology* **106**, S11–S24.

Brown D. and Smith M. (1984) Crop substitution in the estimation of the economic benefits due to ozone reduction. *J. Env. Econ. Manag.* **11**(4), 347–362.

Davies C. (2001) *Air Pollution and Agricultural Insect Pests in Urban and Peri-Urban Areas of India: A Case Study of Varanasi.* PhD thesis, Imperial College of Science Technology and Medicine.

Garrod G. and Willis K. (1999) *Economic Valuation of the Environment.* Edward Elgar, Cheltenham.

Heck W.W., Taylor O.C. and Tingey D.T. (1988) *Assessment of Crop Loss from Air Pollutants.* Elsevier Science, London.

Hunter G. (1978) *Agricultural Development and the Rural Poor.* Declaration of Policy and Guidelines for Action, Overseas Development Institute, London.

Innes J.L. (1993) *Forest Health: Its Assessment and Status,* CAB International.

Lanian S. (1993) *Valuing the Unknown: Cost Benefit Analysis and Air Pollution.* Oxford Institute for Energy Studies, Oxford.

Mackenzie J. and El-Ashry M. (1989) *Air Pollution's Toll on Forests and Crops.* Yale University Press, New Haven.

Marshall F.M., Wildig Z., Stonehouse J., Bell J.N.B., Ashmore M.R. and Batty K. (2000) *The Impacts and Policy Implications of Air Pollution on Crops in Developing Countries.* Final Technical Report, Department for International Development, Environment Research Programme. (R6992) Imperial College of Science Technology and Medicine, London.

Marshall F., Ashmore M. and Hinchcliffe F. (1997) *A Hidden Threat to Food Production: Air Pollution and Agriculture in the Developing World.* International Institute for Environment and Development, London.

Mougeot L.J.A. (1994) *Urban Food Production: Evolution, Official Support and Significance.* Cities Feeding People Report No. 8, International Development Research Centre, Ottawa.

Spash C.L. (1997) Assessing the economic benefits to agriculture from air pollution control. *J. Eco. Surv.* **11**(1), 47–69.

Subject Index

Plant Species Index

winter wheat 132, 133, 143, 247, 312, 323